# Lecture Notes in Computer Science 1315

Edited by G. Goos, J. Hartmanis and J. van Leeuwen

Advisory Board: W. Brauer   D. Gries   J. Stoer

T0241845

# Springer
*Berlin*
*Heidelberg*
*New York*
*Barcelona*
*Budapest*
*Hong Kong*
*London*
*Milan*
*Paris*
*Santa Clara*
*Singapore*
*Tokyo*

Gerald Sommer   Jan J. Koenderink (Eds.)

# Algebraic Frames for the Perception-Action Cycle

International Workshop, AFPAC'97
Kiel, Germany, September 8-9, 1997
Proceedings

 Springer

Series Editors

Gerhard Goos, Karlsruhe University, Germany

Juris Hartmanis, Cornell University, NY, USA

Jan van Leeuwen, Utrecht University, The Netherlands


Volume Editors

Gerald Sommer
Christian-Albrechts-Universität zu Kiel
Institut für Informatik und Praktische Mathematik
Preußerstr. 1-9, D-24105 Kiel, Germany
E-mail: gs@informatik.uni-kiel.de

Jan J. Koenderink
University of Utrecht, Helmholtz Institute
Princetonplein 5, 3584 CC Utrecht, The Netherlands
E-mail: J.J.Koenderink@fys.ruu.nl

Cataloging-in-Publication data applied for

**Die Deutsche Bibliothek - CIP-Einheitsaufnahme**

**Algebraic frames for the perception action cycle** : international
workshop ; proceedings / AFPAC '97, Kiel, Germany, September 8 -
9, 1997. Gerald Sommer ; Jan J. Koenderink (ed.). - Berlin ;
Heidelberg ; New York ; Barcelona ; Budapest ; Hong Kong ;
London ; Milan ; Paris ; Santa Clara ; Singapore ; Tokyo : Springer,
1997
  (Lecture notes in computer science ; Vol. 1315)
  ISBN 3-540-63517-3


CR Subject Classification (1991): I.3.5, I.2.9-10, I.5, I.2.6

ISSN 0302-9743
ISBN 3-540-63517-3 Springer-Verlag Berlin Heidelberg New York

© Springer-Verlag Berlin Heidelberg 1997
Printed in Germany

Typesetting: Camera-ready by author
SPIN 10545824    06/3142 – 5 4 3 2 1 0    Printed on acid-free paper

# Preface

This volume presents the proceedings of the International Workshop on Algebraic Frames for the Perception and Action Cycle, AFPAC'97 held in Kiel, Germany, September 8–9, 1997. The topics of this workshop were trends in conceptualization, design, and implementation of the rapidly growing field of visual sensor based robotics and of autonomous systems. Special emphasis was placed on mathematical problems of integrating relevant disciplines like robotics, computer vision, signal theory, and neural computation.

The growing interest in the design of autonomous artificial systems with the competence to interact with the world according to a set of behaviors is not only based on their growing technical feasibility but is stimulated and accompanied by the formation of a socio-ecological theory of biological systems as a metaphor for the everyday intelligence of these systems, on every stage of their evolution. In that frame, the paradigm of behavior based systems gives grounds for hope that we will be able in the near future to design artificial systems with such properties as robustness, stability, or adaptability. While behavior is an observable property of a system and can be used to define the aim of our efforts, the design of the perception-action cycle (PAC) is the way to reach that aim. Insofar, the purpose of the workshop was to define some relevant frames for the embedding of PAC. These frames have to be powerful enough to unify separately developed mathematical models of the contributing disciplines and to overcome well identified limitations. These frames are of algebraic nature and have to support stochastic approaches for gaining geometry based experience and for representation of geometric-topological knowledge from the environment.

The contributed papers of the workshop were carefully reviewed by at least two members of the program committee. The members of the program committee were invited to present their overall view on important aspects of the workshop. Regrettably, O. Faugeras and Y. Zeevi could not follow that invitation. Many thanks for their careful reviewing of the papers. We want to give our thanks also all the other reviewers. On the basis of their enthusiasm, all the authors of this volume contributed with important aspects to the topic of the workshop. Our thanks go to all the authors, both invited and contributing, for the high quality of their work and for their cooperation.

We thank the Christian-Albrechts-Universität Kiel for hosting the workshop and the industrial and institutional sponsors for financial support. Springer-Verlag Heidelberg and especially Alfred Hofmann are gratefully acknowledged for publishing the AFPAC proceedings in the LNCS series. Last but not least, the workshop could not take place without the extraordinary commitment of the local organizing committee.

Kiel and Utrecht, June 1997                    Gerald Sommer and Jan J. Koenderink

**Conference Chairs:**
Gerald Sommer, Germany
Jan J. Koenderink, The Netherlands

**Program Committee**
Y. Aloimonos, USA
O. Faugeras, France
G. M. Granlund, Sweden
D. Hestenes, USA
J.J. Koenderink, The Netherlands
H. Ritter, Germany
A. Shashua, Israel
G. Sommer, Germany
L. Van Gool, Belgium
Y.Y. Zeevi, Israel

**Organizing Committee**
G. Sommer (Chair)
E. Bayro-Corrochano (Program)
J. Bruske (Accommodation, Publicity)
M. Hansen (Local Arrangement)
U. Mahlmeister (Technical Support)
F. Maillard (Secretary)
J. Pauli (Finance)

**Invited Speakers**
Y. Aloimonos, USA
G. M. Granlund, Sweden
D. Hestenes, USA
K. Kanatani, Japan
J.J. Koenderink, The Netherlands
S.J. Maybank, UK
H. Ritter, Germany
A. Shashua, Israel
G. Sommer, Germany
L. Van Gool, Belgium

**Sponsors**:
Sun Microsystems, COSYCO GmbH/Datacube, Siemens AG, Isatec GmbH, Sparkasse Kiel, TELEROB GmbH, Deutsche Forschungsgemeinschaft, Förderverein der TF Kiel, Wissenschaftsministerium SH, Institut für Praktische Mathematik und Informatik der Universität Kiel

# Table of Contents

## PAC Systems

## Low Level and Early Vision

## Recognition of Visual Structure

## Processing of the 3D Visual Space

# Representation and Shape Perception

# Inference and Action

# Visual and Motor Signal Neurocomputation

# Algebraic Aspects of Designing Behavior Based Systems[*]

Gerald Sommer

Cognitive Systems Group, Institute of Computer Science
Christian-Albrechts-University, Kiel, Germany
gs@informatik.uni-kiel.de

**Abstract.** We address in this paper the design of behavior based systems from a bottom–up viewpoint. Although behavior is an observable property of a system, and therefore immediately causes a top–down model, the approach has to be inverted to support the learning of equivalence classes of the perception–action cycle. After introducing the paradigm in the frame of a socio–ecological theory of biological systems, we discuss the natural science problems to be solved for successful design of behavior based systems by a bootstrap of perception–action cycles. The necessary fusion of robotics with computer vision, neural computation, and signal theory needs a common theoretical framework. This framework consists of a global algebraic frame for embedding the perceptual and motor categories, a local algebraic framework for bottom–up construction of the necessary information, and a framework for learning and self–control, based on the equivalence of perception and action. Geometric algebra will be identified as the adequate global algebraic frame, and the Lie theory will be introduced as the local algebraic frame. We will demonstrate several applications of the frames in early visual processing. Finally, we will finish our discussion with the fusion of local approaches and the global algebraic frame with respect to both the formulation of an adequate multidimensional signal theory and the design of algebraic embedded neural processing. In both cases we will discuss the relation to the non–linear Volterra series approach, which, in our framework, will be reduced to a linear one.

## 1 Introduction

In this paper, we want to promote the consideration of algebraic aspects in the process of fusion of disciplines such as computer vision, robotics, neural computation, and signal theory, which have been developed separately until now. We conclude the necessity of following this line from two contradicting roots. In principle, the paradigmatic frame and the technical resources are available to develop vision based robots or autonomous self–navigating systems. However, the conceptions of handling phenomena of spatio–temporal geometry are very limited in the contributing disciplines and there exist, at least partly, deep gaps between the mathematical languages used. This judgement may surprise but

---

[*] This work was supported by DFG grant So 320–2–1.

will be substantiated later on. Another point of shortcoming is the following. The mathematical framework should support a constructive way of capturing spatio–temporal geometric phenomena of the world, while we often use a descriptive approach. Because each non–pathological algebraic structure can be approached either from top–down or from bottom–up we have the possibility at changing our viewpoint. That means that the problems to be solved have to be formulated from the systems point of view which has to gain structural concepts from seemingly non–coherent phenomena in the learning stage and has to match phenomena of the world to isomorphic representations of them in the matured stage. In other words, we have to support learning by experience by means of separation of variant from invariant phenomena and have to reduce the explicit formulation of categories by programming. In that way algebra will meet stochastics insofar as e.g. geometric entities as points, lines or planes are conceptions of the top–down approach which can only be approximated from bottom–up by recognizing the corresponding algebraic properties (symmetries) in the manifold of sensory data. From this follows that the bottom–up approach does not result in the so–called signal–symbol gap.

In the last decade great progress has been made in the conceptualization of computer vision (paradigmatic extension to active and purposive vision, use of projective geometry), signal theory (wavelet theory, local spectral features), and in the availability of dedicated and powerful computer systems. But we made only minor progress in adequate representation of multidimensional signals from a local approach, in the unification of spatio–temporal phenomena (patterns and motion), in recognition of projective geometric relations on signal level, in integration of neural computation and computer vision, in neural learning theory, in designing robust and adaptive systems, and e.g. in architectural design of visual systems. All these problems in a certain sense are related to the algebraic frames used to formulate the solutions.

The fusion of perception and action in the frame of the so–called perception action cycles (PAC) is the paradigmatic starting point for the design of behavior based technical systems. Behavior is the observable expression of the competence gained in realizing PAC. But while behavior corresponds the top–down approach, the design of PAC is our view of bottom–up approach. The hypothesis of this direction of research consists in a possible bootstrap of robust and adaptive systems by equipping systems with the ability to organize such cycles and the cooperation/competition between by themselves. In contrast to programmed solutions, the result of such self-organization may not be provably correct from a designers point of view, but the degree of success is observable and may be analyzed by statistical means.

Biological systems, if they are plants, ants or human beings, are successful behavior based systems, although most of them do not know about algebra. The question is, do we as system designers know enough on that topic or do the mathematicians have the right algebras in their desks. Nobody knows the answer. But we know the sluggishness of the human society and the preference of the technicians for simple, linear, and in consequence suboptimal solutions.

Therefore, linear algebra of vector spaces is the frame commonly used to embed almost all our problems. It is our opinion that we have indeed powerful mathematical languages for embedding PAC. But they are either not known to the community or are ignored because of the burden of both to make them useful or to pick them up.

D. Hestenes [34] for a period of thirty years has been promoting the use of a geometric interpreted version of Clifford algebra in physics. He calls it geometric algebra. We will follow him in using this algebra as a global frame of embedding PAC and even calling it geometric algebra. On the other hand the bottom–up approach of PAC necessitates a local algebraic embedding of perception and action to recognize and generate patterns of certain symmetry. This local frame is Lie theory [61].

It seems to us that the problems of putting real phenomena in space–time into a rich mathematical frame is rooted in the same manner in algebra as those of physics. May be, a special slot of scale (macroscopic phenomena) has to be considered and we have not to regard electrodynamics, quantum effects or relativity, although their metaphoric potentials are enormous. But our central questions are

- *In which manner the perceptible and experienced world can be struc-tured most successfully?*
- *What has to be the functional architecture of such systems?*

To answer these questions, there is a need of more complex mathematics to formulate things more simply and to make them work with limited resources, taking into account both the complexity and the constraints of real world.

Indeed, geometric algebra and Lie theory are more complex than linear algebra of vector spaces. But our research group started two years ago successful work in overcoming shortcomings in disciplines contributing to the design of PAC systems by using extended possibilities of representing geometric relations in this frame. Nevertheless our work is in infancy. This may not wonder in comparison to the long lasting role of vector algebra in engineering and science.

The outline of the paper is as follows. In section two we will give a sketch of the behavior based systems paradigm. In section three we will outline our vision of the theoretical framework of embedding the design of such systems. Section four is dedicated to a short discussion of exemplary use of the framework of geometric algebra in early vision and learning of geometry. The paper will be accompanied by three special contributions of our group, dedicated to special problems and presented at the same workshop.

Our view will be biased by the aspects of visual sensory recognizing the world. This is based on the fact that we are predominantly interested in behavior based design of visual systems, and we have to make clear with this respect the role of action in building perceptual categories and their use in PAC. Another view may be perceptually supported autonomous systems, especially visual based robotics. Both viewpoints have to be understood as the two sides of one medal and are fused in the paradigm of behavior based systems.

# 2 Behavior Based Systems as Paradigm

In this section, we want to summarize the evolution and the essential features of the paradigm of behavior based systems. With these respects we will focus on the problems of computer vision and their overcoming by extension of the scope. On that base we will draw a sketch of natural science problems which in our opinion have to be coped with to develop technical systems based on that paradigm. The engineering science problems will not be dealt with.

## 2.1 Two Metaphors of Intelligence

The term behavior is borrowed from ethology and stands for the basic capabilities of biological systems which guarantee survival of the species and/or the individual. Famous ethologists as K. Lorenz [40] and N. Tinbergen [66] considerably contributed to the change of the metaphoric view of intelligence, respectively brain theory [57, 24]. Besides, results of molecular and evolutionary genetics on a completely other level of living systems brought into consideration that information processing is an inherent capability of all biological systems, on which level ever [25]. Not only from biology, ethology, and psychology [27] but also from the growing knowledge of physics on complex, non-linear, and dynamic self-organizing systems [32], the behavior based paradigm of "intelligent" systems is superseding the paradigm of knowledge based systems.

Both paradigms of system design are outcomes of quite different metaphors on understanding intelligent achievements or intelligence by themselves (see table 1). Common to both metaphors is only that there are biological systems (man), which are interpreting what they observe and are trying to use the gained models for construction of technical systems with comparable performance to that of living systems. All other aspects are fundamentally different.

The computational theory of mind is rooted in the results of logics, linguistics, and the brilliant von Neumann architecture of computers. Their advocates viewpoint is a top-down one and expresses the dominance of description of observed phenomena.

To oppose that metaphor to the socio-ecological theory of biological systems, the following summary of the implicit assumptions of the computational theory of mind will be given:

1. *The world can be formally modeled using terms of our language or categories as representations (symbols) of equivalence classes.*
2. *Information is an intrinsic entity of the world as energy or matter. Information processing is a process as conversion of energy in physical processes. It can be done on an abstract level of symbols and independently of the material base of the system.*
3. *Intelligence is that of human beings and those may use it for top-down design of tasks. The problem solutions should be provably correct and themselves they can be interpreted as achievement of intelligence of the computing system.*

*4. On that base, any partial aspect of the world can be considered in isolation to construct a complete solution from a set of partial ones of any problem the designer may formulate.*

| Metaphor | Computational Theory of Mind | Socio–Ecological Theory of Biological Systems |
|---|---|---|
| basic roots | logics, linguistics<br>von Neumann computer<br>architecture | cybernetics<br>molecular and evolut. genetics<br>ethology<br>evolutionary theory of knowledge<br>synergetics |
| basic paradigms | information processing<br>(symbol processing,<br>connectionism) | information selection and<br>construction |
| empiric aim | understanding intelligence<br>of human beings | understanding competence<br>of biological systems |
| disciplines | computer science<br>artificial intelligence<br>cognitive science<br>(computational neuroscience) | artificial life<br>neural Darwinism<br>vision based robotics<br>synthetic psychology |
| engineering roots | computer engineering<br>computer vision<br>robotics | mechatronics<br>computer vision<br>(active, animate, qualitative, purposive)<br>reactive robotics |
| engineering paradigms | knowledge based system<br>design | behavior based system<br>design |
| normative aim | construction of intelligent<br>machines | construction of autonomous<br>systems |

**Table 1.** The change of metaphors

Although the power of facts created by universal computer machines is impressive, on that base the hopes of computer vision and robotics regarding the development of engines which are robust and adaptive in their performance could not be realized.

The socio–ecological theory of biological systems stands in very contrast to these conceptions. Although their advocates also start with the observation of phenomena of living systems (behavior), they have to invert their viewpoint into a bottom–up approach by asking what principles are running so that living systems can show stabile behavior.

The implicit assumptions of the socio–ecological theory of biological systems are:

1. *The formalized models of the world are idealized conceptions of the reality and insofar they are not useful, neither to understand living systems nor to realize comparable performance.*

2. *Instead, any behavior is based on a sufficient approximation of categories by equivalence classes with respect to the resources of the system and its purpose in dependence of the situation.*

3. *Information is a prerequisite for behavior. It is a result of an active process of interaction of the system with its environment, and it is the result of purposive construction, gathering and selection. If we call this process also information processing, it happens on real (sensoric mapped) data from the total complexity of the world.*

4. *Equivalence classes stand for the gained order of the system's internal degrees of freedom and without having a language there is no need to conceptualize categories.*

5. *Competence should be understood of having equivalence classes as a prerequisite of behavior. Insofar, competence is a kind of real intelligence and much more general than this because each living system on each level of organization, reaching from phages to vertebrates and from macromolecules to the body, will need it.*

6. *Competence cannot be programmed but has to develop by purposively constrained self-organization of the internalized representations of phenomena of the environment.*

7. *Competence is only provable by the success of behavior. It may be based on knowledge which has been acquired by the species during phylogenesis, has to be transmitted to the individual by genes, or/and may be learned by individuals during ontogenesis.*

8. *As emergent property of a behavior based system, competence is robust and adaptive if there are no dramatical changes of individual resources or environment.*

9. *Behavior based systems are open systems with respect to their environment, non–linear ones with respect to the situation dependence of their responses to sensory stimuli, and dynamic ones as they change their behavior as answer on the same signals in different situations.*

If we want to follow that metaphor [52] to design behavior based systems, the result would be in any case a kind of autonomous system. It seems to become visible that we not only have to make systems running but have to work out fundamental new principles of design. It is far from being sufficient to change from the symbolic to the subsymbolic level of information processing. This change of the computational paradigm remains in the computational theory of mind metaphor if not the essence of behavior based systems will be considered. This essence is to close the loop between system and environment, the perception–action cycle, by using the afferent and efferent channels [35].

## 2.2 Evolution of the Paradigms of Visual Information Processing

Remarkable stimuli for the yet ongoing process of redefinition of the paradigmatic base of artificial intelligence came from deep conflicts recognized in computer vision one decade ago. The proceedings of computer vision conferences of the last years make obvious important contributions to shape the conception of behavior based visual architectures. Progress has also been reached in robotics.

D. Marr [43] formulated the computational theory of vision on the base of the information processing paradigm of cognitive science using the symbol processing paradigm of Newell and Simon [48]. Assuming that vision is an inherent task in the world and resting on a general theory, he postulated the famous three–step procedure for the top–down design of a visual system:

1. **computational theory**: formulation of the general task and the way to find the solution by considering necessary constraints and boundary conditions,
2. **representation and algorithm**: specification of a formal procedure,
3. **implementation**: assumption of independence of the procedure with respect to the hardware at hand.

This approach resulted in insights into important relations between the spatio–temporal macroscopic structures of the world and hypothetical vision tasks. Examples are the role of projection operator and shape–from–X tasks. A dramatical consequence of Marr's theory has been the thesis that vision mainly could be understood as reconstruction of the world from sensory data. The sensory part of the visual system together with geometry and illumination would constitute a general operator of sensory imaging. Perception would be defined by application of the inverse operator. Visual perception as ill–posed inverse problem should be regularized to become well–posed [65] by adding to the sensory data constraints regarding geometry and physics of imaging, and knowledge with respect to the imaged scene. This conception fitted very well to the knowledge based approach of artificial intelligence. Nevertheless, it failed with respect to realize recognition and to construct general useful and robust systems.

To give a summary of characterizations of knowledge based vision the following drawbacks will be noticeable:

1. *The mysterious gap between signals and symbols cannot be closed in the frame of the paradigm.*
2. *The dominant role of symbolic processing versus signal processing totally underestimates the role of early visual processes.*
3. *The top–down control of bottom–up directed data flow allows no return from symbols to signals in case of erroneous interpretations.*
4. *Recognition has to be done by matching prototypes as world models because equivalence classes cannot be adequate modeled.*
5. *The explicit formal representation of models is limited to a simple world and therefore restricts application fields. Other restrictions follow from a time–consuming search–based matching process.*
6. *The use of a maximum of knowledge to solve a visual task contradicts the needs of economy of resources.*

The contemporary spectrum of alternative paradigms was initiated by R. Bajcsy [5], Y. Aloimonos [1], and D. Ballard [6] who remembered that a visual system is a sensory, perceptual, and a motor one.

1. **Active Vision**: [2, 39] The active control of the outer degrees of freedom of the oculomotor system enables the vision system to break down the complexity of the task. If vision tasks are coupled to an oculomotor behavior, the coupling of algorithmic solutions to behavioral strategies will result.
2. **Animate Vision**: [6] Similar to Active Vision also animate vision supports the division of the vision process into two phases: the preattentive and the attentive stages. While preattentive vision is interpreted as fast and data driven bottom–up processing, attentive vision is top–down controlled and related to the next term.
3. **Purposive Vision**: [3] The active system is controlled by the purpose of vision and that is action. Purpose is understood as the driving power of the system to interact with the environment. This indeed is in almost agreement with the behavior based approach.
4. **Qualitative Vision**: [4] The opportunism of purposive vision calls for using minimal efforts in realizing any visual task. That means gathering of sufficient hints in minimum time with respect to the task. That also means the use of minimal models of the world in very contrast to knowledge based vision.

Interesting questions of research are of such kind:

*Which knowledge of the structure of the world is necessary to perform purposive vision in a limited range of time using oculomotor strategies.*

But two fundamental differences to the behavior based paradigm, projected to visual systems, remain. These are the unsolved recognition problem and the problem of the origin of categories of behavior. Although the coupling of visual tasks with oculomotor behavior and purpose introduced an important strategic component, the recognition problem only gained some redefinition but no general solution. Now, indeed, recognition is decomposed into partial tasks connected with the oculomotor behavior and defined by the mismatch between the task and the fusion of partial solutions. Learning and acquisition of competence until now only in exceptional cases is integrated with active vision [67].

Another aspect, not yet well understood, is the mutual support of visual and motor categories. A reasonable hypothesis, drawn from cognitive psychology and ethology, leads to the conjecture that vision as isolated process is a too hard task. Visual used categories cannot be learned and are not defined by vision alone, but can be interpreted as projections of multiple defined (and learned) categories onto the visual system.

## 2.3 Natural Science Problems

We have to our disposal now stereo–camera heads, miniaturized robots, radio Ethernet, and powerful computers. This allows to think of designing new systems, we never had before. We may be encouraged simply to do it and we should.

But there are a lot of serious and fundamental problems to be solved in advance if we want to classify such systems as behavioral ones.

Replying a debate on the paradigmatic changes in computer vision, Aloimonos concludes [2] that the task to be solved may be summarized by:

*Find a general solution for special problems.*

Indeed, the great challenge will be to understand the general principles underlaining the success of living systems in performing their perception/action tasks. Only if we find some sufficient approximation to the answers, we will be able to equip technical systems with the resources to develop the wanted competences.

As nature brought forth rather different levels of behavior, each with different amount of directness and indirectness of behavior, we should start with the simplest categories of systems. That means, although we want to have systems with human like competence, this should not be our goal in the moment. Cognitive processes as indirect behavior can be our concern if we sufficiently understand the functionality of more direct behaviors as e.g. signal–motor mappings.

If also nature is constrained by the principles of evolution, which leads some people to state that nature is a tinker, nature within these constraints most effectively uses those general principles.

The most important features of behavior based systems are:

1. **situatedness**: The system is managing its tasks in the total complexity of concrete situations of the environment.
2. **corporeality**: The system can only gain experiences with respect to the environment by means of the specific resources of its body, including mind.
3. **competence**: The system's order of inner degrees of freedom is an expression of the captured order of the environment and causes order of the system's outer parameters and actions.
4. **emergence**: The system's competence emerges from disorder if the purposive rooted and physical constrained actions are fitting well the general and invariant aspects of the situations in the environment.

With respect to the purpose of the system, any behavior has the properties of usefulness, efficiacy, effectiveness, and suitability. It corresponds the equilibrium with respect to the purpose between the system and its environment. Its robustness with respect to minor distortions of this equilibrium has to be completed by adaptivity with respect to more heavy distortions. A general natural science theory of the principles underlaining behavior based systems will be related to the theory of non–linear dynamic systems. This is a theory of the dynamics of large (physical) systems and far from being the theory of vision, which Marr asked for. Although all the features of behavior based systems are well fitted by the growing up theory of non–linear dynamics, their metaphorical use in practice is limited yet [49].

As the most important conclusion from situatedness with respect to the limited resources of a system, not the knowledge of detailed models of the world but of useful relations between the system and its environment has to be considered. The set of situative important relations and the amount of knowledge has to be minimal because actions should be suitable and effective.

Information cannot be simply gathered and used as Gibsonian invariants [28]. A perceptual system is no pure reactive system (not only controlled by data and instincts as proposed by Brooks [11]), just as it is no pure cognitive system (not only controlled by expectations or knowledge). Instead, perception is bottom–up driven within the limits of corporeality and top–down controlled [19] by the purpose, projected to the situations. This implies that the pragmatic aspect of information is mainly determined by the purpose, the semantic aspect is constructed, respectively selected, by physical experience, and the syntactic aspect is mainly matter of perception.

Recognition in this frame is a purposive constraint matching to equivalence classes whose meaning is based on using corporeality. Consequently, there is in the stage of competence no problem of heaving too less invariants but selecting the right ones. This principle has to be supported by an architecture with sufficient purposive constraint dynamics.

In the stage of incompetence, learning will be the process, which will result in the mentioned equilibrium. Actual learning paradigms hardly can be understood in the frame of emergent systems. The used least mean square minimization as linear approach lacks the attractor properties of non–linear dynamic systems. The most promising approach is reinforcement learning because it supports evaluation of interpretation of sensory data by action most naturally [12]. Using this approach, the important contribution of the knowledge bias for fast learning will become obvious [31]. The resulting question is, how to partition the necessary knowledge base to learned and acquired or programmed contributions.

Our contemporary understanding of multiple supported categories is in agreement with implicit representations as a result of optimization of the perception–action cycle by self–supervised learning. Insofar, the mapping of the non–linear spatio–temporal relations between sensory and motor signals, respectively vice versa, by means of the paradigm of artificial neural nets, is a promising way of semantically based coupling of perception and action. But semantics is based on pragmatics and pragmatics is strongly related to purpose. In the frame of reinforcement strategy of learning, the pragmatics is submitted to the system as cost function or confidence measure. Appearance based vision by self–supervised learning [53] is starting to become useful for the design of bottom–up constructed perception–action cycles.

Such PAC not necessarily must be an elementary one. A more complex behavior is not a linear sum of a set of primitive ones. Therefore, the top–down partitioning of behaviors is of limited value. Moreover, any complex behavior should emerge from a set of primitive ones as a result of adaption to new situations. Although some experiments could be interpreted as such emergence [18], the construction of relations between primitive cycles is not well understood yet [10]. This is also the case for the design of hierarchically structured behavior based systems. Inverse relations of dependence between primitive behaviors (necessary for survival) and higher–order behaviors (necessary for the task) cause different total behaviors [11, 44].

# 3   The Theoretical Framework

The realization of behavior based systems by bootstrap of perception–action cycles necessitates the fusion of robotics, computer vision, neural computation, and signal theory in a unified, respectively compatible framework. That framework should allow to embed all the tasks constituting the PAC system with sufficient flexibility, dynamics, and completeness.

The behavior based system has to experience any perception–action cycle and not to report or to reason on it. Insofar, the explicit symbolic level is restricted to the programming of the dynamic frame of PAC and to the interface between system and user. Within PAC implicit representation of knowledge is dominating a certain amount of explicitly formulated basic rules as instincts.

The level of signal processing with respect to afferents and efferents, including representation of equivalence classes, has to be able to realize all aspects of PAC. Within such frame no signal–symbol gap will exist. Symbols are not necessary as representation of equivalence classes for PAC systems.

The central problem of the existence of a behavior based system will be to cope with all spatio–temporal phenomena of the environment which are of relevance with respect to the objective of the system. Concerning visual information processing, these phenomena are spatio–temporal structures, including those which are caused by the actions of the system.

Within such scenario, situations are embedded in the Euclidean space–time. Spatio–temporal phenomena are fused but may be projected to spatial or temporal ones. This will be supported by the use of oculomotor behavior. With respect to the coupling of perception and action, the most important task will be to recognize and to generate patterns of spatio–temporal geometry with a certain degree of symmetry. These patterns represent equivalence classes of that property and, as expression of the competence of the system, support a statement of mathematical equivalence of perception and action:

> – *similar visual patterns cause similar motor actions*
> – *similar motor actions cause similar visual patterns.*

This unity of perceptual and motor equivalence classes enables the system to self–control and to learn from actions by using oculomotor behaviors.

The theoretical embedding of the perception–action cycle will be constituted by an algebraic framework and by the frame of learning and using the knowledge on implicit (neural) representations. The algebraic framework will be a dual one because it has to support both the forming and representation of experience from global phenomena of the environment and the process of local generation or verification of global pattern concepts. In this section, we will refrain from presentation of the frame of learning and neural information processing. With this respect, we developed a type of neural net, called Dynamic Cell Structure – DCS [14], which in the context of behavior based system design could be successfully proved [15]. Its main idea is the optimally topology preserving adaption to the manifold by self–supervised vector quantization.

## 3.1 The Global Algebraic Frame

The global frame has to represent the perceptual relevant phenomena of the Euclidean space–time $E_4$. These are resting or moving objects of any dimension less than or equal three and their relations between. The classical mathematical framework enables modeling of either resting objects by means of either analytical geometry using entities of dimension zero (points), one (lines), two (planes), respectively three (cubes), or differential geometry using entities as curves (1D) and surfaces (2D), or modeling of objects in motion within the frame of kinematics as rigid body movement in $E_3$ using these entities. Normally, the movement of rigid bodies is described by rigid displacement of a set of points. The entities of motion concepts are geometric transformations as translation, rotation, and scaling. From these entities complex patterns of motion are constructed. This decoupling of space–time can be done by a competent system using the oculomotor behaviors (e.g. fixation and tracking). But fixation to infinity while moving is also useful and will result in patterns from $E_4$.

Both object and motion concepts are determined by the correlations in the data of their patterns. These correlations have certain aggregated global symmetry properties which are important to define and which represent the equivalence classes. But the construction of global symmetry from local correlations is a bottom–up process and therefore, it is matter of the local algebraic frame. Nevertheless, the global symmetries have to be represented to form isomorphic representations to the observed phenomena. If Pellionisz wrote [56] *"the brain is for geometrical representation of the external world"* or if Koenderink [36] and von der Malsburg [42] stated that the brain is a geometry engine, the importance of representation of the environment is underlined. But this representation is implicitly constructed and has not to be complete in a mathematical sense but complete with respect to purpose, situatedness, and corporeality of the system. The algebraic frame should allow to support this flexibility.

The global symmetries and the metric properties of objects in Euclidean space $E_3$ are distorted in a systematic way by global projective transformations between objects and the observer's visual sensory system. Nevertheless, the stratification of the space [26] in connection with the transformation group of each shell (projective, affine, Euclidean) [41] allows recognition of the corresponding invariants. The necessary amount of effort to use the invariants of different shells can be strongly modified by oculomotor behavior. In this way, oculomotor constraint recognition from stratified space is in accordance with the situation dependent use of minimal resources [21].

The global algebraic frame should support all these aspects of geometric mappings and should also support the effective control of actions. This will be possible by realizing geometric transformations in the same framework. But even as invariance is important for recognition, this is valid also for actions. One aspect of invariance with this respect is the independence of transformations from a fixed coordinate system. This aspect is often neglected in the design of behaviors.

The linear vector space of real and complex numbers is not satisfactory to represent all the mentioned phenomena in the requested quality. This is known for a long time and therefore, tensor algebra has been intensively studied in neural science [56], robotics [20], and computer vision [30]. Pellionicz [56] argues that neural coupling of sensory receptions with motor actions has to consider the special transformation properties of these signals (contravariance of motor signals and covariance of sensory signals) to gain a metric internalization of the world. The arguments of using tensor algebra in the frame of signal processing [30] mainly refer to the poorness of the representation power of vectors and scalars with respect to multidimensional signals. While the first argument seems not so obvious, the second one corresponds also our experience and is of greatest importance for the bottom–up design of the PAC.

Indeed, the Hilbert vector space is representing only sets of points (using vectors) with the result that all correlations which specify symmetries of higher dimensional entities with great effort have to be reconstructed by the analysis of the occupation of the vector space and by construction of subspaces. In the linear frame, besides addition and multiplication of vectors, there is only the poor operation of the scalar product, which even shrinks vectors to a scalar, and which in addition is defined only as bilinear operation. What we want to have are more rich structures in the vector space, which represent in any way higher order entities as planes or volumes as expression of their correlation. In tensor algebra and in vector algebra, there are constructs of such type as outer product tensor or cross product of vectors to extend the algebraic limitations.

**Vector Space Structuring by Geometric Algebra** In our opinion, the most systematic way to endow a vector space with such entities is based on Clifford algebra [58] and can be most intuitive related to geometric conceptions in the version promoted by Hestenes [33]. This is the geometric algebra. Only to give an impression of the richness of defining structure in vector spaces by definition of partial product spaces from the contributing vectors, we will give a short summary of the subspace conception of the algebra [64].

The geometric algebra $G(\mathbf{A}) = G_n$ is the algebra of multivectors $\mathbf{A}$,

$$\mathbf{A} = \mathbf{A}_o + \mathbf{A}_1 + ... + \mathbf{A}_n$$

with $\mathbf{A}_k$, $k \leq n$, as homogeneous $k$–vectors, i.e. multivectors of grade $k$. The geometric algebra $G_n$ results from a vector space $V_n$ by means of endowing it with a so–called geometric product. This geometric product causes a mapping of $V_n$ onto $G_n$, which themselves is a linear multivector space of dimension $2^n$. Any $n$ linear independent vectors $\mathbf{a}_1, ..., \mathbf{a}_n \in V_n$ are therefore transformed to the multivector $\mathbf{A}$. The basis of this multivector space is constituted by $n+1$ sets of $\binom{n}{k}$ linear independent $k$–blades $\mathbf{M}_k$ which themselves constitute the basis of linear subspaces $G_k(\mathbf{A})$ of dimension $\binom{n}{k}$ of all $k$–vectors in $G_n$.

The geometric product of 1–vectors (i.e. normal vectors) $\mathbf{a}, \mathbf{b} \in V_n$,

$$\mathbf{ab} = \mathbf{a} \cdot \mathbf{b} + \mathbf{a} \wedge \mathbf{b}$$

has the important property that it maps vectors $\mathbf{a}, \mathbf{b}$ into both a scalar as a result of the symmetric inner product $\alpha_0 = \mathbf{a} \cdot \mathbf{b}$ as well as a bivector as a result of the antisymmetric outer product $\mathbf{A}_2 = \mathbf{a} \wedge \mathbf{b}$, $\alpha_0, \mathbf{A}_2 \in G_n$.

Any two homogeneous multivectors $\mathbf{A}_r, \mathbf{B}_s \in G_n$ are mapped by the geometric product into a spectrum of multivectors of different grade, ranging from grade $|r - s|$ as a result of the pure inner product $\mathbf{A}_r \cdot \mathbf{B}_s = \mathbf{C}_{|r-s|}$ until grade $r + s$ as a result of the pure outer product $\mathbf{A}_r \wedge \mathbf{B}_s = \mathbf{C}_{r+s}$. This corresponds the partitioning of $G_n$ into the subspaces $G_k(\mathbf{A})$. Thus, any $k$–blade $\mathbf{M}_k \in G_k(\mathbf{A})$ can be geometrically interpreted as the uniquely oriented $k$–dimensional vector space $V_k = G_1(\mathbf{A}_k)$ of all vectors $\mathbf{a}$, satisfying $\mathbf{a} \wedge \mathbf{A}_k = 0$, respectively of all $k$ linear independent vectors $\mathbf{a}_1, ..., \mathbf{a}_k$, spanning $V_k$ and constituting a factorization of the $k$–blade $\mathbf{M}_k = \mathbf{a}_1 \mathbf{a}_2 ... \mathbf{a}_k$. Therefore, any $\mathbf{A}_k$ is understood as projection of $\mathbf{A}$ onto $G_k$ and, on the other hand, can be formulated as a linear superposition of all $k$–blades, constituting the basis of $G_k$.

Because $\mathbf{A}_n = \lambda I$ with $I$ as unit pseudoscalar or direction of $V_n$, there follows an intrinsic duality principle of geometric algebra. This duality is based on $I_k I_{n-k} = I$ for any $k$–blade, respectively $(n - k)$–blade. From the dual $\mathbf{A}^* = \mathbf{A}I^{-1}$ of any multivector $\mathbf{A}$ follows that there is a simple change of the base of any multivector with respect to the dual blade.

The property of orientation of $G_k$ obviously transmits to the multivectors $\mathbf{A}_k \in G_k$. Therefore, the multivectors of the Euclidean spaces $E_2$ or $E_3$ result in directed numbers with the algebraic properties of complex numbers, respectively quaternions. Additionally, certain algebraic restrictions or extensions will result in other number conceptions as dual or double numbers [60]. They altogether possess nice algebraic properties which can be either interpreted in the frame of geometry or kinematics. Besides, each multivector has a quantitative or an operational interpretation. The reason for that duality of interpretation of multivectors lies in the fact that the Clifford algebra unifies Grassmann algebra with quaternion algebra of Hamilton. This makes the geometric algebra so attractive for our fusion of disciplines because the same number may be operand or operator.

Moreover, geometric algebra does not only subsume the mathematics of metric vector spaces but also that of projective geometry. For instance the qualitative incidence operations meet and join of entities, homogeneous coordinates, and the operation called projective split which relates vector spaces of different dimension in a simple way can be used.

## 3.2 The Local Algebraic Frame

With the local algebraic frame we will understand several aspects of supporting the bottom–up approach of PAC. That means with respect to the conception that the system uses (visual) behavior to reduce the amount of data, to reduce the complexity of the task, and to control the gathering of hints in a purposive manner. All these strategic aspects can be subsumed by attention to a local patch of signals to get a partial contribution to resolve conflicts or to solve the actual task. Global interpretations result from local contributions by fusion.

Even the same problem the system will have with the control of actions. Of course, these constructive processes are covered in biological systems by high parallel processing and are accelerated by special equipment as the sensory geometry in the human retina. Besides, in the highly trained stage, there is a process of shortcutting to global interpretations. These special effects will be disregarded here.

In the following, we will discuss some problems which are related to the local algebraic frame, with respect to irregular sampling using extended sampling functions and to the choice of basis functions for signals and operators. We will propose an attentive, purposive, and progressive visual architecture, and the use of the canonical local basis generated by Lie group invariants.

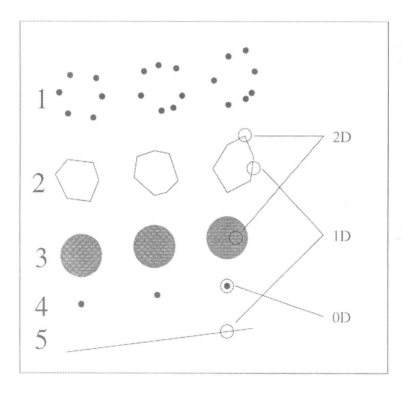

**Fig. 1.** Several conceptions of structure, dependend on purpose and/or aperture of the sampling operator.

**Sampling with Finite Aperture** The first problem concerns the term "local". In mathematics exists the conception of the entity point as interpretation of an ideal location, defined by its coordinates. With respect to our problem of designing a real technical system by understanding something of living systems, we have to define a locus as a finite extended patch and the extension depends on both situation and task. Thus, a local patch of (visual) attention has to be scalable. The scaling function of wavelet theory or blob hierarchy [37] plays the role of a regularization operator by integration on a finite set of signals. But we have to consider two aspects. Each structure has at least one intrinsic scale. Because of the hierarchy of concepts of structure, there are normally some. The observer should be able to adapt the scale of his local patch of interest with the chance of getting one or several unambiguous interpretations. In figure 1, there are several levels of interpretation. In dependence of the purpose of the observer and of the aperture function used, rather different interpretations can be found. The canonical coordinate frame of pointwise regular sampling in linear signal theory (sampling theorem) does no longer fit such strategy of sampling with extended operators. Now, estimation theory has to be considered to get high significance using bloblike sampling without loosing resolution as metric property or failing in estimation of dimension as topologic property.

**Local Intrinsic Dimensionality of Data** In the example of figure 1, also the local dimension of the structure dramatically changes if both different apertures or different conceptions are used. A visual sensory system will interpret the data of local patches with respect to their local dimension because it strongly correlates with local symmetry. A set of measured signals (1), e.g. as a result of corner detection, may be interpreted as examples, taken from entities as in (2). By assuming that these entities are disturbed by noise, a more reasonable interpretation may be given by (3). From step (2) to (3), a change of the dimensionality conception happened. While in (2) a sequential process of small aperture induces an one-dimensional contour, in (3) a two-dimensional patch is assumed. In (3) all locations, within the patches may have the same meaning. Therefore, as a result of vector quantization, the points of (4) may suffice to represent the manifold. In (5), on a global level these representations again may be fused by an one-dimensional conception. As an implicit assumption of this discussion, we used the definition of the local intrinsic dimension of an entity as the number of degrees of freedom which suffice the chosen criterion. While in (2) the sequential process induces an one-dimensional contour, its embedding in the plane as a polygon necessitates the assumption of maximal intrinsic dimension two (see also chapter 4). In (2), the contour is constituted by fusion of multiple one–dimensional components. Therefore, any patch covering a corner can only represent the detected local geometry in a two–dimensional base, as the outer product of two one–dimensional bases. The interpretation of (noised) data with respect to their intrinsic local dimension has to separate between the intrinsic dimension of the manifold and the dimensional aspect, induced by noise. Both are dependent of the used aperture. By vector quantization, the dimensionality orthogonal to that of the manifold may be surpressed. In [13] an interesting approach of the estimation of the local intrinsic dimensionality is presented which is based on the KLT of vector quantized optimally topology representing sets of sampling points. Interestingly, this approach relates pattern recognition with signal theory.

**Global Partial Reconstruction using Wavelet Nets** The third problem is related to the global reconstruction from local hints. The linear signal theory can from algebraic reasons only support regular sampling in a compact way. Wavelet theory elegantly relates the distance of regular sampling with the scaling factor of the wavelet functions. In the wavelet transform at each sampling position, the signal will be mapped to a complete set of functions. From these projection coefficients, the original signal may be reconstructed. But with respect to the economy of resources, this strategy is very dumb. In contrast to that, irregular sampling makes possible the adaption of sampling positions to interesting structures. The choice of positions of interest may be based on the signal structure and/or on the interest of the observer. In our group, a drastic modification of wavelet transform in direction of a wavelet net has been developed [55]. The wavelet basis functions are coupled to irregular sampling points of the wavelet net. Scale and orientation are continuous adaptively controlled. In very contrast to the wavelet transform, there is a need for a minimum of irregular sampling positions and a minimum of basis functions at the sampling points. Indeed, only one wavelet per sampling point is used. This one has to be globally optimized with respect to the wavelets at the other points. As result of global optimization, from a purposive controlled set of sampling points an image of partial reconstruction raises up from an extremely sparse code. The purpose of the task not only influences the locations of the sampling points, but also the scheme of regularization in the optimization process.

In figure 2, we demonstrate the result of putting about 30 sampling points on the right eye (top left) or 90 sampling points on both eyes and on the mouth (top right) as most prominent regions of a face. The results of reconstruction with the same sets of wavelets, coupled to these points, is shown in the bottom part of the figure. The left code needs 98 bytes and the right one needs 196 bytes for representation on a level of $256 \times 256$ pixels.

This example of global fusion from irregular distributed local hints demonstrates a philosophy of purposive minimal decomposition of the visual signal by a limited set of basis functions. Obviously, the goal is not complete global reconstruction but in the sense of qualitative vision the reconstruction of the necessary structures to suffice the task.

**Optimal Filter Control by Steerability** In the fourth problem we consider the filtering process in the attentive stage, i.e. if the operator is fixated at a keypoint of interest. In traditional signal processing, if the translation DOF is frozen, there is no other possibility than projecting the signal energy to the operator. But remembering the remaining degrees of freedom, that means rotation and scale, there should be the possibility of a dense sampling with respect to these DOF. Because coarse sampling may seriously mismatch the structure,

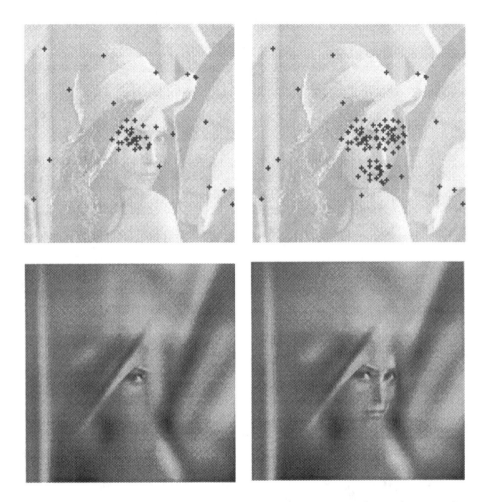

**Fig. 2.** Partial global reconstruction from local hints as purposive and progressive vision task.

dense sampling would be wanted. As a way to prevent the infinite amount of effort with respect to such processing, the steering filter scheme can be used. This consists in (now in contrast to the last problem) decomposition of the wanted filter into a small set of basis functions. These basis functions have the property of exact or approximate reconstruction of the wanted filter. Instead of filtering with an infinite set of (e.g. oriented) filters, only the projection of the local signal to the small set of basis functions (e.g. 10) will be done.

From that very limited set of projection coefficients, the response of the optimal adapted filter will be reconstructed by interpolation. Figure 3 shows a set of ten basis functions of an edge detection filter. Both the basis functions and the interpolation functions can be computed either from Lie group theory [46] or using SVD [47]. In the Lie group approach, the eigenfunctions of the

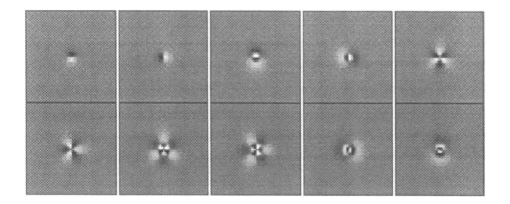

**Fig. 3.** Ten basis functions of an edge detection filter with orientation as DOF.

generating operator of the considered Lie group are the basis functions and the eigenvalues are the interpolation functions of the steering filter problem. In the SVD approach, the basis functions will be given by the right singular values, whereas the interpolation functions will be given by the left singular values.

Steering filters is a very powerful principle for local analysis of multidimensional signals. Coupled with the quadrature filter concept, the steering of filters results in signatures of the local energy and of the local phase as functions of the considered DOF.

**Attentive, Purposive and Progressive Visual Architecture** In problem number five, we want to propose a visual architecture for attentive, purposive, and progressive recognition.

This architecture has to follow the general needs of visual perception in a PAC system:

1. **fast** with respect to the time scale of the special PAC,
2. **flexible** with respect to the change of intention or purpose,
3. **complete** with respect to the task,
4. **unambiguous** with respect to the internalized conceptions or categories,
5. **selective** with respect to the importance of data,
6. **constructive** with respect to learning and recognition of equivalence classes.

From an algebraic point of view the Lie group approach is the canonical way for exactly local signal analysis (in multiple parameters of deformation). Because the generating operator of the considered Lie group generates local symmetry, this will be an invariance criterion for a bottom–up approach of grouping aggregated patterns from irreducible invariants. A scheme of an attentive visual system architecture will be shown in figure 4.

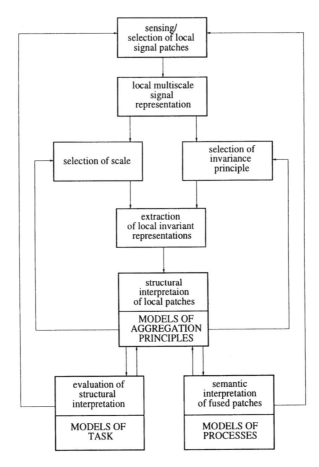

**Fig.4.** The cycles of an attentive processing scheme.

Noticeable is the organization of several cycles of processing. While the outer cycle indicates to the gaze control aspect, the inner cycle allows the consideration of several invariance principles for grouping of aggregated invariants from irreducible ones in dependence of the evaluation process with respect to the known aggregation results (coherent patterns of e.g. faces) and with respect to the purpose. This bottom–up process is top–down modulated or controlled. This is purposiveness. The recognition should also be progressive in the sense that the process should be stopped if a measure of confidence or evidence is sufficient high. This principle of economy of time requires support by the matching process. The template in this frame is the equivalence class which is only a purposively constrained approximation of the ideal one. With other words, the template should be decomposable to allow a gradual increase of matching results. The principle of progressive recognition is inherent both to the wavelet net approach of reconstruction and to the steering filter approach of recognition. In the first case, the sampling points are weighted by a factor of importance. Therefore, the reconstruction can be stopped if wanted. In the case of steering filters, a large set of basis functions may be computed offline. This ordered set will be sequentially used. In figure 5, the proposed and partially realized visual architecture of purposive progressive recognition will be shown.

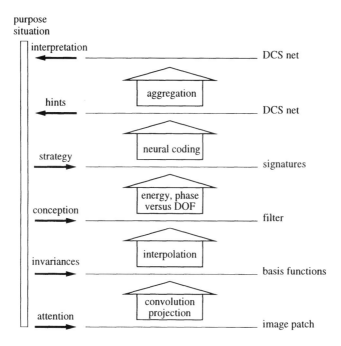

**Fig.5.** Layered architecture for purpose, progressive visual recognition.

On the left side, we see the communication with the control unit of purpose and task. To the right, we see a hierarchy of structure levels which will be passed by bottom–up driven recognition. The procedure starts with a mapping of the signal to irreducible invariants which correspond to Lie group eigenvectors as symmetry primitives and which are the basis functions of filters. These filters may be any complex templates as aggregates of the irreducible invariants, and indeed they are steerable filters [47]. Their output can be evaluated or/and associated in a cascade of DCS nets [14] for recognizing more and more complex patterns. This process of aggregation is modulated from top–down.

This bottom–up scheme, embedded in the cyclic scheme of fig. 4, seems to be too slowly with respect to the summarized needs. But first, the projection to the basis functions and several other steps of processing can be done in a parallel procedure. Second, in the non–competent phase, respectively in the attentive phase, this sequential scheme may be adequate. In the competent phase, respectively in the preattentive phase, the pathway of recognizing the grandmother should be engraved to shortcut quickly the grandmother cells with her visual pattern.

**Lie Group Approach** The Lie group approach is the general method to design the perception–action cycle as bottom–up process. It may constitute the local algebraic frame for both recognition and forming patterns in space–time from a differential viewpoint. Because local patterns have local symmetry of low dimension, the task of recognizing or forming of smooth patterns by a sequential process will be locally supported. Within that frame, the equivalence between recognition and action can be seen most obviously. To follow a chosen conception means selection of the corresponding Lie group and the change of conception in case of any events means change of the Lie group, or at least their parameters. As an example, in fig. 1, case 2, the polygons result from a sequence of the translation group and the rotation group.

Although Lie algebra and Lie group theory for a long time is known in the community, its breakthrough is missed. There is a lot of relevant papers in computer vision [29, 63]. But in robotics [17, 51] and neural computation [54] the attention is very limited yet.

Both the local and the global algebraic frames can be fused with success [23] because every Lie algebra corresponds a bivector algebra and every Lie group can be represented as a spin group.

## 4   PAC and Geometric Algebra

The special problems of PAC with respect to the geometric algebra are

- multi–dimensionality of visual signals from space–time,
- need of fast multilink actions in an invariant manner,
- nonlinearity of perception–action mapping,
- nonlinearity of recognition of higher–order correlations,
- need of a switch between strategies, conceptions, qualitative and quantitative aspects.

After working for two years on that field of geometric algebra, we can summarize the momentary expected benefits from geometric algebra:

- flexible change of interpretation frames of different dimension for multi–dimensional local signals [16],
- flexible change of interpretation frames between entities of different dimension of the Euclidean space [7],
- flexible change between projective, Euclidian, and motor space (of kinematics) [9],
- transformation of nonlinear to linear problems [22],
- enrichment of representation and learning capabilities of neural nets [8],
- effective recognition of higher–order correlations.

We often observe a reduced complexity of symbolic coding of a problem. Of course, the numeric complexity often expands on computers which do not know the type of quaternions or so. Nevertheless, the answer depends on the problem, e.g. in conversion of nonlinear iterative solutions to linear straightforward ones.

In the following, we want to introduce into a special problem, which is of central importance for the design of PAC systems. This problem concerns the complete recovery of local geometry and the capturing of geometric relations in neural nets. We will show, how nonlinear processing of signals may be reduced to a linear one, without loss of information.

The designer of PAC systems has to guarantee that in principle all structures could be recognized, if necessary. This was not possible till now, without loosing the nice properties of the linear signal theory. The reason is based on the fact that until now we had no adequate signal theory for multidimensional signals. Textbooks refer to that problem with the statement that a two–dimensional local phase is not defined. Indeed, the deep reason is that, within the frame of complex numbers (the Fourier domain), there is no possibility to express all the symmetries of a two–dimensional signal. While one–dimensional signals with respect to a fixed position can be decomposed into an even and an odd component, two–dimensional signals can have any combination of both in each dimension.

Using a linear integral transformation as the Fourier transform, in the domain of complex numbers only even and odd symmetries can be represented. Thus, the Fourier transform in complex domain is only adequate to one–dimensional signals, although, as we know, the two–dimensional Fourier transform in complex domain is well defined.

In the case of using energy and phase as local spectral features, this limitation becomes obvious. To give an example, the recognition of an L–junction with steering filters [45] results in two peaks of the signature of energy. These have to be identified in a second step of non–linear processing, that means by application of a threshold to the energy. In contrast to that, complete linear processing would result in only one peak if the steering filter could represent all the symmetries of two–dimensional signals.

The geometric algebra offers the right frame for embedding the considered problem of representation of the Fourier transform of multidimensional signals in a higher–order domain of numbers without loosing the property of linearity. Indeed, a line as one–dimensional ideal structure (case 4 in fig. 1) may be represented as a vector, while any two–dimensional structure as e.g. the L–junction and any constant 2D–path (cases 2, respectively 3 in fig. 1) are represented by bivectors as the result of the outer product of vectors. The even subalgebra $G_3^+$ of $G_3$ equals the algebra of quaternions. Therefore, quaternions constitute the domain of embedding a two–dimensional Fourier transform without loss of information.

We show in [16] that even the analytic signal, the Hilbert transform, and the Gabor filter can systematically be embedded in this domain. Of course, the two–dimensional phase is now constituted by three components, each resulting from the combination of the real part with one of three separate imaginary parts. In figure 6, middle row, we show the four components of the quaternionic Gabor filter with (from left to right) real, i–imaginary, j–imaginary, and k–imaginary parts. These quaternionic Gabor filter components are estimators of symmetry conceptions, adequate to differential geometric curvatures, from raw data. Insofar, they are adequate to the bottom–up approach, embedded into the global algebraic frame. By combination of these components, the well known complex Gabor filter results and, in addition, several components, which are able to respond to the other symmetries, which are missed else.

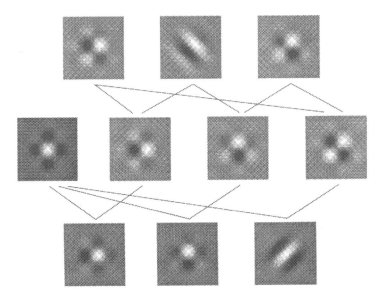

**Fig.6.** The quaternionic Gabor function (middle row) and their compositions to the symmetrie concepts of the plane.

The problem of recognition of structure in spatial domain is based on the recognition of the correlations with the corresponding symmetry. The Volterra series approach of filter design [62] recently was used to formulate non–linear filters for responding to higher–order correlations. In [38] the higher–order statistics of two–dimensional signals was estimated using a second–order Volterra filter in frequency domain to estimate the local intrinsic dimensionality of two–dimensional signals. In their approach, the authors had to extend the Fourier transform to 4D to respond the second–order Volterra approach.

Of course, the quaternionic quadrature filters are linear ones in contrast to the Volterra filters.

To summarize, the embedding of the analysis of multidimensional signals remains the linearity of operators, respectively remains the use of Fourier transform, but needs higher–order numbers. Any switch to lower dimensions is possible by combination of higher order numbers to lower order ones, just as we requested for a flexible vision system, embedded into the global algebraic frame.

We will finish our journey on linear signal processing by considering the functionality of neural nets. Of central importance in neural nets is the so–called linear associator, that means the functionality of a neuron, which linearly associates its input with the weights. This simple scalar product is based on the conception of linear vector algebra just as other linear methods of signal processing or recognition. The idea is to substitute the linear associator by a so–called geometric associator. That means to design a geometric algebra based neuron [8].

There have been several trials in the past, to algebraically extend neurons to complex numbers or quaternions. In [8], the general frame of geometric algebra with its multivector entities has been used with respect to MLP and RBF nets. In MLP nets, the central problem is to define an activation function which remains the locality of the backpropagated error function. This problem could be solved. Another aspect is the adequate representation of the input data to the used algebra. With this respect, we chose an outer product polynomial extension in accordance with the multiplication rules of the geometric algebra. Not completely to our surprise, both the convergence of learning as well as the generalization capability of the nets gained profit from that extension. Indeed, the used polynomial preprocessing is with some respect comparable to the well known design of higher–order nets (HONN, [50]), respectively to the Volterra connectionist model (VCM, [59]). But in these approaches, both the processing within the neurons and the polynomial extension were not algebraically correctly embedded, although they handle the polynomial extension also for multiple linear processing. The proposed so–called geometric neuron operates as multi–linear neuron on multivector entities. We hope to demonstrate in nearest future not only its capability of capturing correlations, but its useful application in real geometric problems.

# 5  Summary and Conclusions

In this paper, we discussed the socio–ecological metaphor as a base of the design of behavior based systems. But in contrast to the well known methodology of top–down designing a system from the model which we develop in our mind, this direct mapping between the functional behavior of the system and its model does not work in the context of behavior based systems. Instead, the model of functionality which we gained by observation of the appearance of the behavior has to be inverted into a bottom–up approach for the design of the PAC. This bottom–up approach implies both the importance of signal processing and learning of the competence.

Therefore, we concluded that both a global frame and a local one for algebraic embedding are requested to fulfill all the needs of the fusion of computer vision, neural computation, signal theory, and robotics. These global frames have been motivated and identified as geometric algebra and Lie theory. Finally, we discussed several aspects of the application of both frames. We demonstrated the usefulness with respect to recovery of geometry from signals by filtering and neural net processing.

As a result of this discussion of the task of designing behavior based systems, both the chances for general scientific progress as well as the amount of work, which has to be done, will become obvious. Only by concentration of the power and the experience of different disciplines, and only by considering the theoretical roots of the task, and growing support of basic research, we will successfully proceed in next future. This is in contrast to the actual interests of politics and industry, which want to support short–term development instead of long–term research.

# References

1. Aloimonos Y., Weiss I. and Bandopadhay A.: Active vision. Int. J. Comp. Vis. 7: 333–356. 1988
2. Aloimonos Y. (Ed.): Active Perception. Lawrence Erlbaum Ass., Hillsdale. 1993
3. Aloimonos Y.: Active vision revisited. In: [2], pp. 1–18
4. Aloimonos Y.: What I have learned. CVGIP: Image Understanding 60 (1): 74–85. 1994
5. Bajcsy R.: Active perception. Proc. IEEE 76 (8): 996–1005. 1988
6. Ballard D.H.: Animate vision. Artif. Intelligence 48: 57–86. 1991
7. Bayro–Corrochano E. and Lasenby J.: A unified language for computer vision and robotics. In: Proc. AFPAC'97. Kiel. G. Sommer and J.J. Koenderink (Eds.). LNCS Vol. XXX. Springer–Verlag. Heidelberg. 1997
8. Bayro–Corrochano E. and Buchholz S.: Geometric neural networks. In: Proc. AFPAC'97. Kiel. G. Sommer and J.J. Koenderink (Eds.). LNCS Vol. XXX. Springer–Verlag. Heidelberg. 1997
9. Bayro–Corrochano E., Daniilidis K. and Sommer G.: Hand–eye calibration in terms of motion of lines using geometric algebra. In: Proc. 10th Scandinavian Conference on Image Analysis. Lappeenranta. Finland. 1997
10. Benediktsson J., Sveinsson J., Ersoy O. and Swain P.: Parallel consensual neural networks. IEEE Trans. Neural Networks 8 : 54–64. 1997
11. Brooks R.A.: A robust layered control system for a mobile robot. IEEE Trans. Robotics and Automation 2 (1): 14–23. 1986
12. Bruske J., Ahrns I. and Sommer G.: Neural fuzzy control based on reinforcement learning and Dynamic Cell Structures. In: Proc. EUFIT, Vol. 2. 710–714. Aachen. 1996
13. Bruske J. and Sommer G.: An algorithm for intrinsic dimensionality estimation. In: Proc. CAIP'97, Kiel, Sept. 1997. LNCS Vol. XXX. Springer–Verlag. Heidelberg. 1997
14. Bruske J. and Sommer G.: Dynamic cell structure learns perfectly topology preserving map. Neural Computation 7 (4): 845–865. 1995
15. Bruske J., Hansen M., Riehn L. and Sommer G.: Adaptive saccade control of a binocular head with dynamic cell structures. In: Proc. ICANN'96. 215–220. LNCS Vol. 1112. Springer–Verlag. 1996
16. Bülow T. and Sommer G.: Multi–dimensional signal processing using an algebraically extended signal representation. In: Proc. AFPAC'97. Kiel. G. Sommer and J.J. Koenderink (Eds.). LNCS Vol. XXX, Springer–Verlag. Heidelberg. 1997
17. Chevallier D.P.: Lie algebras, modules, dual quaternions and algebraic methods in kinematics. Mech. Mach. Theory 26 (6): 613–627. 1991
18. Connell J.A.: A behavior–based arm controller. IEEE Trans. Robot. & Automation 5 (6): 784–791. 1989
19. Cutting J.E.: Perception with an Eye for Motion. MIT Press. Cambridge. 1986
20. Danielson D.A.: Vectors and Tensors in Engineering and Physics. Addison–Wesley Publ.. Redwood City. 1992

21. Daniilidis K.: Fixation simplifies 3D motion estimation. Comp. Vis. and Image Understanding. 1996. to appear

22. Daniilidis K. and Bayro–Corrochano E.: The dual quaternion approach to hand–eye calibration. In: Proc. 13th ICPR, Vienna, Vol. A. 318–322. ICPR. 1996

23. Doran C., Hestenes D., Sommen F. and Van Acker N.: Lie groups as spin groups. J. Math. Phys. 34 (8): 3642–3669. 1993

24. Eccles J.C.: Evolution of the Brain: Creation of the Self. Routledge. London and New York. 1989

25. Eigen M. and Schuster P.: The Hypercycle — a Principle of Natural Self–Organization. Springer-Verlag. Heidelberg, New York. 1979

26. Faugeras O.: Stratification of three–dimensional vision: projective, affine, and metric representations. J. Opt. Soc. Am. A 12 (3): 465–484. 1995

27. Furth H.G.: Piaget and Knowledge. Theoretical Foundations. Prentice Hall, Inc. Englewood Cliffs. 1969

28. Gibson J.J.: The Ecological Approach to Visual Perception. Houghton Mifflin. Boston. 1979

29. Gool van L., Moons T., Pauwels E. and Oosterlinck A.: Vision and Lie's approach to invariance. Image and Vision Computing 13(4): 259–277. 1995

30. Granlund G.H. and Knutsson H.: Signal Processing for Computer Vision. Kluwer Academic Publ. 1995

31. Hailu G. and Sommer G.: Learning from reinforcement signal. submitted. 1997

32. Haken H.: Synergetics. An Introduction. Springer-Verlag. Berlin. 1977

33. Hestenes D. and Sobczyk G.: Clifford Algebra to Geometric Calculus: A Unified Language for Mathematics and Physics. D. Reidel. Dordrecht. 1984

34. Hestenes D.: New Foundations for Classical Mechanics. Kluwer Academic Publ. Dordrecht. 1986

35. Koenderink J.J.: Wechsler's vision. Ecological Psychology 4 (2): 121–128. 1992

36. Koenderink J.J.: Embodiments of geometry. In: A. Aertsen (Ed.). Advanced Neural Computers. 303–312. Elsevier Science Publ. Amsterdam. 1990

37. Koenderink J.J., Kappers A. and van Doorn A.: Local operations: the embodiment of geometry. In: G.A. Orban, H.H. Nagel (Eds.): Artificial and Biological Vision Systems. 1–23. Springer–Verlag. Berlin. 1992

38. Krüger G. and Zetzsche C.: Nonlinear image operators for the evaluation of local intrinsic dimensionality. IEEE Trans. Image Processing 5 (6): 1026–1042. 1996

39. Landy M.S., Maloney L.T. and Pavel M. (Eds.): Exploratory Vision. The Active Eye. Springer-Verlag. New York. 1996

40. Lorenz K.: Evolution and Modification of Behavior. Merhuen. London, 1966

41. Luong Q.T. and Vieville T.: Canonic representations for the geometries of multiple projective views. In: J.O. Eklundh (Ed.) Proc. ECCV'94. 589–599. LNCS Vol. 800. Springer–Verlag. Berlin. 1994

42. Malsburg von der C.: Considerations for a visual architecture. In: R. Eckmiller (Ed.). Advanced Neural Computers. 303–312. Elsevier Science Publ. Amsterdam. 1990

43. Marr D.: Vision. W.H. Freeman. San Francisco. 1982

44. Mataric M.J.: Integration of representation into goal–driven behavior–based robots. IEEE Trans. Robot. & Automation 8 (3): 304–312. 1992

45. Michaelis M. and Sommer G.: Junction classification by multiple orientation detection. In: J.O. Eklundh (Ed.). Proc. ECCV'94. 101–108. LNCS Vol. 800. Springer–Verlag. Berlin. 1994

46. Michaelis M. and Sommer G.: A Lie group approach to steerable filters. Patt. Rec. Lett. 16: 1165–1174. 1995
47. Michaelis M.: Low Level Image Processing Using Steerable Filters. PhD Thesis. Technical Faculty. University of Kiel. Germany. 1995
48. Newell A. and Simon H.A.: Computer science as empirical enquiry: symbol and search. Commun. of the ACM 19: 113–126. 1976
49. Nicolis J.S.: Chaotic Dynamics Applied to Biological Information Processing. Akademie–Verlag, Berlin. 1987
50. Pao Y.H.: Adaptive Pattern Recognition and Neural Networks. Addison–Wesley. 1989
51. Park F.C., Bobrow J.E. and Ploen S.R.: A Lie group formulation of robot dynamics. Int. J. Robotics Res. 14 (6): 609–618.1995
52. Paton R. (Ed.): Computing with Biological Metaphors. Chapman and Hall. London. 1994
53. Pauli J., Benkwitz M. and Sommer G.: RBF networks for appearance–based object detection. In: Proc. ICANN'95, Vol. 1. 359–369. ICANN, Paris. 1995
54. Pearson D.W.: Linear systems equivalent to artificial neural networks via Lie theory. Neurocomputing 8: 157-170.1995
55. Pelc A.: Purposive Progressive Representation of Images using Wavelet Nets. Master Thesis (in German). Cognitive Systems Group. Computer Science Institute. University of Kiel. Germany. 1997
56. Pellionisz A. and Llinas R.: Tensor network theory of the metaorganization of functional geometries in the central nervous system. Neuroscience 16 (2) : 245–273. 1985
57. Popper K.R.: Objective Knowledge. The Clarendon Press. Oxford, 1972
58. Porteous I.R.: Clifford Algebras and the Classical Groups. Cambridge University Press. 1995
59. Rayner P.J.W. and Lynch M.R.: A new connectionist model based on a nonlinear adaptive filter. In: Proc. IEEE Int. Conf. Acoust., Speech & Signal Process. Vol. 2. 1191–1194. 1989
60. Rooney J.: On the three types of complex numbers and planar transformations. Environment and Planning B5: 89–99. 1978
61. Sattinger D.H. and Weaver O.L.: Lie Groups and Algebras with Applications to Physics, Geometry, and Mechanics. Springer–Verlag. New York. 1993
62. Schetzen M.: Nonlinear system modeling based on the Wiener Theory. Proc. IEEE 69 (12): 1557-157
63. Segman J. and Y.Y. Zeevi: Image Analysis by wavelet–type transforms: group theoretical approach. J. Math. Imaging and Vision 3: 51–77. 1993
64. Sommer G., Bayro–Corrochano E. and Bülow T.: Geometric algebra as a framework for the perception–action cycle. In: Proc. Int. Workshop Theor. Foundations of Computer vision. Dagstuhl. March 1996. Springer–Verlag. Wien. 1997
65. Tikhonov A.N. and Arsenin V.Y.: Solutions of Ill–Posed Problems. Wiley. New York. 1977
66. Tinbergen N.: The Study of Instinct. Oxford University Press. New York, 1951
67. Workshop on Visual Behaviors. Seattle. Washington. June 19, 1994, Proc.: IEEE Computer Society Press. Los Alamitos. 1994

# From Multidimensional Signals to the Generation of Responses

Gösta H Granlund

Computer Vision Laboratory, Department of Electrical Engineering,
Linköping University, S-581 83 Linköping, Sweden

**Abstract.** It has become increasingly apparent that perception cannot be treated in isolation from the response generation, firstly because a very high degree of integration is required between different levels of percepts and corresponding response primitives. Secondly, it turns out that the response to be produced at a given instance is as much dependent upon the state of the system, as the percepts impinging upon the system. The state of the system is in consequence the combination of the responses produced and the percepts associated with these responses. Thirdly, it has become apparent that many classical aspects of perception, such as geometry, probably do not belong to the percept domain of a Vision system, but to the response domain.

There are not yet solutions available to all of these problems. In consequence, this overview will focus on what are considered crucial problems for the future, rather than on the solutions available today. It will discuss hierarchical architectures for combination of percept and response primitives, and the concept of combined percept-response invariances as important structural elements for Vision. It will be maintained that learning is essential to obtain the necessary flexibility and adaptivity. In consequence, it will be argued that invariances for the purpose of vision are not geometrical but derived from the percept-response interaction with the environment. The issue of information representation becomes extremely important in distributed structures of the types foreseen, where uncertainty of information has to be stated for update of models and associated data.

## 1 Introduction

A fundamental problem is how to assemble sufficiently complex models and the computational structures required to support them. In order for a system modeling a high structural complexity to be manageable and extendable, it is necessary that it exhibits modularity in various respects. This implies, for example, standardized information representations for interaction between operator modules. One way to satisfy these requirements is to implement the model structure in a regular fashion as a *hierarchy*, although we should bear in mind that the communication need not be restricted to adjacent layers of such a hierarchy.

We can distinguish between two different types of hierarchies:

— *Scale hierarchies*

— *Abstraction hierarchies*

Most of the work on pyramids so far has dealt with size or scale, although they have indirectly given structural properties. They will not be dealt with in this paper, but descriptions can be found in [6,15,21].

Granlund introduced an explicit abstraction hierarchy [4], employing symmetry properties implemented by Gaussian envelope functions, in what today is commonly referred to as Gabor functions or wavelets [2]. An *abstraction hierarchy* implies that the image can be considered as an expansion into *image primitives*, which can be viewed as conceptual building blocks forming the image. In this concept lies the assumption of a hierarchy, such that building blocks at a lower level form groups which constitute a single building block at a higher level. Building blocks at the two levels are viewed as having different *levels of abstraction*.

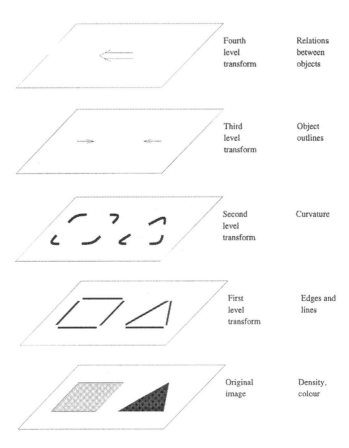

**Fig. 1.** Conceptual representation of an image as an abstraction hierarchy.

As an example we can look at Figure 1, which suggests a particular set of abstraction levels. At the lowest level we assume the image itself, describing a distribution of density and possibly color. At the second level we have a description of line and edge elements. At the third level we may have a description of curvature, or convexity and concavity. At the fourth level we may have outlines. At the fifth level we may have relations between objects, and continue to higher levels as appropriate.

It is necessary to omit the important discussion of mathematical representations for information in this document. It turns out that for 2-D information a *vector* representation is advantageous, while for 3 dimensions and higher, *tensors* will do the work. These representations allow the use of certainty statements for all features, which can be updated with respect to models and data. For further details, reference has to be made to [6].

In the same way that context information affects our interpretation of a more local event, information at a higher level can be used to control the processing at a lower level. This is often referred to as *top-down* processing. A hierarchical structure allowing this is illustrated intuitively in Figure 2.

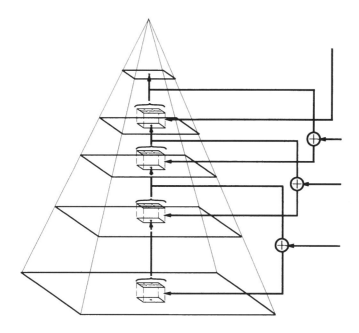

**Fig. 2.** A hierarchical structure with bottom-up and top-down flow of information.

In such a structure, processing proceeds in a number of stages, and the processing in one stage is dependent upon the derived results at higher levels. This leads to a model based analysis, where models are assembled from combinations of primitives from several levels. An important property is that these models do not remain constant, but adapt to the data, which can be used for adaptive filtering in multiple dimensions [8]. This is a very important issue which, however, goes beyond the objectives of this document, and reference has to be made to [6].

# 2 Similarity Representations for Linked Structures

Most information representation in vision today is in the form of arrays. This is advantageous and easily manageable for stereotypical situations of images having the same resolution, size, and other typical properties equivalent. Increasingly, various demands upon flexibility and performance are appearing, which makes the use of array representation less obvious.

The use of actively controlled and multiple sensors requires a more flexible processing and representation structure, compared to conventional methodology. The data which arrives from the sensor(s) can be viewed as patches of different sizes, rather than frame data in a regular stream and a constant array arrangement. These patches may cover different parts of the scene at various resolutions. Some such patches may in fact be image sequence volumes, at a suitable time sampling from a particular region of the scene, to allow estimation of the motion of objects. The information from all such various types of patches has to be combined in some suitable form.

The conventional array form of image information is in general impractical as it has to be searched and processed when some action is to be performed. It would be desirable to have the information in some suitable, partly interpreted form to fulfill its purpose to rapidly evoke actions. The fact that the information has to be in some interpreted form, implies that it should be represented in terms of content or *semantic* information, rather than in terms of array values. As we will see, content and semantics implies relations between units of information or symbols. It is consequently attractive to represent the information as *linked objects*. The discussion of methods for representation of objects as linked structures will be the subject of most of this document, and we can already now observe how some important properties of such a representation relate to that of conventional array representations:

- The array implies a given size frame, which can not easily be extended to incorporate a partially overlapping frame
- Features of interest may be very sparse over an array, leaving a large number of unused positions in the array

There is a great deal of literature available on the topic of object representation using classical methods[7], which however will not be reviewed here. Most of these methods treat objects with respect to geometric properties expressed in some coordinate system. They relate to rules of interpretation, which should be input into the system. This is probably appropriate for the representation of low level properties. For higher level properties the requirements are different.

## 2.1 Continuity, Similarity and Semantics

In the world around us, things generally appear different, whether they are or not. A particular object will appear different seen from different angles. Still we can recognize most objects at arbitrary positions, orientations, distances, etc. An object which persists in appearing different from anything else we know, can not be matched to any known class, which is a common purpose of recognition. There have to be aspects which are sufficiently familiar, for us to start a process of recognition. For that reason we will be interested in the *simultaneous appearance of similarity and difference* of properties of objects. This is related to notions which will be termed *invariance* and *equivariance* for future discussions.

Along the lines spelled out earlier, we can view the detection of invariance as the result of a test establishing similarity, while equivariance is the result of a test establishing a distance measure.

The representation of information in a cognitive system is crucial for effective performance. Traditionally, the representation is in the form of natural language, which has the following less desirable properties:

- Discrete and discontinuous: Similarity is established by matching, and the result is MATCH or NO MATCH
- Non-metric: It is not possible to establish a degree of similarity or distance between symbols

As an example we can take the words:

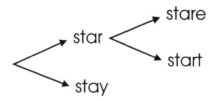

**Fig. 3.** Example of words having a small distance in terms of an ASCII letter metric, but large distances in content or meaning

Establishing a similarity measure e.g. using their ASCII numerical value would not be useful. Such a representation can not be used for efficient processing of semantic or cognitive information.

We can conclude that a suitable representation for semantic information requires:

*Continuous representation of similarity in content*

In the preceding case with words, we can observe that we deal with two types of adjacency:

– Time or position adjacency between words
– Content adjacency or similarity in meaning between words

It is apparent that both of these distance measures have to be represented in the description, although this is not the case in the example above. It is fairly obvious what the space or time distance represents, but what about the similarity in property? We will adress this question in the next section.

## 2.2 Channel Representation of Information

A representation of similarity assumes that we have a distance measure between items. For an advanced implementation of a linkage structure, we assume that information is expressed in terms of a *channel representation*. See Figure 4.

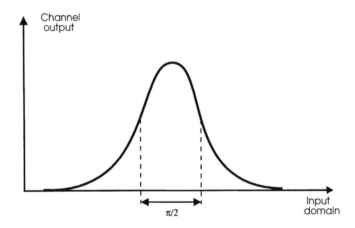

**Fig. 4.** Channel representation of some property as a function of match between filter and input pattern domain

Each channel represents a particular property measured at a particular position of the input space. We can as a first approximation view such a channel as the output from some band pass filter sensor for some property. If we view the channel output as derived from a band pass filter, we can first of all establish a measure of *distance* or *similarity* in terms of the parameters of this filter. See Figure 4. For a conventional, linear simple band pass filter, the phase distance between the flanks is a constant $\pi/2$. Different filters will have different band widths, but we can view this as a standard unit of similarity or distance, which is supplied by a particular channel filter.

This resembles the function of biological neural feature channels. There are in biological vision several examples available for such properties; edge and line detectors, orientation detectors, etc. Such a channel filter has the characteristic that it is local in some input space, as well as local in some property space.

It may map from some small region in the visual field, and indicate, say, the existence of a line at some orientation.

We can view every channel as an originally independent fragment of some space. It should now be observed that we have two different types of distances:

– *Distance in input space*
– *Distance in property space*

The distance in input space may be the distance between two different positions of a line within the receptive field of an orientation detector, where the line has a constant orientation.

The distance in property space may be the difference between two different orientations of a line, located centrally within the receptive field of an orientation detector.

A variation of position or a variation in orientation will both of them give a variation of the output according to Figure 4, and a priori, we cannot distinguish between these two situations, having a single and simple stimulus acting upon a single orientation detector. Either a line at the proper orientation can be moving over the spatial region of the detector, or a line at the proper position can be rotating over the detector, or most likely a combination of both.

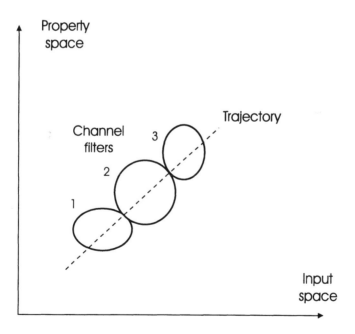

**Fig. 5.** Visualization of channels in input space as well as property space

This leads us to consider the input space and the property space as two orthogonal subspaces, which in the general case both will contribute to the output in some linear combination. See Figure 5. The distance represented by the channel filter will be in a linear combination from both of these spaces. Distance is a property which is well defined in a multidimensional space. Distance does not allow us to order events, but to define a sequence of events, represented by nodes which are joined by links. Every such link will represent a different one-dimensional projection from the multidimensional space under consideration, than a joining link.

The fact that we can view the phase distance between two adjacent channel peaks as $\pi/2$, implies that we can view the two subspaces as orthogonal in the metric defined by the joining band pass filter. Still these subspaces are parts of some larger common vector space.

This means that the subspaces which relate to each filter output are different, and cannot really be compared in the same two-dimensional projection plane, as suggested in Figure 5. Each subspace can for intuitive visualization be represented as a vector, which is orthogonal to its nearest neighbor subspaces. This is illustrated in Figure 6. As can be seen from Figure 6, the vectors are tapering off from the center. This indicates that while adjacent subspaces are orthogonal, we cannot say much about the relation between vector subspaces at a larger distance. What we can assume is that the subspaces "bend" into other parts of the common vector space, which makes them disappear from the horizon of any given vector subspace. This can be viewed as a curvature of the local space around a particular subspace, or as a windowing effect. As such, it may well be a necessary introduction of locality providing a stabilizing effect for the resulting networks, much like lateral inhibition.

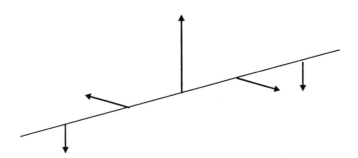

**Fig. 6.** Representation of channels as orthogonal subspaces

## 2.3 Implications of Multiple Measurements

From the previous section it follows that similarity is measured and valid along a single, one-dimensional subspace only, given the output from one single channel. For a particular object, there will be different distance measures to another particular object, in terms of different properties. In addition, any two successive links may not necessarily represent distances along the same, one-dimensional subspace, as we have no way to track what filters are involved where. There is consequently no way to order objects unambiguously for two different reasons:

1. There is no way to order items which are defined in a multi-dimensional space, which is the *Curse of Multi-dimensionality* [6]
2. It is not possible to sort objects with respect to a particular property, as a similarity between subspaces of different filters can never be established

The potential possibility to sort objects with respect to similarity, to produce a list is consequently not available. The fact that we have different measures of distance between two objects implies that we can represent the objects as points in a sufficiently high dimensional, common space. See Figure 7.

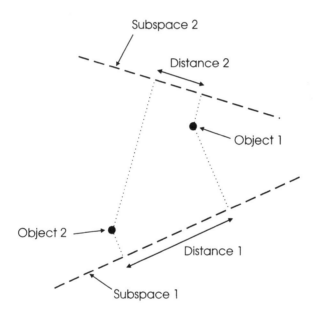

**Fig. 7.** Distance between two objects measured with two different probes, implying projections upon two different subspaces

### 2.4 Representation Using Canonical Points

It is postulated that we do not observe the world continuously although it may appear so to us. Rather observations and representations are made in particular, discrete points. We call these *canonical points*.

It is postulated that canonical points relate to certain phases of the features output from the filters involved. It is postulated that canonical points correspond to phases $0, 180$ and $\pm 90$ degrees in outputs from these filters. Parenthetically, these values do as well correspond to the discrete eigensystems which are derived from observation operators used for continuous fields in quantum mechanics [10].

It is furthermore postulated that a representation at these characteristic phases gives us exactly the sampling resolution required to provide a sufficiently good description. This can be viewed as a variable sampling density controlled by the content of the image.

It is obvious that there has to be some discretization in the representation of objects and events, implying a certain limited resolution. What is stated here is that this resolution is directly related to the forms and scales of objects and events themselves, mediated by the measurement probes or filters involved. These canonical points will imply different things dependent upon the level and phenomenon under consideration, but in general be points of symmetry, etc. of objects. Canonical points represent what we will abstractly denote an *object*, which in everyday language can be a feature, an edge, a line, an object, an event, a position, a view, a sequence, etc. Every feature or object is provided at some level of the processing hierarchy by something equivalent to a filter. The implementation of this is apparent for low-level features, but we can find equivalent interpretations at higher levels.

# 3 Representation and Learning of Object Properties

Traditionally it has been assumed that knowledge could actively be input into systems by an operator through specification of rules and instances, an assumption underlying the classical AI approach. It has become increasingly apparent, however, that knowledge cannot be represented as prespecified, designated linkages between nodes, as suggested in common AI symbolic representation. The difficulties of the classical symbolic representation are that:

1. It requires that the available types of relations are predefined and already existing in the system, and that an external system keeps track of the development of the system such as the allocation of storage, and the labeling of information.
2. It requires an entity which can "observe labels and structure", and take action on this observation.

These external, centralized functions make it impossible to have the system itself organize its information.

## 3.1 Object Invariants Formed by Percept-Response Combinations

Over the years there has been an increasing interest in research on invariants, [22,11,12,14]. Most of the methods proposed treat invariants as geometric properties, the rules for which should be input into the system. Theoretical investigation of invariance mechanisms is certainly an important task, as it will give clues to possibilities and limitations. It is not likely, however, that more advanced invariants can be programmed into a system. The implementation of such invariance mechanisms in systems will have to be made through learning.

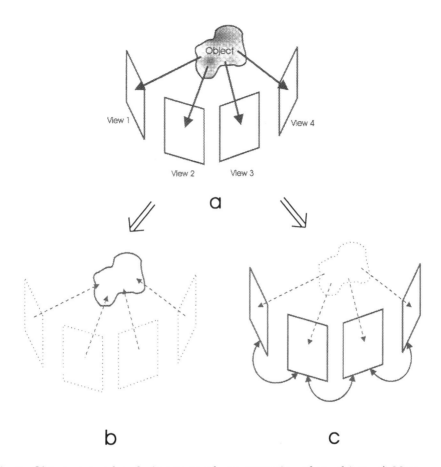

**Fig. 8.** Object-centered and view-centered representation of an object. a) Measurements produce information about different views or aspects of an object. b) Object-centered representation: The views are used to reconstruct a closed form object representation. c) View-centered representation: The views are retained as entities which linked together form a representation of the object

An important application of invariant representation is for object description. There are traditionally two major lines of approach which have been used for object description: *object-centered* and *view-centered* representation. See Figure 8.

An object-centered representation employs the information from a number of views to produce a composite geometrical object [7]. See Figure 8b. The image appearance of an object is then obtained using separate projection mappings. An important idea is that matching can be done more easily as the object description is independent of any viewpoint-dependent properties. A view-centered representation, on the other hand, combines a set of appearances of an object, without trying to make any closed form representation [27,24,1]. See Figure 8c.

We will not make any exhaustive review of the properties of these two approaches, or compare their relative advantages here, but only give some motivation for our choice of what is closer to the view-centered representation. A drawback of the object-centered representation for our purpose is that it requires a preconceived notion about the object to ultimately find, its mathematical and representational structure, and how the observed percepts should be integrated to support the hypothesis of the postulated object. Such a preconceived structure is not well suited for selforganization and learning. It is also a more classical geometrical representation, rather than a response domain related representation.

A view-linked representation on the other hand, has the advantage of potentially being self-organizable. There are also indications from perceptual experiments, which support the view-centered representation.

Continuing the discussion of the preceding section, it is postulated that we can advantageously represent objects as invariant combinations of percepts and responses. We will start out from the view centered representation of objects, and interpret this in the light of invariant combinations of percepts and responses.

In order for an entity to have some compact representation, as well as to be learned, it has to exhibit invariance. This means that there has to exist some representation which is independent of the frame in which it is described. The representation must not depend on the different ways it can appear to us. As discussed in the last section, the variation in appearance of views has to be directly related to responses we can make with respect to it.

There are different ways to interpret the combination of views to form an object:

$$View + Change\ in\ Position = Invariant$$
$$View + View\ Linkage \qquad = Invariant$$
$$View + View\ Linkage \qquad = Object$$

The combination of views with information concerning the position of these views, which is equivalent to the combination of percepts and responses, should produce an entity which allows an interpretation independently of the angle of observation. This is again equivalent to our notion of an object, as something which is not itself affected by the angle from which we view it.

As an example at a higher level, we can take a robot navigating in a room. The combination of detected corners and objects in the room, and the motion responses which are linking these corners together, constitutes an invariant representation of the room. The fact that a combination is an invariant, will make it interesting as a data object to carry on for further computations.

It is furthermore postulated that the invariance mechanism for the representation of an object as a combination of views and the responses involved, implies a form of *equivalence between structures* in the feature domain and in the response domain. We may say that for the domain of an object, we have a "balance", or an equivalence between a particular set of features and a particular response.

To emphasize, they are equivalent precisely because the combination of them forms an invariant; an entity whose variation is not perceivable in the combined percept-response domain interface surrounding this description. An invariant inevitably implies the balanced combination of a percept and a response. Thus a given response in a particular system state is equivalent or complementary to a particular percept.

Unless we have to postulate some external organizing being, the preceding must be true for all interfaces between levels where invariants are formed, which for generality must be for all levels of a system. This must then be true for the interface of the entire system to its environment as well. What this implies is that the combination of percepts that a system experiences and the responses it performs, constitute an invariant viewed from the wider percept-response domain. This in turn implies that the entire system appears as an invariant to the environment in this wider domain. To avoid misunderstanding, it has to be emphasized that a system can only be observed externally from its responses, and the effects in this subdomain are as expected not invariant, otherwise the system could not affect its environment.

## 3.2 Response as the Organizing Mechanism for Percepts

A vision system receives a continuous barrage of input signals. It is clear that the system cannot attempt to relate every signal to every other signal. What properties make it possible to select a suitable subset for inclusion to an effective linkage structure? We can find two major criteria:

1. Inputs must be sufficiently close in the input space where they originate, the property space where they are mapped and/or in time-space. This is both an abstract and a practical computational requirement: It is not feasible to relate events over too large a distance of the space considered. This puts a requirement upon the maps of features available, namely the requirement of *locality*.

2. A response or response equivalent signal has to be available, for three different reasons:
   - The first reason is to provide an indication of motive; to ascertain that there are responses which are associated to this percept in the process of learning.
   - The second reason is to provide a limitation to the number of links which have to be established.
   - The third reason is to provide an output path to establish the existence of this percept structure. Without a response output path, it remains an anonymous mode unable to act into the external world.

From the preceding we postulate that:

*The function of a response or a response aggregate within an equivalence class is to produce a set of inputs on its sensors, which similarly can be assumed to belong to a common equivalence class, and consequently can be linked.*

In consequence we propose an even more important postulate:

*Related points in the response domain exhibit a much larger continuity, simplicity and closeness than related points in the input domain. For that reason, the organisation process has to be driven by the response domain signals.*

Signal structure and complexity is considerably simpler in the response domain than in the percept domain, and this fact can be used as a focusing entity on the linkage process, where the system's own responses act as organizing signals for the processing of the input. There is a classical experiment by Held and Hein, which elegantly supports this model [9]. In the experiment, two newborn kittens are placed in each of two baskets, which are hanging in a "carousel" apparatus, such that they are tied together to couple the movements of the kittens. One of the kittens can reach the floor with its legs, and move the assembly, while the other one does not reach the floor and is passively towed along. After some period of time, the kitten which can control its movements develops normal sensory-motor coordination, while the kitten which is passively following the movements fails to do so until being freed for several days. The actively moving animal experiences changing visual stimuli as a result of its own movements. The passive animal experiences the same stimulation, but this is not the result of self-generated movements.

It is apparent that there is no basis for any estimation of importance or "meaning" of percepts locally in a network, but that "blind and functional rules" have to be at work to produce what is a synergic, effective mechanism. One of these basic rules is undoubtedly to register how percepts are associated with responses, and the consequences of these. This seems at first like a very limited repertoir, which could not possibly give the rich behavior necessary for intelligent systems. There is a traditional belief that percepts are in some way "understood" in a system, after which suitable responses are devised. This does however require simple units to have an ability of "understanding", which is not a reasonable demand upon structures. This is a consequence of the luxury of our own capability of consciousness and verbal logical thinking; something which is not available in systems we are trying to devise and in fact a capability which may lead us astray in our search for fundamental principles. Rather, we have to look for simple and robust rules, which can be compounded into sufficient complexity to deal with complex problems in a "blind" but effective way.

Driving the system using response signals has two important functions:

– *To simplify, learn and organize the knowledge about the external world in the form of a linked network*
– *To provide action outputs from the network generated*

It is necesssary that the network structure generated has an output to allow activation of other structures outside the network. This output is implemented by the linkage to response signals, which are associated with the emergence of the invariance class. If no such association were made, the network in question would have no output and consequently no meaning to the structure outside.

Driving a learning system using response signals for organization, is a well known function from biology. Many low level creatures have built in noise generators, which generate muscle twitches at an early stage of development, in order to organize the sensorial inputs of the nervous system. More generally, it is believed that noise is an important component to extend organization and behavior of organisms [13].

There are other important issues of learning such as representation of purpose, reinforcement learning, distribution of rewards, evolutionary components of learning, etc, which are important and relevant but have to be omitted in this discussion [16–19].

## 3.3 Object Properties – Part Percept – Part Response

In the tradition developed within the Vision Community, vision has been the art of combining percepts in a way that will describe the external world as well as possible for purposes of interacting with it. There has been an increasing awareness, however, that perception cannot be treated solely as a combination of perceptual attributes, in isolation from the response generation. As an example, it appears that many classical aspects of perception, such as geometry, most likely do not exclusively belong to the percept domain of a Vision system, but include the response domain. This is supported by recent reseach about the motor system, and in particular the cerebellum [25].

Invariance mechanisms are central in the description of properties for recognition and analysis. It can be seen as an axiom or a truism that only properties which are sufficiently invariant will be useful for learning and as contributions to a consistent behavior.

To start with, I would like to postulate that the following properties are in fact response domain features, or features dominated by their origin in the response domain:

- Depth
- Geometric transformations
- Motion
- Time

In the percept domain, a plane is not a distinguishable entity. We may perceive the texture of the plane, or we may perceive the lines which limit the plane, and these may be clues, but they do not represent the system's model of a plane. A learning system trying to acquire the *concept of plane*, has to associate perceptual and contextual attributes with a translational movement of the response actuator. This response movement can be a translation laterally along the plane, or it can be a movement in depth to reach the plane. This appears to be the invariance property of a plane, and this invariance property is not located in the percept domain but in the response domain. Similarly, the depth to that plane or any other structure will correspond to the movements required to reach it. For modeling and prediction, actual movements do not have to be effected, but the equivalent implementation signals can be shunted back into the structure, in a way discussed in the section on representation of time.

Similarly, it is believed that projective transformations are to a large extent response domain features. The reason is that they describe how the external world changes its appearance as a function of our movements in it. The primary step in that modeling is to relate the transformations to ego-motion. A secondary step is for the system to generalize and relate the transformations to a relative motion, be it induced by the system itself or any other cause. This is an important example of equivalence, but also an example of invariance. The system can learn the laws of geometrical transformation as a function of its own responses, and then generalize it to any situation of relative motion of the object in question.

An analogous analysis can be made with respect to motion, and the learning of this representation. Motion is the response domain representation of the material provided by various visual clues.

In the same way, the representation of *time* is postulated to be residing on the response side of the structure as well. What this means is that time is represented in terms of elements of responses. This makes it possible to give a network representation of the phenomenon of time, a representation which does not involve unit delays of some form, but is related to the scale of the time phenomena in a particular context. The network linkages allow us to represent time units in an abstracted way for predictive simulation as well, without actually implementing the responses. A further discussion of the representation of time will be given in Section 4.1.

## 3.4   Response and Geometry

In the proposed system response and geometry are equivalent. Responses always imply a change of geometry, and geometry is always implemented as the consequence of responses or something directly associated with a response. An associated change of geometry can be implemented as a response. We can view responses as the means to modify geometry in the vicinity of the system. What logically follows is that object and scene geometry is in fact represented as invariant combinations of objects and responses.

Relative position is a modular, scaled property, which is probably uni-directional. In addition, it is directly related to a particular movement. Relative position is directly related to a particular displacement, in the sequential representation of things. There is also a simultaneous parallel representation of things. In our model terms, it implies the shunting linkage between the two nodes, without an external action.

Geometry, position and response are all relative properties, which are defined by a description which is valid within a window of a certain size. This window corresponds to the band pass representation of some property filter.

# 4 Representation in Linked Structures

The conventional way to represent association in neural network methods is to use a covariance matrix. There are however some disadvantages with such a matrix structure for the representation:

- The matrix structure and size has to be determined before the learning process starts
- It is a centralized representation, which in turn assumes a centralized computational unit
- To sufficiently well define the matrix and its components, generally requires a large number of training samples
- A conventional covariance matrix does track the relation between points mapped, but it does not track typical dynamical sequences
- There will generally be a large number of zeroes for undefined relations

As a consequence, a closed form matrix organization is not feasible for self-organizing, extendable representation structures.

Rather, an attractive representation should be oriented towards *sparse representation*, and not be organized in terms of spatial coordinates, nor in terms of feature coordinates. It is also postulated that a fundamental property of effective representation structures is the ability of *effective representation of instances*. Many procedures in neural network methodology require thousands of training runs for very simple problems. Often, there is no apparent reason for this slow learning, except that the organisation of the learning does not take into account the dynamics of the process, and considers every data point as an isolated event. We know that biological systems often require only one or two examples for learning per item. The reason is that the succession of states is a very important restrictive mechanism for compact representation as well as fast learning. The system must consequently be able to learn from single or few instances as a base for the knowledge acquisition.

As a consequence of the preceding, it is postulated that it is more important to keep track of transitions between association states, than the actual association states themselves as static points.

For that reason it is postulated that the basis of the representation is one-dimensional trajectories linking these associated states, which have earlier been referred to as canonical points. The essential properties of *similarity* or *distance* can be represented by linkages implemented by operators stating this distance in a general form. The use of discrete points is a way to resolve the problem of scaling, in that it allows the linkage of items, regardless of distance between the objects in the external physical space. The canonical points are linked by a trajectory, which is basically one-dimensional, but may fork into alternative trajectories. The linkage can be given an intuitive representation according to Figure 9.

Two or more canonical points linked together can itself be represented by a canonical point in the actual resolution and the actual set of features. It can be

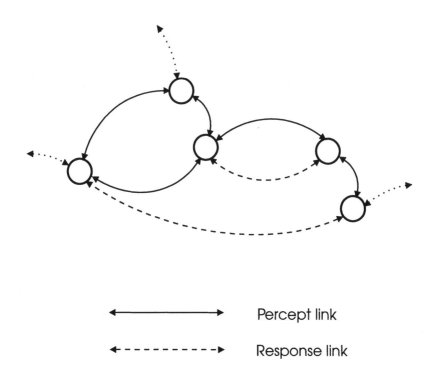

Percept link

Response link

**Fig. 9.** Intuitive illustration of linkage structure

viewed as an interval over which only one thing is happening at the level under consideration. It can, however, also be an aggregate of, or a sequence of canonical points at some level. We can view such a path as a single canonical point at a low resolution. As an example, we can take the walking along a corridor between two major landmarks. Between these two major landmarks there may well be other minor landmarks, but at a different level of resolution.

Another good argument for the representation of structures as locally one-dimensional trajectories is the *The Curse of Multi-dimensionality* [6]. This states that objects or events can not be unambiguously ordered in a space of dimensionality two or higher. The ordering sequence in multi-dimensional spaces is consequently either dependent upon rather arbitrary conventions set up, or they can be provided by contextual restrictions inferred by the spatial arrangements of items in the multi-dimensional space. This "curse" is of fundamental importance, and it has a number of effects in related fields. It is the reason that polynomials of two or more variables cannot, in general be factored. It is the reason why stability in recursive filters is very difficult to determine for filters with a dimensionality of two or higher. It is finally the reason why grammars or rule-based structures are inherently one-dimensional, although their effects may span a higher dimensionality as a result of combination, concatenation or symmetric application. This is what relates to the original issue.

An experimental system has been built up to test response driven learning of invariant structures using channel representation, with successful results [23].

## 4.1 Representation of Time

A crucial issue is how time should be represented in a percept/response hierarchy. The first likely choice would be to employ delay units of different magnitudes. This is a probable mechanism for low-level processing of motion. To use such delays at higher levels, implementing long time delays, has a number of problems as will appear in the ensuing discussion.

We postulate that:

*Time is represented only on the response side of the pyramid. Time is in fact represented by responses, as time and dynamics are always related to a particular action of physical motion. The linkage which is related to a particular time duration is mediated by actuator control signals expressing this duration. This allows long time intervals to be implemented as the duration of complex response actions.*

This gives us a consistent representation of displacement and of time. Time must be given a representation which is not a time sequence signal, but allows us to treat time like any other linked variable in the system. This is e g necessary as time sequence processes are to be compared. The model obtained is completely independent of the parameter scaling which generated the model. As there is not always a correspondence in time between percepts and the responses which should result, the equivalence relation must contain time as a link, rather than to match for equivalence or coincidence between the percept and the response for every time unit.

An important property of this representation is that it allows us to generate *predictive models* which allow simulations of actions in faster than real time. It is postulated that this is implemented as a direct shunting of response control signals, replacing those normally produced at the completion of a response action. See Figure 10. It is well known that there are such on-off shunts for output response signals in the nervous system, which are activated e g during dreaming. It is also believed that memory sequences can be processed at a much higher speed than real time, e g as they are consolidated into long term memory during REM sleep.

Another benefit is that something which is learned as a time sequential process, can later be represented as a completely parallel, time-delay independent model.

It appears that such an organization procedure goes in two steps:

1. The *establishment* of a model employs knowledge about the band pass adjacency between different features to build a model having the appropriate structure.
2. The *use* of a model assumes that features input to the model will exhibit the same adjacency pattern as before, although it is not tested for.

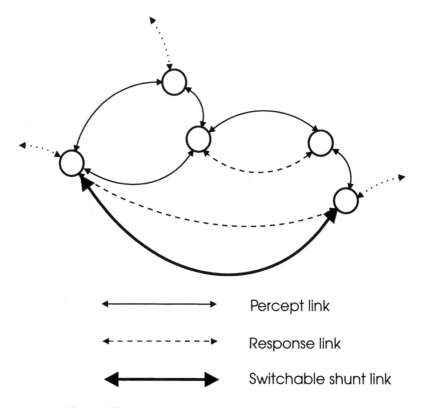

**Fig. 10.** Time representation for fast predictive modeling

The fact that adjacency is not crucial in the second case implies that a time sequential activation of features, in the same way as in the learning process, is no longer necessary to activate the model. Features can in fact be applied in parallel. While responses are inherently time sequential signals, we can still represent them in a time independent form as described earlier. This implies that we can activate model sets of responses in parallel. The execution of such response model sets then has to be implemented in some separate response sequencing unit. It appears likely that the *cerebellum* may have such a function [25].

Signals representing phenomena having dynamic attributes, such as responses, can be substituted by other signals giving equivalent effects. These equivalent signals will then take on the meaning of time, although they may be generated in a totally different fashion. A particular time delay is by no means necessary in the activation of the structure. This is why the dynamics of a movement can be translated into spatial distance. This is another illustration of the advantages of this coordinate free representation.

The preceding representation of time gives us in conclusion:

1. A flexible representation of time, which is scalable to the durations required and not dependent upon fixed delay units.

2. A representation of time, which is not itself represented as a time delay, but as a linkage like all other variables in the structure.

3. A linkage which can be routed over the response system for generation of actions, or be connected directly back for fast simulation and prediction.

# 5 The Extended Percept-Response Pyramid

The ultimate purpose of vision, or in fact all aspects of information processing, is to produce a response, be it immediate or delayed. The delayed variety includes all aspects of knowledge acquisition. This response can be the actuation of a mechanical arm to move an object from one place to another. The system can move from one environment to another. It can be the identification of an object of interest with reference to the input image, a procedure we customarily denote classification. Another example is enhancement of an image, where the system response acts upon the input image (or a copy of it) in order to modify it according to the results of an analysis. In this case, the input image or the copy is a part of the external world with respect to the system, upon which the system can act.

A major problem in the implementation of an effective vision structure is that the channel between the analysis and the response generation parts traditionally is very narrow. This implies that the information available from the analysis stage is not sufficiently rich to allow the definition of a sufficiently complex response required for a complex situation. It has also become increasingly apparent that perception cannot be treated in isolation from the response generation, firstly because a very high degree of integration is required between different levels of percepts and corresponding response primitives. Secondly, it turns out that the response to be produced at a given instance is as much dependent upon the state of the system, as the percepts impinging upon the system. The state of the system is in consequence the combination of the responses produced and the percepts associated with these responses. Thirdly, it has emerged that many classical aspects of perception, such as geometry, probably do not belong to the percept domain of a Vision system, but to the response domain.

In view of this, we want to propose a different conceptual structure, which has the potential of producing more complex responses, due to a close integration between visual interpretation and response generation [5], as illustrated in Figure 11.

This structure is an extension of the computing structure for vision, which we have developed over the years [6]. As discussed earlier, the input information enters the system at the bottom of the processing pyramid, on the left. The interpretation of the stylized Figure 11 is that components of an input are processed through a number of levels, producing features of different levels of abstraction. These percept features of different levels, generated on the left hand side of the pyramid, are brought over onto the right hand side, where they are assembled into responses, which propagate downward, and ultimately emerge at the bottom on the right hand side. A response initiative is likely to emerge at a high level,

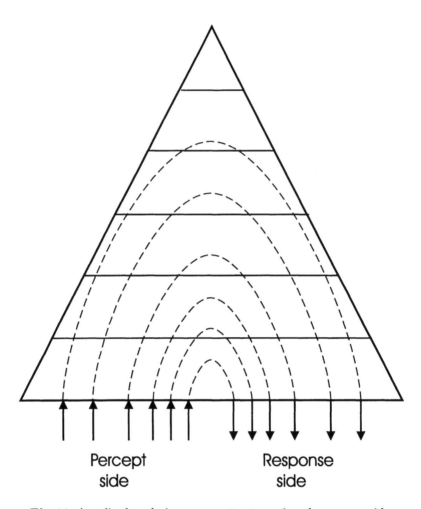

**Fig. 11.** A stylized analysis-response structure viewed as a pyramid .

from where it progresses downward, through stages of step-by-step definition. This is illustrated intuitively as percepts being processed and combined until they are "reflected" back and turned into emerging responses.

The number of levels involved in the generation of a response will depend on the type of stimulus input as well as of the particular input. In a comparison with biological systems, a short reflex arch from input to response may correspond to a skin touch sensor, which will act over interneurons in the spinal cord. A complex visual input may involve processing in several levels of the processing pyramid, equivalent to an involvement of the visual cortex in biological systems.

A characteristic feature of this structure is that the output produced from the system leaves the pyramid at the same lowest level as the input. This arrangement has particular reasons. We believe that processing on the percept side going upward in the pyramid, usually contains differentiating operations upon data which is a mixture between input space and property space. This means that variables in the hierarchical structure may not correspond to anything which we recognise at our own conscious level as objects or events. In the generation of responses on the right hand side, information of some such abstract form is propagated downward, usually through integrating operations. Only as the emerging responses reach the interface of the system to the external world, do they have a form which is in terms of objects as we know them. In conclusion, this is the only level at which external phenomena make sense to the system; be it input or output.

This mechanism has far-reaching consequences concerning programming versus learning for intelligent systems. Information can not be "pushed" directly into the system at a higher level, it must have the correct representation for this particular level, or it will be incomprehensible to the system. A more serious problem, is that new information will have to be related to old information, on terms set by the system and organized by the system. It will require the establishment of all attribute links and contextual links, which in fact define the meaning of the introduced item. It is apparent that information can only be input to a system through the ordinary channels at the lowest level of a feature hierarchy system. Otherwise it cannot be recognized and organized in association with responses and other contextual attributes, which makes it usable for the system.

In biological systems, there appear to be levels of abstraction in the response generation system as well, such that responses are built up in steps over a number of levels [20,26]. Arguments can be made for the advantage of fragmentation of response generation models, to allow the models to be shared between different response modes.

The function of the interior of the response part of the pyramid in Figure 11 can be viewed as a more general response action command entering from the top of the structure. This command is then modified by processed percept data input entering from lower levels, to produce a more context specific response command. This is in turn made even more specific using more local, processed lower-level input data.

A typical response situation may be to stretch out the hand towards an object to grasp it. The first part of the movement is made at high speed and low precision until the hand approaches the object. Then the system goes into a mode where it compares visually the position of the hand with that of the object, and sends out correcting muscle signals to servo in on the object. The grasping response can now start, and force is applied until the pressure sensors react. After this, the object can be moved, etc.

There is a great deal of evidence that this type of hierarchical arrangement is present also at higher levels of the cortex, where a response command is modified and made specific to the contextual situation present. The processing of the cerebellar structure performs some such coordinating, context sensitive response modification [25,3]. The structure discussed seems capable of building up sufficiently complex and data-driven responses.

So far the discussion may have implied that we would have a sharp division between a percept side and a response side in the pyramid. This is certainly not the case. There will be a continuous mixture of percept and response components to various degrees in the pyramid. We will for that purpose define the notion of *percept equivalent* and *response equivalent*. A response equivalent signal may emerge from a fairly complex network structure, which itself comprises a combination of percept and response components to various degree. At low levels it may be an actual response muscle actuation signal which matches or complements the low level percept signal. At higher levels, the response complement will not be a simple muscle signal, but a very complex structure, which takes into account several response primitives in a particular sequence, as well as modifying percepts. The designation implies a complementary signal to match the percept signal at various levels. Such a complex response complement, which is in effect equivalent to the system state, is also what we refer to as *context*.

A response complement also has the property that an activation of it may *not necessarily* produce a response at the time, but rather an activation of particular substructures which will be necessary for the continued processing. It is also involved in knowledge acquisition and prediction, where it may not produce any output.

**Acknowledgements**

The author wants to acknowledge the financial support of WITAS: The Wallenberg Laboratory for Information Technology and Autonomous Systems, as well as the Swedish National Board of Technical Development. These organisations have supported a great deal of the local research and documentation work mentioned in this overview. Considerable credit should be given to the staff of the Computer Vision Laboratory of Linkoeping University, for discussion of the contents as well as for text and figure contributions to different parts of the manuscript.

# References

1. D. Beymer and T. Poggio. Image Representations for Visual Learning. *Science*, 272:1905–1909, June 1996.
2. D. Gabor. Theory of communication. *J. Inst. Elec. Eng.*, 93(26):429–457, 1946.
3. J.-H. Gao, L. M. Parsons, J. M. Bower, J. Xiong, J. Li, and P. T. Fox. Cerebellum Implicated in Sensory Acquisition and Discrimination Rather Than Motor Control. *Science*, 272:545–547, April 1996.
4. G. H. Granlund. In search of a general picture processing operator. *Computer Graphics and Image Processing*, 8(2):155–178, 1978.
5. G. H. Granlund. Integrated analysis-response structures for robotics systems. Report LiTH-ISY-I-0932, Computer Vision Laboratory, Linköping University, Sweden, 1988.
6. G. H. Granlund and H. Knutsson. *Signal Processing for Computer Vision*. Kluwer Academic Publishers, 1995. ISBN 0-7923-9530-1.

7. W. E. L. Grimson. *Object Recognition by Computer: The Role of Geometric Constraints*. MIT Press, Cambridge, MA. USA, 1990.

8. L. Haglund, H. Knutsson, and G. H. Granlund. Scale and Orientation Adaptive Filtering. In *Proceedings of the 8th Scandinavian Conference on Image Analysis*, Tromsö, Norway, May 1993. NOBIM. Report LiTH–ISY–I–1527, Linköping University.

9. R. Held and A. Hein. Movement–produced stimulation in the development of visually guided behavior. *Journal of Comparative and Physiological Psychology*, 56(5):872–876, October 1963.

10. R. I. G. Hughes. *The structure and interpretation of quantum mechanics*. Harvard University Press, 1989. ISBN: 0-674-84391-6.

11. L. Jacobsson and H. Wechsler. A paradigm for invariant object recognition of brightness, optical flow and binocular disparity images. *Pattern Recognition Letters*, 1:61–68, October 1982.

12. K. Kanatani. Camera rotation invariance of image characteristics. *Computer Vision, Graphics and Image Processing*, 39(3):328–354, Sept. 1987.

13. L. C. Katz and C. J. Shatz. Synaptic activity and the construction of cortical circuits. *Science*, 274:1133–1138, November 15 1996.

14. J. J. Koenderink and A. J. van Doorn. Invariant properties of the motion parallax field due to the movement of rigid bodies relative to an observer. *Opt. Acta 22*, pages 773–791, 1975.

15. J. J. Koenderink and A. J. van Doorn. The structure of images. *Biological Cybernetics*, 50:363–370, 1984.

16. T. Landelius. Behavior Representation by Growing a Learning Tree, September 1993. Thesis No. 397, ISBN 91–7871–166–5.

17. T. Landelius and H. Knutsson. A Dynamic Tree Structure for Incremental Reinforcement Learning of Good Behavior. Report LiTH-ISY-R-1628, Computer Vision Laboratory, S–581 83 Linköping, Sweden, 1994.

18. T. Landelius and H. Knutsson. Behaviorism and Reinforcement Learning. In *Proceedings, 2nd Swedish Conference on Connectionism*, pages 259–270, Skövde, March 1995.

19. T. Landelius and H. Knutsson. Reinforcement Learning Adaptive Control and Explicit Criterion Maximization. Report LiTH-ISY-R-1829, Computer Vision Laboratory, S–581 83 Linköping, Sweden, April 1996.

20. R. A. Lewitt. *Physiological Psychology*. Holt, Rinehart and Winston, 1981.

21. L. M. Lifshitz. Image segmentation via multiresolution extrema following. Tech. Report 87-012, University of North Carolina, 1987.

22. J. L. Mundy and A. Zisserman, editors. *Geometric Invariance in Computer Vision*. The MIT Press, Cambridge, MA. USA, 1992. ISBN 0–262–13285–0.

23. K. Nordberg, G. Granlund, and H. Knutsson. Representation and Learning of Invariance. In *Proceedings of IEEE International Conference on Image Processing*, Austin, Texas, November 1994. IEEE.

24. T. Poggio and S. Edelman. A network that learns to recognize three-dimensional objects. *Nature*, 343:263–266, 1990.

25. J. L. Raymond, S. G. Lisberger, and M. D. Mauk. The Cerebellum: A Neuronal Learning Machine? *Science*, 272:1126–1131, May 1996.

26. G. M. Shepherd. *The Synaptic Organization of the Brain*. Oxford University Press, 2nd edition, 1979.

27. S. Ullman and R. Basri. Recognition by linear combinations of models. *IEEE Transactions on Pattern Analysis and Machine Intelligence*, 13(10):992–1006, 1991.

# Perception and Action Using Multilinear Forms*

Anders Heyden, Gunnar Sparr, Kalle Åström

Dept of Mathematics, Lund University
Box 118, S-221 00 Lund, Sweden
email: {heyden,gunnar,kalle}@maths.lth.se

**Abstract** In this paper it is shown how multilinear forms can be used in the perception-action cycle. Firstly, these forms can be used to reconstruct an unknown (or partially known) scene from image sequences only. Secondly, from this reconstruction the movement of the camera can be calculated with respect to the scene, which solves the so called hand-eye calibration problem. Then action can be carried out when this relative orientation is known. The results are that it is sufficient to either use bilinear forms between every successive pair of images plus bilinear forms between every second image or trilinear forms between successive triplets of images. We also present a robust and accurate method to obtain reconstruction and hand-eye calibration from a sequence of images taken by uncalibrated cameras, based on multilinear forms. This algorithm requires no initialisation and gives a generic solution in a sense that is clearly specified. Finally, the algorithms are illustrated using real image sequences, showing that reconstruction can be made and that the hand-eye calibration is obtained.

## 1  Introduction

The problem of reconstruction of an unknown scene from a number of its projective images has been studied by many researchers. Firstly, calibrated cameras where considered, see [13], making it possible to reconstruct the scene up to an unknown similarity transformation (Euclidean plus scale). The drawback of this approach is that the camera need to be calibrated before the image sequence is captured and when this have been done it is neither possible to change focus nor to zoom. Another drawback is that the algebra gets complicated because orthogonal matrices have to be used, giving 6 algebraic constraints on each such matrix.

Another approach is to model the projective transformation as an affine transformation, giving an affine reconstruction of the scene, see [21]. This approach gives a very simple algebraic formulation and the reconstruction and hand-eye calibration can be obtained using linear methods. The drawback is that the approximation, giving an affine camera, is often not accurate, e.g. when there are perspective effects present. When some action has to be made, the camera often has to move closer to the scene, giving large perspective effects, which makes this approximation very bad in perception-action tasks. Other similar attempts are orthographic projections, see [21] and paraperspective, see [15].

---

* This work has been supported by the Swedish Research Council for Engineering Sciences (TFR), project 95-64-222

Recently, methods requiring no camera calibration have become popular, [2,14,19]. These methods are based on projective geometry and give a reasonable algebra. The first result obtained is that it is only possible to reconstruct the object up to an unknown projective transformation, see [18,2,14,5]. This drawback is not always important, especially in perception-action tasks, because this kind of reconstruction, together with the head-eye calibration is often sufficient in order to take some action, e.g. obstacle avoidance, positioning, picking up objects, navigation, etc.

The first algorithms used the so called fundamental matrix, see [2], which is obtained from the bilinear form. Later on, higher order multilinear forms where introduced; trilinearities between triplets of views, see [16,6,3], and quadrilinearities between quadruples of views, see [22,8,3]. It soon became apparent that there was no need to go beyond quadrilinearities and after a while it was discovered that the trilinearities was sufficient to use. Recently, the algebraic and geometric properties of the bilinearities and the trilinearities have been exploited, see [10], and it has been shown that both geometrically and algebraically, the bilinearities are sufficient to use. These multilinear forms can be used both to reconstruct the scene and to obtain hand-eye calibration.

There has also been intensive research on more robust methods, based on the multilinear forms, see [23,20,11]. These iterative methods requires no initialisation step and gives optimal reconstruction in a sense that will be specified later. The advantage of these methods is that they use all images and all available point matches with equal priority, i.e. no image are selected as a reference image. They can easily be extended to recursive algorithms, taking more and more images into account as they become available, see [1].

In this paper, we will present reconstruction methods based on multilinear forms and iterative robust methods. It will be shown, both theoretically and in experiments that an accurate reconstruction and hand-eye calibration are obtained. In this way the perception-action problem is treated.

## 2 Camera Model

The image formation system (the camera) is modeled by the equation

$$\lambda \begin{bmatrix} x \\ y \\ 1 \end{bmatrix} = \begin{bmatrix} \alpha_x & s & x_0 \\ 0 & \alpha_y & y_0 \\ 0 & 0 & 1 \end{bmatrix} [\, R \mid -Rt\,] \begin{bmatrix} X \\ Y \\ Z \\ 1 \end{bmatrix} \quad \Leftrightarrow \quad \lambda \mathbf{x} = K[\, R \mid -Rt\,] \mathbf{X} \ . \quad (1)$$

Here $\mathbf{X} = [\, X\, Y\, Z\, 1\,]^T$ denotes object coordinates in extended form and $\mathbf{x} = [\, x\, y\, 1\,]^T$ denotes extended image coordinates. The scale factor $\lambda$, called the *depth*, accounts for perspective effects and $[\, R \mid -Rt\,]$ represents a rigid transformation of the object, i.e. $R$ denotes a $3 \times 3$ rotation matrix and $t$ a $3 \times 1$ translation vector. Finally, the parameters in $K$ represent intrinsic properties of the image formation system: $\alpha_x$ and $\alpha_y$ represent magnifications in the $x$- and $y$-directions in the light sensitive area, $s$ represents the skew, i.e. nonrectangular arrays can be modelled, and $(x_0, y_0)$ is called the principal point and is interpreted as the orthogonal projection of the focal point onto the image plane. The parameters in $R$ and $t$ are called *extrinsic parameters* and the parameters in

$K$ are called the *intrinsic parameters*. Observe that there are 6 extrinsic and 5 intrinsic parameters, totally 11, the same number as in an arbitrary $3 \times 4$ matrix defined up to a scale factor. If the extrinsic as well as the intrinsic parameters are unknown (1) can compactly be written

$$\lambda \mathbf{x} = P \mathbf{X} \ . \tag{2}$$

Since there is a freedom in the choice of coordinate system in the object, it is a general practice to chose $P_1 = [\, I \,|\, 0 \,]$, which will be done in the sequel. However, three degrees of freedom remains, since a projective change of coordinates by the matrix

$$\begin{bmatrix} 1 & 0 & 0 & 0 \\ 0 & 1 & 0 & 0 \\ 0 & 0 & 1 & 0 \\ a & b & c & d \end{bmatrix}$$

does not change $P_1 = [\, I \,|\, 0 \,]$. This ambiguity in the determination of the camera matrices corresponds to the fact that it is only possible to reconstruct the scene up to an unknown projective transformation. This ambiguity in representation will be called the *projective ambiguity*. It is furthermore convenient to use the notation

$$P_2 = [\, A \,|\, t \,], \quad P_3 = [\, B \,|\, u \,], \quad P_4 = [\, C \,|\, v \,] \tag{3}$$

for the first three camera matrices.

In the sequel we will assume that we have $n$ points (with known correspondences) in $m$ different images and that the intrinsic parameters are allowed to vary between the different imaging instants. Image coordinates in image $i$ are denoted by $\mathbf{x}_i$ and the camera matrix for image number $i$ will be denoted by $P_i$, i.e.

$$\lambda_i \mathbf{x}_i = P_i \mathbf{X}, \quad i = 1, \dots, m \ . \tag{4}$$

## 3 Multilinear Forms

Consider the equations in (4). These equations can be written

$$M u = 0 \ , \tag{5}$$

with

$$M = \begin{bmatrix} P_1 & \mathbf{x}_1 & 0 & 0 & \dots & 0 \\ P_2 & 0 & \mathbf{x}_2 & 0 & \dots & 0 \\ P_3 & 0 & 0 & \mathbf{x}_3 & \dots & 0 \\ \vdots & \vdots & \vdots & \vdots & \ddots & \vdots \\ P_m & 0 & 0 & 0 & \dots & \mathbf{x}_m \end{bmatrix}, \quad u = \begin{bmatrix} \mathbf{X} \\ -\lambda_1 \\ -\lambda_2 \\ -\lambda_3 \\ \vdots \\ -\lambda_m \end{bmatrix} . \tag{6}$$

Since $M$ has a nontrivial nullspace, it follows that

$$\text{rank}(M) \leq m + 3 \ . \tag{7}$$

The matrix $M$ in (6) contains one block with three rows for each image. Observe that all determinants of $(m + 3) \times (m + 3)$ submatrices of $M$ in (6) are *multihomogeneous* of degree $(1, 1, 1, \ldots, 1)$, that is of the same degree in every triplet of image coordinates.

**Definition 1.** The subdeterminants of size $(m + 3) \times (m + 3)$ from $M$ in (6) are called the *multilinear constraints*.

The multilinear constraints obtained from submatrices containing all rows corresponding to two images and one row from each of the other images are called the *bilinear constraints*. The bilinear constraints between image $i$ and image $j$ can be written as a product of $x$-, $y$- and $z$-coordinates in the other images and

$$b_{i,j}(x_i, y_i, z_i, x_j, y_j, z_j) := \det \begin{bmatrix} P_i & \mathbf{x}_i & 0 \\ P_j & 0 & \mathbf{x}_j \end{bmatrix} = 0 \ . \tag{8}$$

Since the first factors consists of projective coordinates, some combination of these projective coordinates has a nonvanishing product and the bilinear constraints are equivalent to the constraint in (8), which is sometimes called the *epipolar constraint*. These constraints can be written

$$\mathbf{x}_i^T F_{i,j} \mathbf{x}_j = 0 \ , \tag{9}$$

where $F_{i,j}$ is called the *fundamental matrix* between images $i$ and $j$.

Given at least 8 corresponding points in 2 images, it is possible to solve linearly for the entries in $F_{i,j}$ by using 9. Then it is possible to calculate the camera matrices $P_i$ and $P_j$ up to the projective ambiguity.

The multilinear constraints obtained from submatrices containing all rows corresponding to one image, two rows each from two other images and one row from each of the other images are called the *trilinear constraints*. The trilinear constraints between image $i$, $j$ and $k$ can be written as a product of $x$-, $y$- and $z$-coordinates in the other images and $6 \times 6$ subdeterminants of

$$\begin{bmatrix} P_i & \mathbf{x}_i & 0 & 0 \\ P_j & 0 & \mathbf{x}_j & 0 \\ P_k & 0 & 0 & \mathbf{x}_k \end{bmatrix} \ . \tag{10}$$

Again the first factors consists of projective coordinates, and some combination of these projective coordinates has a nonvanishing product, thus the trilinear constraints are equivalent to the constraints expressed by subdeterminants from (10).

The trilinear constraints are often expressed using the so called *trilinear tensor*, $\alpha$, with components defined by

$$\alpha_{i,j,k} = t_i(B)_{j,k} - u_j(A)_{i,k} \ , \tag{11}$$

using the notations in (3), see [17,6,8]. There turns out to be four linearly independent (in $\alpha_{i,j,k}$) trilinear constraints, obtained from subdeterminants of (10). This makes it possible to obtain four linear constraints on the coefficients of the trilinear tensor for each corresponding point in 3 images. Thus having 7 corresponding points in 3 images makes it possible to linearly recover the 27 components of the trilinear tensor and from them the camera matrices (up to the previous mentioned ambiguity).

The multilinear constraints obtained from submatrices containing two rows corresponding to each of three images and one row from each of the other images are called the *quadrilinear constraints*. The quadrilinear constraints between image $i$, $j$, $k$ and $l$ can be written as a product of $x$-, $y$- and $z$-coordinates in the other images and $7 \times 7$ subdeterminants of

$$\begin{bmatrix} P_i & \mathbf{x}_i & 0 & 0 & 0 \\ P_j & 0 & \mathbf{x}_j & 0 & 0 \\ P_k & 0 & 0 & \mathbf{x}_k & 0 \\ P_l & 0 & 0 & 0 & \mathbf{x}_l \end{bmatrix}. \tag{12}$$

Again the first factors consists of projective coordinates, and some combination of these projective coordinates has nonvanishing product, thus the quadrilinear constraints are equivalent to the constraints expressed by subdeterminants from (12).

The quadrilinear constraints are usually expressed using the *quadrifocal tensor*, $\beta$, with components, $\beta_{i,j,k,l}$, expressed in $A$, $t$, $B$, $u$, $C$ and $v$. There turns out to be 16 linearly independent (in $\beta_{i,j,k,l}$) quadrilinear constraints, see [22], obtained from subdeterminants of (12). This makes it possible to obtain 16 linear constraints on the coefficients of the quadrilinear tensor for each corresponding point in 4 images. Thus having 6 corresponding points in 4 images makes it possible to linearly recover the 81 components of the quadrilinear tensor and from them the camera matrices (up to the previous mentioned ambiguity).

Because of the large number of components of the trifocal and quadrifocal tensor (27 and 81 respectively) simplifications that reduces these numbers are often needed in calculations. One such simplification is the use of *reduced affine coordinates* in the images and in the object, see [9]. In this case three corresponding points are chosen as an affine basis in each image and all other coordinates used are affine coordinates with respect to these three basis points. Using this simplification one obtains the *reduced fundamental matrix*, the *reduced trifocal tensor*, and the *reduced quadrifocal tensor* containing, 6, 15 and 27 components respectively.

## 4 Dependencies between Multilinear Forms

It is obvious that (7) describes all available constraints on the camera matrices in an image sequence. These constraints can be expressed by bilinear, trilinear and quadrilinear constraints. All higher order multilinear constraints are just products of these three different types.

**Theorem 2.** *All constraints on the camera matrices in a sequence of images can be expressed in the bilinear, trilinear and quadrilinear constraints between all combinations of 2, 3 and 4 images respectively.*

Obviously, this large amount of multilinear constraints is unsatisfactory to work with. In order to reduce the number, assume that the trilinear constraints between every successive triplet of images is known, in the form of the trilinear tensor. Then the each triplet of camera matrices can be calculated up to the projective ambiguity. Then it is a simple task to adapt the whole sequence of camera matrices to each other and obeying all multilinear constraints.

**Theorem 3.** *All constraints on the camera matrices in a sequence of images can be expressed by the trilinear constraints between every successive triplet of images.*

The interesting question now is if we can reduce the representation to bilinear constraints. Introduce the notation

$$P_i = Q_i[I \,|\, t_i] \ . \tag{13}$$

Then the three bilinear constraints between images $i$, $j$ and $k$ can be expressed by

$$\mathbf{x}_i^T F_{i,j} \mathbf{x}_j, \quad \mathbf{x}_i^T F_{i,k} \mathbf{x}_k, \quad \mathbf{x}_j^T F_{j,k} \mathbf{x}_k \ , \tag{14}$$

with

$$F_{i,j} = Q_{j,i}^T T_{t_{j,i} \times}, \quad F_{i,k} = Q_{k,i}^T T_{t_{k,i} \times}, \quad F_{j,k} = Q_{k,j}^T T_{t_{k,j} \times} \ , \tag{15}$$

where the notation

$$Q_{i,j} = Q_i^{-1} Q_j, \quad t_{i,j} = Q_i(t_i - t_j), \quad T_{u \times} v = u \times v$$

have been used. Thus from the three bilinearities, $t_{i,j}$, $t_{i,k}$ and $t_{j,k}$ can be recovered up to scale. Since

$$Q_1^{-1} t_{12} - Q_1^{-1} t_{13} + Q_2^{-1} t_{23} = 0$$

a relative scale between these two vectors can be obtained (if they are not coplanar). Then the relative scale between $t_i$, $t_j$ and $t_k$ can be calculated and the only ambiguity present in the determination of the camera matrices $P_i$, $P_j$ and $P_k$ is the projective ambiguity, see also [12]. Note that this relative scale is naturally obtained from the trilinearities in (10) written as

$$\mathrm{rank} \begin{bmatrix} Q_i & Q_i t_i & \mathbf{x}_i & 0 & 0 \\ Q_j & Q_j t_j & 0 & \mathbf{x}_j & 0 \\ Q_k & Q_k t_k & 0 & 0 & \mathbf{x}_k \end{bmatrix} \leq 6 \ . \tag{16}$$

This shows that geometrically, the bilinearities between view $i$ and $i + 1$ and between view $i$ and $i + 2$ is sufficient to represent the multiple view geometry.

**Theorem 4.** *All constraints on the camera matrices in a sequence of images can be expressed geometrically by the bilinear constraints between image $i$ and $i + 1$ and between image $i$ and $i + 2$.*

Observe that this theorem is true under the hypothesis of general motion, i.e. three successive camera centers are not collinear.

Turning to the algebraic point of view, the problem is not so easy. The first difficulty is that the ideal defined by the trilinearities, $\mathcal{I}_t$, is not the same as the ideal defined by the bilinearities, $\mathcal{I}_b$, observed in [4]. For instance for three views three arbitrary points on the trifocal lines (the lines in the images connecting the epipoles from the other two images) in each image obeys all bilinear constraints, but not necessarily all trilinear ones. This corresponds to the fact that trilinearities can distinguish between different points on the trifocal plane, but the bilinearities can not. However, even if the ideals are different it is possible to calculate $\mathcal{I}_t$ from $\mathcal{I}_b$ in the following algebraic sense, see [10].

**Theorem 5.** *The bilinear ideal is reducible and has a primary decomposition, $\mathcal{I}_b = \mathcal{I}_t \cup \mathcal{I}_{tp}$, where $\mathcal{I}_{tp}$ is the ideal corresponding to an arbitrary point on each trifocal line.*

This theorem, together with the previous one shows that both geometrically and algebraically the bilinearities are sufficient to describe image sequences. However, it is not the best way to treat these problems numerically.

## 5  Reconstruction using a Subspace Approach

In order to reconstruct a scene from an image sequence, one can use the bilinear (or even trilinear) constraints to build up the sequence of camera matrices and the find the reconstruction by triangulation. A numerically better way is to user all available multilinear constraints in (7). Such a reconstruction algorithm will be outlined in this section.

### 5.1  Notations

Introduce the notation $(x_{i,j}, y_{i,j})$ for the coordinates of point number $j$ in image number $i$ and

$$\mathbf{x}_i = \begin{bmatrix} x_{i,1} & x_{i,2} & x_{i,3} & \dots & x_{i,n} \\ y_{i,1} & y_{i,2} & y_{i,3} & \dots & y_{i,n} \\ 1 & 1 & 1 & \dots & 1 \end{bmatrix}, \quad i = 1, \dots, m , \tag{17}$$

$$\mathbf{X} = \begin{bmatrix} X_1 & X_2 & X_3 & \dots & X_n \\ Y_1 & Y_2 & Y_3 & \dots & Y_n \\ Z_1 & Z_2 & Z_3 & \dots & Z_n \\ 1 & 1 & 1 & \dots & 1 \end{bmatrix}, \tag{18}$$

where $(X_j, Y_j, Z_j)$ denotes the coordinates of object point number $j$. Describe the depths, $\lambda_{i,j}$ in (1), of point $j$ in image $i$ by the diagonal matrices

$$\Lambda_i = \operatorname{diag}(\lambda_{i,1}, \lambda_{i,2}, \lambda_{i,3}, \dots, \lambda_{i,n}), \quad i = 1, \dots, m . \tag{19}$$

### 5.2  Subspace Formulation

Using these notations (4) can be written

$$\mathbf{x}_i \Lambda_i = P_i \mathbf{X}, \quad i = 1, \dots, m . \tag{20}$$

Denote the linear subspace in $R^n$ spanned by the rows in $\mathbf{x}_i$ and by the rows in $\mathbf{X}$ by $\mathcal{D}_i$ and $\mathcal{D}$ respectively. Then (20) can be interpreted as

$$\Lambda_i \mathcal{D}_i \subset \mathcal{D} \ , \tag{21}$$

where the diagonal matrix $\Lambda_i$ is interpreted as an operator that acts on a linear subspace by componentwise multiplication of the diagonal elements. One advantage of this formulation in (21) is that it is independent of the chosen coordinate system in the images, since $\mathcal{D}_i$ is the same subspace for every choice of affine coordinates in the images.

The advantage of the introduced notations in (17), (18) and (19) is that (1) can be written for $m$ images of $n$ points:

$$\begin{bmatrix} \mathbf{x}_1\Lambda_1 \\ \mathbf{x}_2\Lambda_2 \\ \vdots \\ \mathbf{x}_m\Lambda_m \end{bmatrix} = \begin{bmatrix} P_1 \\ P_2 \\ \vdots \\ P_m \end{bmatrix} \mathbf{X} = \mathbf{PX} \ . \tag{22}$$

For convenience, introduce the notation $\mathbf{x}_i\Lambda_i = \hat{\mathbf{x}}_i$. Then, multiplying each side with its transpose and dividing with the number of images, (22) can be written

$$\frac{1}{m}\sum_{i=1}^{m}\hat{\mathbf{x}}_i^T\hat{\mathbf{x}}_i = \mathbf{X}^T\Big(\frac{1}{m}\sum_{i=1}^{m}P_i^TP_i\Big)\mathbf{X} = \mathbf{X}^T\frac{1}{m}\mathbf{P}^T\mathbf{PX} \ . \tag{23}$$

The corresponding equation for the subspace analogy follows directly from (21):

$$\Lambda_1\mathcal{D}_1 + \Lambda_2\mathcal{D}_2 + \Lambda_3\mathcal{D}_3 + \ldots + \Lambda_m\mathcal{D}_m \subseteq \mathcal{D} \ , \tag{24}$$

where $S_1 + S_2 = \{\, x + y \,|\, x \in S_1, y \in S_2 \,\}$ denotes the sum of the two subspaces $S_1$ and $S_2$. Our goal now is to use (22), (24) to design an algorithm for calculating $\mathbf{X}$, $\Lambda_i$ and $P_i$ from the image data $\mathbf{x}_i$. Then $\mathbf{X}$ gives us the reconstruction (structure, perception) and $P_i$ give us the camera matrices (motion; obtained as the nullspaces of $P_i$, action). Introduce the matrices

$$\mathbf{T} := \frac{1}{m}\sum_{i=1}^{m}\mathbf{T}_i, \quad \mathbf{T}_i = \hat{\mathbf{x}}_i^T(\hat{\mathbf{x}}_i\hat{\mathbf{x}}_i^T)^{-1}\hat{\mathbf{x}}_i \ , \tag{25}$$

where $\mathbf{T}_i$ is the projection matrix onto $\Lambda_i\mathcal{D}_i$. The purpose of the factor $1/m$ in (25) is just to make the size of the entries in $\mathbf{T}$ more independent on the number of images. Observe that $\mathbf{T}$ is independent on the chosen basis for $\hat{\mathbf{x}}_i$ since

$$(A\hat{\mathbf{x}}_i)^T(A\hat{\mathbf{x}}_i(A\hat{\mathbf{x}}_i)^T)^{-1}A\hat{\mathbf{x}}_i = \hat{\mathbf{x}}_i^TA^T(A^T)^{-1}(\hat{\mathbf{x}}_i\hat{\mathbf{x}}_i^T)^{-1}A^{-1}A\hat{\mathbf{x}}_i = \hat{\mathbf{x}}_i^T(\hat{\mathbf{x}}_i\hat{\mathbf{x}}_i^T)^{-1}\hat{\mathbf{x}}_i \ ,$$

which implies that $\mathbf{T}_i$ is independent on the chosen coordinate system in the image. Now (24) implies that

$$\sum_{i=1}^{m}\mathbf{T}_iv \in \mathcal{D}, \quad \forall v \quad \Rightarrow \quad \text{rank}\,\mathbf{T} \leq 4 \ , \tag{26}$$

since $\mathcal{D}$ is a 4 dimensional subspace. This can be viewed as another way of describing the multilinear constraints, which is more suitable for numerical calculations.

Let $\sigma_i$, $i = 1, \ldots, m$ be the singular values of $\mathbf{T}$ and introduce the proximity measure $\mathcal{P} = \sigma_5$, which measure the degree of 4-dimensionality of the sum of the subspaces $\mathcal{D}_i$. Then the reconstruction problem can be formulated by a variational formula

$$\min_{\{\lambda_{i,j}\}} \mathcal{P} . \tag{27}$$

In the noise free case the minimum value is equal to 0. When noise is present in the measurements, minimising this variational formula gives a reconstruction that is independent of the chosen coordinate systems and the ordering of the points. This reconstruction can be obtained from the range of $\mathbf{T}$, which is equal to $\mathcal{D}$. We remark that the proximity measure can not be interpreted in terms of distances in the images, since distances have no meaning in this coordinate independent approach. Instead it is an abstract measure of the dimension of a subspace.

## 5.3 An Algorithm

Reconstruction can be made from an iterative algorithm consisting of the following steps:

1. Start by putting $\lambda_{i,j} = 1$.
2. Calculate $\mathbf{T}$ from (25).
3. Calculate the singular value decomposition of $\mathbf{T}$, i.e. $T = U^T S U$ and the proximity measure $\mathcal{P}$. If $\mathcal{P}$ is sufficiently small stop.
4. Let $\mathbf{X}$ denote the first four rows of $U$, which will be used as an approximation of the object.
5. Use (20) to estimate $\Lambda_i$ from $\mathbf{x}_i$ and $\mathbf{X}$.
6. Goto 2.

The criteria on the proximity measure for terminating have to be chosen appropriately. Step 5 above can done giving a result that is independent of both coordinate systems and the ordering of images and points. Let $T_\mathcal{D}$ and $Q_\mathcal{D} = I - T_\mathcal{D}$ denote the projection matrices onto $\mathcal{D}$ and its orthogonal complement respectively. Then we use (21) in the form

$$\min_{||\Lambda_i||=1} \sum_{j=1}^{3} ||Q_\mathcal{D} \Lambda_i \tilde{\mathbf{x}}_i^j||^2 , \tag{28}$$

where $\tilde{\mathbf{x}}_i^j$ denotes an orthonormal basis for $\mathcal{D}_i$. For details see [11].

We remark that the main difference between this algorithm and the one presented in [23] is that the former one gives a result that is independent of both the ordering of image and points and the choice of coordinate systems in the images, whereas the latter does not. This is due to the use of Euclidean distances between corresponding points and epipolar lines as well as an asymmetry in the use of different images.

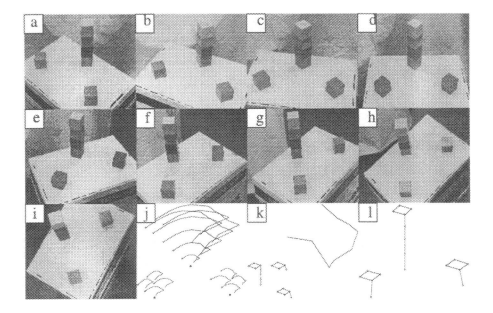

**Figure1.** Illustration of the trilinear approach. a)-i) show nine images of a simple scene. In j), some of the extracted points are shown in the affine coordinate system defined by three basis points. In k), the reconstructed camera motion is shown together with a reconstruction of the extracted corner points. l) highlights the reconstructed object.

## 6 Experiments

In this section two different experiments will be presented.

**Reconstruction using Trilinear Forms**

We illustrate the trilinear approach with an experiment where 9 images have been taken of a simple scene. These are shown in Fig. 1. Some of the corners are extracted and three corners, one from each box on the floor, are used as an affine basis in each image in order to simplify the algebra (see [9,8]). The corner positions have been used to estimate the trilinear forms and therefrom the reconstruction and the motion of the camera. This motion is presented together with the reconstruction of some of the corner points.

**Reconstruction using the Subspace Approach**

Consider the four images of a toy block scene in Figure 2. 32 corresponding points (corners of blocks) were picked out manually and were used as inputs to the algorithm. The results are shown in Figure 2, together with the obtained reconstruction. In order to display the reconstruction a Euclidean representation of all projectively equivalent reconstructions, obtained from the algorithm, has to be chosen. This has been done using a three-dimensional model of the object and selecting the closest possible of the projective reconstructions, in least squares sense. However, this is the only stage where the model has been used.

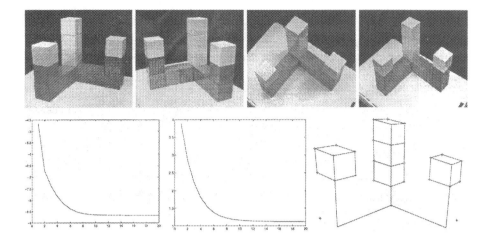

**Figure2.** Illustration of the performance of the algorithm. Above: Four images of the toy block scene. Below: The logarithm of the proximity measure, the estimated standard deviation and the obtained reconstruction.

# 7 Conclusions

In this paper we have shown that multilinear forms can be used in perception and action tasks in computer vision. Firstly, multilinear forms can be directly used to reconstruct the scene, giving the perception, and to calculate the camera motion, giving a necessary input to action tasks. Secondly, a generic algorithm, based on multilinear forms, for reconstruction and calculation of camera motion has been described. The algorithm is generic in the sense that all corresponding points and all images are treated in an equal way, i.e. no points or images are used as base points or reference image respectively. Furthermore, the algorithm gives a result that is independent on the chosen coordinate systems in the images, which is the natural way to deal with uncalibrated cameras. Although, there is no guarantee that this iterative algorithm will converge, experiments show that the convergence is usually very fast.

The algorithms are presented using real images, showing that it is possible to obtain both reconstruction and camera motion. Further investigations would be towards specific action tasks, such as picking up an object or positioning.

## References

1. Berthilsson, R., Heyden, A., Sparr, G., Recursive Structure and Motion from Image Sequences using Shape and Depth Spaces, *to be presented at CVPR'97*, 1996.
2. Faugeras, O., D., What can be seen in three dimensions with an uncalibrated stereo rig?, *ECCV'92, Lecture notes in Computer Science, Vol 588. Ed. G. Sandini, Springer-Verlag*, 1992, pp. 563-578.

3. Faugeras, O., D., Mourrain, B., On the geometry and algebra on the point and line correspondences between N images, *Proc. ICCV'95, IEEE Computer Society Press*, 1995, pp. 951-956.
4. Faugeras, O., D., Mourrain, B., About the correspondences of points between N images, *Proc. IEEE Workshop on Representation of Visual Scenes*, 1995.
5. Hartley, R., I., Projective Reconstruction and Invariants from Multiple Images, *IEEE Trans. Pattern Anal. Machine Intell.*, vol. 16, no. 10, pp. 1036-1041, 1994.
6. Hartley, A linear method for reconstruction from lines and points, *Proc. ICCV'95, IEEE Computer Society Press*, 1995, pp. 882-887.
7. Heyden, A., Reconstruction and Prediction from Three Images of Uncalibrated Cameras, *Proc. 9th Scandinavian Conference on Image Analysis, Ed. Gunilla Borgefors, Uppsala, Sweden*, 1995, pp. 57-66.
8. Heyden, A., Reconstruction from Image Sequences by means of Relative Depths, *Proc. ICCV'95, IEEE Computer Society Press*, 1995, pp. 1058-1063. An extended version to appear in *IJCV, International Journal of Computer Vision*, 1996.
9. Heyden, A., Åström, K., A Canonical Framework for Sequences of Images, *Proc. IEEE Workshop on Representation of Visual Scenes*, 1995.
10. Heyden, A., Åström, K., Algebraic Varieties in Multiple View Geometry, *ECCV'96, Lecture notes in Computer Science, Vol 1065, Ed. B. Buxton, R. Chippola, Springer-Verlag* 1996, pp. 671-682.
11. Heyden, A., Projective Structure and Motion from Image Sequences using Subspace Methods, *Submitted to SCIA'97*, 1996.
12. Luong, Q.-T., Vieville, T., Canonic Representations for the Geometries of Multiple Projective Views, *ECCV'94, Lecture notes in Computer Science, Vol 800. Ed. Jan-Olof Eklund, Springer-Verlag*, 1994, pp. 589-599.
13. Maybank, S., *Theory of Reconstruction from Image Motion*, Springer-Verlag, Berlin, Heidelberg, New York, 1993.
14. Mohr, R., Arbogast, E., It can be done without camera calibration, *Pattern Recognition Letters*, vol. 12, no. 1, 1991, pp. 39–43.
15. Poelman, C., J., Kanade, T., A Paraperspective Factorization Method for Shape and Motion Recovery, *ECCV'94, Lecture notes in Computer Science, Vol 801. Ed. Jan-Olof Eklund, Springer-Verlag*, 1994, pp. 97-108.
16. Shashua, A., Trilinearity in Visual Recognition by Alignment, *ECCV'94, Lecture notes in Computer Science, Vol 800. Ed. Jan-Olof Eklund, Springer-Verlag*, 1994, pp. 479-484.
17. Shashua, A., Werman, M., Trilinearity of Three Perspective Views and its Associated Tensor, *Proc. ICCV'95, IEEE Computer Society Press*, 1995, pp. 920-925.
18. Sparr, G., An algebraic-analytic method for affine shapes of point configurations, *proceedings 7th Scandinavian Conference on Image Analysis*, 1991, pp. 274-281.
19. Sparr, G., A Common Framework for Kinetic Depth, Reconstruction and Motion for Deformable Objects, *ECCV'94, Lecture notes in Computer Science, Vol 801. Ed. J-O. Eklund, Springer-Verlag* 1994, pp. 471-482.
20. Sparr, G., Simultaneous Reconstruction of Scene Structure and Camera Locations from Uncalibrated Image Sequences, *proceedings 13th International Conference on Pattern Recognition*, 1996, pp. 328-333.
21. Tomasi, C., Kanade, T., Shape and Motion from Image Streams under Orthography: A Factorization Method, *IJCV, 9(2):137-154*, 1992.
22. Triggs, B., Matching Constraints and the Joint Image, *Proc. ICCV'95, IEEE Computer Society Press*, 1995, pp. 338-343.
23. Sturm, P., Triggs, B., A Factorization Based Algorithm for Multi-Image Projective Structure and Motion, *ECCV'96, Lecture notes in Computer Science, Vol 1065. Ed. B. Buxton and R. Cipolla, Springer-Verlag* 1996, pp. 709-720.

# Local Image Operators and Iconic Structure

Jan J. Koenderink and Andrea J.van Doorn

Universiteit Utrecht, Helmholtz Instituut,
Princetonplein 5, 3584CC Utrecht, The Netherlands

**Abstract.** We study the image structure revealed by a coherent set of local image operators (*e.g.*, edge finders, ...). An observation specifies an equivalence class of images of infinite cardinality. A given set of image operators thus induces a parcellation of all images into such equivalence classes. Taking tolerances into account leads to a further growth of the equivalence classes. We analyze the qualitatively different image structures that can be classified by the set of image operators, that is the discriminatory power of the image operators.

## 1  Local Iconic Structure

In this paper we are concerned with the type of image structure that is relevant in determining the response of image operators such as "edge finders", *etc.* This is essentially the image structure in the immediate neighborhood of a point, the size of the neighborhood being determined by the support of the edge finder. When we run our full battery of local image operators at a point we sample the image structure at that point *exhaustively* (for otherwise we didn't really run our full battery). Obviously there will still exist an infinity of images that would yield identical data. We denote the equivalence class of images that give rise to a single datum an "icon". Then our method determines the "(local) iconic structure" at the point. Important questions are how many icons (or icon classes) there are, and how many images are compatible with a given iconic structure. This is a very fundamental issue of image processing.

Typical icon classes are "edge", "corner", various types of "textons" (texture elements), *etc.* Notice that an icon is not simply a property of the image, but a property of the image *as sampled* by our battery of operators. Thus it has a purely *operational* meaning. "Image structure" is merely a vague, general term, whereas the notion of iconic structure makes it precise and amenable to analysis.

The observer *never* sees all detail that is potentially available in a scene because of the finite resolution. (The same holds for photographic records of course.) Thus "mystery" is an ever present aspect of visual perception[19]. The question has to be where mystery starts, not whether one has to consider mystery at all.

Due to this (fundamental!) lack of resolution the observer obtains a *generalized impression* of the available optical structure. Thus one may not see the individual leaves on a tree or bricks in a wall, but merely "foliage" or "brick work". Nevertheless the foliage may appear "oak–like" and the brick work may likewise appear in many varieties. Thus there exists the necessity to be able to describe, represent or render optical structure at any reasonable level of detailing. "Generalization" ought to be an important science, in reality it is typically ignored[1].

It should be noted that the concept of "finite resolution" is fundamental because *any* observation is done at some finite resolution. To speak of image structure apart from the issue of resolution is strictly senseless.

## 2  Scale

### 2.1  Scale Space

By now the structure of "scale–space" has been well established[3,4]. If you entertain no prior assumptions that are specific to the scene, then there turns out to exist a unique way in which you may implement "blurring", that is lowering of resolution. That method is uniform diffusion of luminance over the visual field, or—equivalently—convolution with Gaußian kernels. Only this operation is translation and rotation invariant and yields "causal" resolution degrading. By "causal" we mean that luminance maxima invariably are lowered by blurring whereas luminance minima are invariably raised. Intuitively that means that lowering resolution will always lead to a *loss of detail*, never a gain.

### 2.2  Local Jets

In image processing tasks one needs to *differentiate* the optical structure. This is inherently a dangerous and ill defined operation: In effect no natural image will be "differentiable" in the $1^{st}$ place. One may differentiate blurred images though[2] because these are smooth and differentiation poses no problems at all. Thus differential operations can only be defined in the context of scale–space.

One has to be very careful though. Let $L(x, y)$ denote the luminance as a function of the Cartesian coordinates $(x, y)$ of an image. Let

$$G(x, y; s) = \frac{e^{-\frac{x^2 + y^2}{4s}}}{4\pi s} \tag{1}$$

denote a Gaußian kernel with resolution parameter $s$. Then the blurred image $L_s$ is defined by the convolution $L_s = G \otimes L$. Now linearity allows you to write

---

[1] One field that has attempted to build such a science of generalization is cartography. A map should always be a *summary description* for otherwise the map would coincide with the landscape itself and be superfluous. Thus making a map is inherently generalizing the earth's surface. Whether a dot should stand for a house, a city block, or a whole town is an important choice for the cartographer.

[2] At least in the sense that they are differentiable *in principle* whereas real images are not even that. However, any *real* blurred image is a *represented* (or coded) data structure and thus suffers from noise, *e.g.*, truncation noise. Thus *in practice* blurred images shouldn't be differentiated either. The right way to proceed is to regularize the differentiating operator itself, not its target (see below).

$\partial L_s$ in any one of the various forms

$$\partial(G \otimes L) = (\partial L) \otimes G = L \otimes \partial G, \tag{2}$$

where $\partial$ denotes some differentiating operator. Although these expressions are indeed formally equivalent, they are practically as different as can be. The

$1^{st}$ **expression** is the derivative of the blurred image. It is numerically unstable, because the blurred image has to be represented in some way, thus will be corrupted by noise (*e.g.*, truncation or rounding errors). The

$2^{nd}$ **expression** is the blurred differential of the image. This expression makes no sense because the differential of the image is not defined to begin with. The

$3^{rd}$ **expression** is the image convolved with the derivative of the Gaußian kernel. Only this expression makes perfect sense and is numerically stable. This is the only sane way to implement differential operators[5,6,8–15].

Notice that the result is equal to the exact derivative of the blurred image. Thus we may set up operators with finite support that implement exact local derivative operators!

# 3   Metamerism

"Metamerism" is a term borrowed from the science of colorimetry. The setting—in a suitably generalized way—is as follows: We consider a space (see figure 1) of "images" $\mathcal{I}$ (that are scalar fields on some parameter space) and a space of "observations" $\mathcal{O}$. We assume that $\dim \mathcal{O} < \dim \mathcal{I}$, thus an observation does not suffice to identify an image. In colorimetry one considers the space of spectra (the "images") and the space of colorimetric coordinates (the "observations", a typical observation would be a set of three numbers, say the red, green and blue intensities). A colorimetric specification does not identify a spectrum, but a "metamer", that is a $^\infty \mathcal{D}$ family of spectra with the same colorimetric specification (these "look the same color", though they correspond to distinct physical beams). Exactly the same formalism applies to spatial patterns (the images): If we observe the pattern via a battery of local image operators we obtain a specification that is not sufficient to identify the image, but instead specifies a large class of images that may be called the "metamer" corresponding to that observation.

   In practice one often indicates a metamer via a canonical member that is uniquely identifiable. One can do this by putting suitable constraints on these canonical images. This is common practice in image processing where one often sets up processes[1,2] that look for "edges" or "bars" (say). These are canonical members of the metamers (of $\infty$ cardinality). Whatever the true nature of the image is, those processes will only identify edges or bars.

## 3.1  General Notions

Consider the simple case where the observations are obtained via a set of linear operators acting on the image. (The typical "edge finders", "Laplacean operators", *etc.*, fall in this scheme.) Then we may set up a basis for image space such that the $1^{st}$ dim $\mathcal{O}$ basis vectors span observation space, whereas the remaining dim $\mathcal{I}$ – dim $\mathcal{O}$ dimensions lead to null observations. This factors the image space into two components, the "fundamental component space" and the "metameric black space". The observation scheme is (by construction) completely blind for the metameric black images (hence the name). An observation is sufficient to uniquely identify a member of the fundamental component space though. Thus the fundamental components may be used as canonical representatives. In practice this is often unattractive though because the fundamental components may well turn out to possess undesirable features. (For instance, the fundamental components typically have pixel values that are not positive definite, thus they correspond to physically impossible images.)

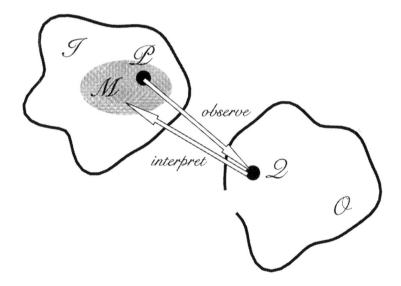

**Fig. 1.** *In this figure $\mathcal{I}$ denotes the space of all images, $\mathcal{O}$ the space of all possible outcomes of observations (in a given paradigm) on these images. Then an observation ($\mathbf{Q}$ say) could be caused by any member of the set $\mathcal{M} \subset \mathcal{I}$ called the metamer specified by the observation $\mathbf{Q}$. Any such member equals the fundamental $\mathbf{F}$ plus a virtual image $\mathbf{B}$ that is totally invisible in the paradigm, a member of the metameric black space $\mathcal{B}$. ($\mathbf{F}$, $\mathbf{B}$ and $\mathcal{B}$ are not represented in this figure.) Since the fundamental $\mathbf{F}$ need not be a physically realizable image (for instance it may have some negative or very large pixel intensities) we often prefer to represent the metamer by some canonical representative $\mathbf{P} \in \mathcal{M}$ that is singled out by some desirable property.*

We obtain a particularly clear description in the scale space setting. Here an obvious choice for the set of linear operators is apparent: The $n^{th}$–order jet $\mathcal{J}^n$ of truncated Taylor expansions. It is comparatively easy to analyze this important case in detail. It is not merely a nice example though: In actual image analysis the jet representation has many practical advantages quite apart from being the obvious choice from $1^{st}$–principles.

## 3.2 The $^1\mathcal{D}$ Case

Here we treat the simplest case, that is $^1\mathcal{D}$. In the $^1\mathcal{D}$ case the $2^{nd}$–order jet $\mathcal{J}^2$ has 3–degrees of freedom. Thus we can map the sample obtained by a local $\mathcal{J}^2$ as a point in $^3\mathcal{D}$ "pattern space". A "pattern" represents an $\infty$ class of images, namely those images that all yield the given sample, thus agree in their local $\mathcal{J}^2$'s. Such a class is called a "metamer" in colorimetry, and the fact that *many* images correspond to a given sample is known as "metamerism". The case of colorimetry is formally identical to $\mathcal{J}^2$ in $^1\mathcal{D}$. We employ the powerful methods developed in colorimetry by Schrödinger[20,21] in the 1920's.

**Metamerism in $\mathcal{J}^2$** We explore this paradigmatic case in some detail. We consider $\mathcal{J}^2$ at the origin $x = 0$. We define the *scale* $s$ of the jet via its "aperture function"

$$A(x; s) = \frac{e^{-\frac{x^2}{8s}}}{\sqrt{8\pi s}}. \tag{3}$$

In this report we proceed by immediately setting $s = 1/4$ because we will only consider the structure at a single scale. Only in the initial few steps will we keep the scale parameter for the sake of clarity. The aperture function roughly "picks out the segment $(-\sqrt{4s}, +\sqrt{4s})$" (thus $(-1, +1)$ for $s = 1/4$) from the infinite $x$–axis. It offers "a window" near the origin on the $x$–axis.

The *Hermite functions* form an orthonormal base whose support is nominally the whole axis, but in practice only a region around the origin. The Hermite functions $\varphi_n(\xi)$ are defined in terms of the Hermite polynomials $H_n(x)$:

$$\varphi_n(\xi) = \frac{H_n(\xi)e^{-\frac{\xi^2}{2}}}{\sqrt{2^n n! \sqrt{\pi}}}, \tag{4}$$

with $\xi = x/\sqrt{4s}$. If we want to expand a function $f(x)$ we first pick the part of $f(x)$ as seen through the aperture, namely $f(x)A(x; s)$ and then project this on the Hermite functions. In effect we compute the scalar products $\langle f(x)A(x; s)|\varphi_n(x)\rangle$, which equals $\langle f(x)|A(x; s)\varphi_n(x)\rangle$. These latter functions $A(x; s)\varphi_n(x)$ turn out to be proportional with the derivatives of the Gaußian kernel

$$G(x; s) = \frac{e^{-\frac{x^2}{4s}}}{\sqrt{4\pi s}}. \tag{5}$$

Thus the coefficients of the expansion of the function $f(x)$ as seen through the aperture $A(x; s)$ (thus $f(x)A(x; s)$) in terms of the orthonormal basis of

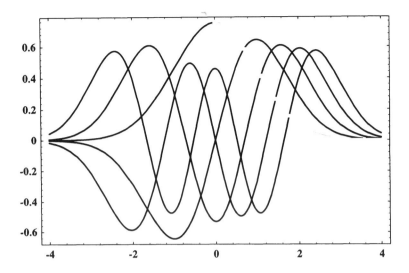

**Fig. 2.** The $1^{st}$ five Hermite functions. Notice the character of these functions: they are oscillatory in a region near the origin but monotonically approach zero for larger distances from the origin. In the piece picked out by the aperture the higher orders are very similar to sine–wave patterns.

Hermite functions equal the convolutions of $f(x)$ itself with the derivatives of a Gaußian kernel, or—what again amounts to the same thing—these expansion coefficients equal the derivatives of the function $f(x)$ at the scale $s$, that is the convolution $f(x) \otimes G(x;s)$ of $f(x)$ with the Gaußian kernel $G(x;s)$.

The first few Hermite functions are depicted in figure 2.

Notice that the functions

$$\psi_n(x;s) = \frac{\varphi_n(\frac{x}{\sqrt{4s}})A(x;s)}{(\sqrt{4s})^n} \tag{6}$$

are proportional with the Gaußian derivatives. In fact you have

$$D_0(x;s) = G(x;s) = \frac{e^{-\frac{x^2}{4s}}}{\sqrt{4\pi s}}$$

$$D_n(x;s) = \frac{\partial^n D_0(x;s)}{\partial x^n} = (-1)^n \sqrt{8\pi s}\sqrt{2^n n! \sqrt{\pi}}\psi_n(x;s). \tag{7}$$

These important functions are depicted in figure **??**. he Pythagorean sum of the $\psi_n$ equals the aperture function:

$$\sqrt{\sum_{n=0}^{\infty} \psi_n(x;s)^2} = A(x;s). \tag{8}$$

In practice the aperture function is already very well approximated if one truncates the sum at a rather low order (see figure 4). The $\psi_n$ don't form an or-

**Derivatives of Gaußian kernel**

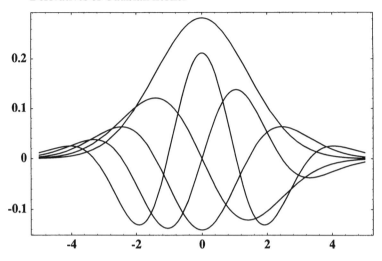

**Fig. 3.** $\mathcal{J}^4$

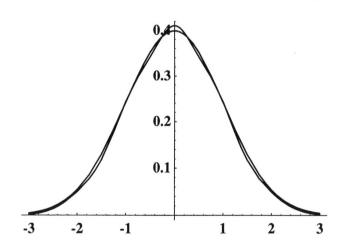

**Fig. 4.** *The Pythagorean sum of the* $1^{st}$ *five* $\psi_n$ *functions plotted together with the aperture function. Notice the already very close correspondence.*

thogonal system. It is much more convenient to split off the aperture function: Then we see that the Gaußian derivatives of a function are just the projections of the windowed function on the Hermite functions. Thus a Taylor expansion at a given scale is nothing but a development of the windowed function in terms of the (orthonormal) basis of Hermite functions of the corresponding scale. The $\psi_n(x; s)$ satisfy the *diffusion equation*

$$\triangle \Xi(x; s) = \frac{\partial \Xi(x; s)}{\partial s} \tag{9}$$

with kernel

$$G(x; s) = \frac{e^{-\frac{x^2}{4s}}}{\sqrt{4\pi s}}. \tag{10}$$

Usually we set $\xi = x/\sqrt{4s}$ as the normalized, scale independent variable. Notice that $G(x; s)$ is proportional to $\varphi_0(\xi)A(x; s)$. (For arbitrary order you have $\psi_n(x; s) = (1/\sqrt{4s})^n\varphi_n(\xi)A(x; s)$.) Also notice that we factor the kernel of the diffusion equation into two similar kernels except for their larger width.

All this is easily generalized to higher dimensions. Notice that the $\psi_n(x; s)$ are the natural image operators in a scale space setting. According to their shape they are generally known as "blurring kernel" ($0^{th}$ order), "edge finder" ($1^{st}$ order), "bar detector" ($2^{nd}$ order), *etc.* However, we will argue in the sequel that such a practice has to be condemned: The image samples obtained via these operators only make sense as an ensemble ($\mathcal{J}^n$), not individually.

Because the $\psi_n(x; s)$ satisfy the diffusion equation, the Hermite functions $\varphi_n(\xi)$ satisfy the partial differential equation[3]

$$\triangle \varphi_n(\xi) + ((2n + 1) - \xi^2)\varphi_n(\xi) = 0. \tag{11}$$

This equation is useful because it serves to describe the Hermite functions as a family and can be used to derive useful properties of this family.

For the $\mathcal{J}^2$ we are concerned with the kernels (from here on we set $s = \frac{1}{4}$):

$$\psi_0(x) = \frac{e^{-x^2}}{\pi^{3/4}2^{1/2}}$$
$$\psi_1(x) = \frac{xe^{-x^2}}{\pi^{3/4}} \tag{12}$$
$$\psi_2(x) = \frac{(x^2 - \frac{1}{2})e^{-x^2}}{\pi^{3/4}}.$$

---

[3] This is the Schrödinger equation of the $^1\mathcal{D}$ harmonic oscillator[18], but there appears to be no relation, we note this merely for curiosity. However, if you are a physicist you may use this fact to good advantage. For instance, the "turning points" of the corresponding classical harmonic oscillator are clearly $\pm\sqrt{2n + 1}$, hence the envelope of the $\varphi_n^2$ peaks at $\pm\sqrt{2n + 1}$ and decays as one recedes farther from the origin. The $\varphi_n$ are like sine–waves between the classical turning points, like exponentially decaying functions outside this region.

In the sequel we will also have to deal with the primitives of these functions, $\Psi_n(x) = \int_{-\infty}^{\xi} \psi_n(\xi)\,d\xi$, specifically:

$$\Psi_0(x) = \frac{1+\mathrm{erf}\,x}{2^{3/2}\pi^{1/4}}$$
$$\Psi_1(x) = \frac{-\mathrm{e}^{-x^2}}{2\pi^{3/4}} \tag{13}$$
$$\Psi_2(x) = \frac{-x\mathrm{e}^{-x^2}}{2\pi^{3/4}}.$$

Suppose we let the operators $(\psi_0, \psi_1, \psi_2)$ act on an image. We obtain the responses $(\varrho_0, \varrho_1, \varrho_2)$. In an operational sense all we may know of an image locally are such triples of samples. If you move an impulse function over the full axis you describe a closed curve in "response space", that is $(\varrho_0, \varrho_1, \varrho_2)$– space. The curve starts and ends at the origin. For various amplitudes of the impulse you obtain scaled versions of this loop. Together these describe a conical surface[4] (See figure 5). We limit the discussion to functions that are nowhere

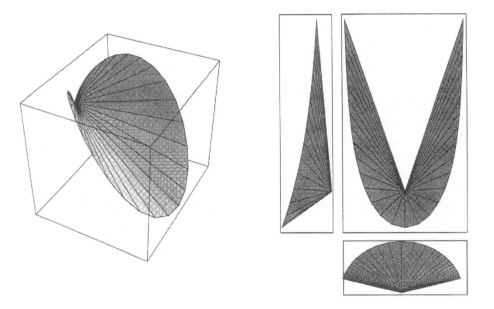

**Fig. 5.** *The cone of impulse patterns. Left: The length of the generators is the intensity for the corresponding impulse. Compare this figure with that on the right in which the length of the generators has been normalized; Right: The general structure should be intuitively obvious from these three views. The parabolic shape of the locus in unit intensity space is readily apparent because the length of the generators has been normalized.*

negative, since we are only interested in *images*. In this case the region *inside*

---

[4] Known as the "spectral cone" in color science. The impulse is then a "monochromatic beam".

this cone corresponds to possible responses, whereas the region outside the cone corresponds to responses that can never occur. The reason is simple enough: Since *every* function can be decomposed into impulses, with positive weight, all responses must lie in the convex hull of the locus of impulse responses. Thus you see that the responses of Gaußian derivatives are not independent: Only certain combinations of responses can occur at all.

In practice it is often useful to consider a lower dimensional representation. Because images that differ only in contrast (overall intensity scaling) are essentially similar, we often use the representation[5]

$$(\zeta_1, \zeta_2) = (\varrho_1/\varrho_0, \varrho_2/\varrho_0). \tag{14}$$

This is a central projection from the origin of $(\varrho_0, \varrho_1, \varrho_2)$–space to the plane $\varrho_0 = 1$. In this plane the cone appears as a parabola[6] (conveniently parameterized by the coordinate $x$), namely $(\zeta_1, \zeta_2) = (x\sqrt{2}, (2x^2-1)/\sqrt{2})$. Instead of using the coordinate $x$ to indicate the impulse functions we also use the phase angle $\lambda = \arctan(-\zeta_2, \zeta_1)$. This is convenient[7] because we deal with a finite parameter, moreover, the implied periodicity turns out to make much sense later. We can parameterize any response in terms of its *intensity* $\zeta_0$, its *phase angle* $\lambda$ or "equivalent position"[8] $\overline{x} = \tan\lambda$ and its *contrast*[9]$\chi = \sqrt{(\zeta_1^2 + \zeta_2^2)}/\sqrt{(4\overline{x}^2 + 1)/2}$. This effectively interprets the (unknown) local image in terms of the superposition of an impulse function (at the equivalent position) and a constant overall intensity[10]. Notice that *any* local image can be represented in this manner. The representation picks out a definite canonical metamer of the image. It is not a particularly fortunate choice in most cases though. This is because the amplitude of the impulse may be very high, whereas the image intensities are typically bounded from above (are in the range 0 to 255 for a typical image say).

If you *add two images* you obtain a $3^{rd}$ one. It is always possible to find an image that will exactly *cancel* the impulse component of any given image: One simply picks an image with the equivalent position such that the line–segment defined by the given image and the $2^{nd}$ one in $(\zeta_1, \zeta_2)$–space includes the origin. This is the case because the addition of images simply corresponds to barycentric addition in $(\zeta_1, \zeta_2)$–space, which again corresponds to vector addition in $(\varrho_0, \varrho_1, \varrho_2)$–space which is obvious because of linearity[11]. The equivalent position $\alpha^\sharp$ that can be used to cancel a given position $\alpha$ may be called its

---

[5] This is known as the "chromaticity diagram" in color science.

[6] Known as the "spectral locus" in color science.

[7] The phase angle corresponds roughly to the "hue" in color science.

[8] The equivalent position corresponds to the "dominant wavelength" in color science.

[9] The "contrast corresponds to "excitation purity" in color science. Notice that the contrast vanishes for the uniform image and reaches the maximum value of unity for the impulse functions.

[10] In color science this was first done by Graßmann. This parameterization is known as "Helmholtz coordinates".

[11] Because $\mathcal{J}^2$ is just a linear transformation from the space of images $\mathcal{I}$ to $(\varrho_0, \varrho_1, \varrho_2)$–space, addition of images corresponds to vector addition in $(\varrho_0, \varrho_1, \varrho_2)$–space. Because $(\zeta_1, \zeta_2)$–space is the projection of $(\varrho_0, \varrho_1, \varrho_2)$–space on the plane $\varrho_0 = 1$ from

"complementary equivalent position". One has simply $\alpha\alpha^\sharp = -1/2$. (Thus there exist weights such that an impulse at position $\alpha$ added to an impulse at position $\alpha^\sharp$ may be equivalent to a uniform intensity over the whole axis.) This will turn out to be an extremely important relation[12] because the structure of $\mathcal{J}^2$ is largely due to it. Notice that the origin is the complementary of both $\pm\infty$: This indicates that the parabolic locus has to be considered as *closed* in matters of complementarity. Indeed, in terms of the phase angle one has simply $\lambda^\sharp - \lambda = \pi$ with no exception. This clarifies the periodic character of the phase angle.

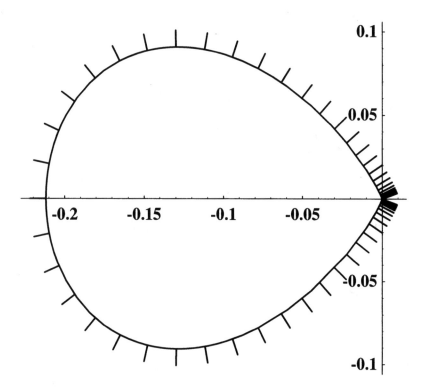

**Fig. 6.** *The locus of left edge patterns. The marks define an uniformly spaced equivalent position scale.*

---

the origin (a perspectivity), addition of images corresponds to a linear interpolation in $(\zeta_1, \zeta_2)$–space. It is easy to show that *barycentric* addition where the "mass" is $\varrho_0$ is the required operation. This is similar to Newton's barycentric rule for his color circle.

[12] In color science $\alpha$ and $\alpha^\sharp$ would be known as "complementary wavelengths. In the case of human color vision many, but not all, wavelengths possess a complementary. In $\mathcal{J}^2$ every position has a complementary one except the origin (which has both $+\infty$ and $-\infty$ as complementary). The case of the origin is singular in another way because the Hermite functions approach zero at both infinities. These complications are of rather minor importance though.

There is a simple relation to the $(\zeta_1, \zeta_2)$–space because the tangent to this curve is just the position on the parabola in $(\zeta_1, \zeta_2)$–space. This moment representation is extremely useful in the discussion of metamers as will be demonstrated later[13].

Another representation that illustrates the periodic character of $\mathcal{J}^2$ is the "moment representation". We consider the "unit left edge patterns"

$$\text{ULE}(x, x_0) = 1, x <= x_0,$$
$$\text{ULE}(x, x_0) = 0, x > x_0. \tag{15}$$

We have

$$(\varrho_0, \varrho_1, \varrho_2) = (\Psi_0(x_0), \Psi_1(x_0), \Psi_2(x_0)) \tag{16}$$

with

$$\Psi_i(x) = \int_{-\infty}^{+\infty} \psi_i(y) \text{ULE}(y, x)\, dy = \int_{-\infty}^{x} \psi_i(y)\, dy. \tag{17}$$

A plot of $(\Psi_1(x_0), \Psi_2(x_0))$ for $x_0 \in (-\infty, \infty)$ shows a closed convex curve (see figure 6).

Suppose that we know that the image intensities are bounded from above (they are already trivially bounded from below too since the intensities can only be positive). For the sake of concreteness[14] we assume that image intensities are in the range $[0, 1]$. Then we have quite a few constraints on the possible responses $(\varrho_0, \varrho_1, \varrho_2)$. For instance, the highest intensity is reached for the uniform white image, it amounts to $1/\sqrt{2\sqrt{\pi}} = 0.5311\dots$. Images of very high contrast $\chi \approx 1$ must be very dim indeed (almost black) because they can only have appreciable pixel intensities in a very narrow region. In fact, it makes sense to ask for the images that have *the highest intensity* for a given phase angle and contrast. (Notice that the phase angle and contrast together specify $(\zeta_1, \zeta_2)$, that is a line through the origin (parameter being the intensity) in $(\varrho_0, \varrho_1, \varrho_2)$–space.) One easily demonstrates (following Schrödinger in his classical paper on color space of the twenties) that such images are *bars*: They are either zero or one and have at most two transition loci. We will refer to such patterns as "optimal images". They are optimal in the sense of being the highest intensity ones for a given phase angle and contrast.

Since *any* image that corresponds to a given phase angle and contrast must be darker than the corresponding optimal image, all such images lie on a line segment between the origin (uniform black image) and the optimal image in $(\varrho_0, \varrho_1, \varrho_2)$–space. This implies that all possible responses lie inside a volume that is bounded by the surface described by the optimal images. We call this the "pattern solid" (see figure 7).

---

[13] In color science this curve is known as the "Luther curve". Luther used this curve to derive many important properties of the object color body.

[14] Notice that this assumption is reasonable in the context of non–selfluminous bodies illuminated by a single light source (think of printed material on your desk under general room light illumination). The bodies have albedos between zero and one and these modulate the maximum available illumination.

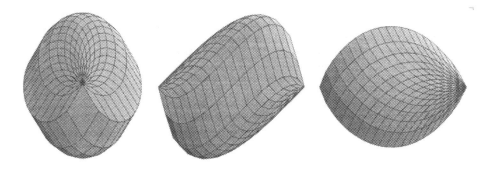

**Fig. 7.** *The pattern solid seen from the three coordinate directions. Notice the central symmetry and the apices (conical points) at the lowest (black) and highest (white) points. The parameterization is by transition positions of the optimal patterns.*

The pattern solid has several symmetric features that are worth noticing. First of all the pattern solid is a convex body. This is obvious because if there were a concavity we could immediately fill it through linear combination of images. However, it is not trivial that the surface of optimal patterns describes the boundary of the convex solid. This is due[15] to the nature of the $\psi_n$. The uniform (gray) images lie on the linear segment between the white and black point. In the center of this segment we find the uniform medium gray (intensity 0.5 of the maximum) image. (Notice that this "uniform gray image" has $\infty$ many articulated metamers. However, $\mathcal{J}^2$ can't "see" these.)

The medium gray image is a *symmetry center* of the pattern solid as you easily demonstrate when you consider the "negatives" of images, that is $I'(x) = I_{max} - I(x)$.

The boundary of the pattern solid is smooth except for the white and black points which are conical. At the black point the impulse cone is tangent to the pattern solid[16], at the white point an inverted copy of the impulse cone is also tangent[17].

Notice that the central symmetry is closely related to our previous notion of "complementarity". Also notice that the optimal patterns are specified by *chords* (line segments between points of the locus of left edge patterns) in the moments representation. Chords of a given direction correspond to images of fixed phase angle. The length of the chord specifies the amplitude of the equivalent impulse. Because the moment function is convex, we find that for any phase angle there exists a unique optimal pattern for which this amplitude is a maximum. Such patterns are in a sense "the most articulated" ones. We will refer to them as the "strong patterns". The strong patterns are optimal patterns for which the transition loci are complementary. This follows from the central symmetry if

---

[15] It is easy to produce examples for which this property doesn't apply.

[16] This is obvious because the optimal patterns become impulses at the black point.

[17] This is also obvious since the optimal patterns approximate $I_{max}$ *minus* some impulse at the white point.

you notice that the strong patterns have the greatest distance from the gray axis: Thus they must lie on the circumscribed cylinder to the pattern solid with generators parallel to the gray axis. We may plot the locus of strong patterns in $(\zeta_1, \zeta_2)$–space (see figure 8). This shows that the region of high intensity images

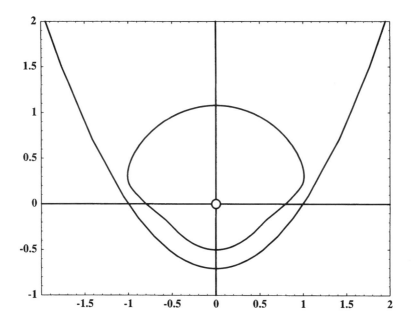

**Fig. 8.** *The locus of impulse functions (parabola) and the locus of strong patterns (the loop encircling the origin) in the $(\zeta_1, \zeta_2)$–plane.*

is very limited. Most of the interior of the parabola in $(\zeta_1, \zeta_2)$–space corresponds to almost black images.

The strong patterns are not all equally intense although their intensity doesn't vary very much. In figure 9 *left* we plot the intensity of strong patterns as a function of their phase angle.

Clearly the patterns with near zero phase angles (these are light bars) have the highest intensity (more than half of the maximally attainable intensity) whereas those with phase angles near $\pm\pi$ are darker (about a third of the maximum intensity). Fairly sharp transitions occur at phase angles near $\pm\pi/2$. In this sense $\mathcal{J}^2$ seems to "prefer light bars".

The phase angle of a strong pattern with transitions at $(\lambda - \pi/2, \lambda + \pi/2)$ is very near to $\lambda$. We show the actual relation in figure 9 *middle*. This is often a useful rule of thumb.

The contrast of the strong patterns depends critically on their phase angle. We plot the dependence in figure 9 *right*. The contrast is almost uniformly low for the bandgap patterns, and uniformly high for the bandpass patterns. Thus

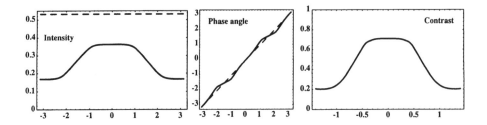

**Fig. 9.** Left: *Intensity of the strong patterns as a function of their phase angle. The intensity of uniform white images is also indicated, it is the upper limit of intensity per se;* Middle: *The phase angle of the strong patterns as a function of their band center* $\lambda$; Right: *Contrast of the strong patterns as a function of their phase angle.*

$\mathcal{J}^2$ again seems to "prefer light bars": they have both the highest contrast and the highest intensity among all strong patterns.

If we want we can now classify any pattern uniquely as the superposition of an attenuated strong pattern with a uniform gray pattern. Such patterns have two transitions at complementary locations and they switch between two levels that are neatly in the range of allowed intensities ($[0, 1]$). We may parameterize them by phase angle (this defines the transition locations), pattern content $p$, white content $w$ and black content $b$, with the relation $p + w + b = 1$. If the intensity of the strong pattern is $I_s$, of the uniform white pattern $I_w$, of the uniform black pattern $I_b = 0$, then the levels are simply $wI_w + bI_b$ and $wI_w + bI_b + pI_s$. This classification is analogous to Ostwald's classification of surface colors[16,17]. It (almost) succeeds in assigning a sensible unique metamer to any observation[7]. The problem is that *very* contrasty images may actually have more contrast than the strong pattern. In that case the pattern, black or white content may become negative. In realistic cases they will turn out to lie in the range $[0, 1]$ though. The problem is easy to fix[18], but we won't concern ourselves with such matters in this report.

Before going on with the discussion we illustrate the structure of the pattern solid into somewhat richer detail. First notice that the optimal patterns occur in four distinct kinds:

$1^{stly}$ we have the light bar patterns,
$2^{ndly}$ we have the dark bar patterns,
$3^{rdly}$ we have the left edge patterns, and
$4^{thly}$ we have the right edge patterns.

The bar patterns form 2–parameter families, whereas the edge patterns form only 1–parameter families. Thus the bar patterns will fill areas on the boundary of the pattern solid, whereas the edge patterns will only describe certain curves on it. We find that the light and dark bar patterns occupy mutually symmetric

---

[18] Instead of the strong patterns one uses "virtual patterns" defined as the points of intersection of the tangent cones to the pattern solid at the black and white apices.

connected areas that meet each other at the curves defined by the edge patterns. We plot the projections of these curves in order to elucidate the structure (figure 10). One of these projections is just the moment curve. We see that the

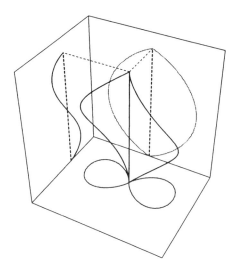

**Fig. 10.** *The edge patterns. The vertical line segment is the gray axis, its end points are the black and white points. The loci of left and right edge patterns are $^3\mathcal{D}$ spirals. We have plotted the projection on the coordinate planes. The projection on the base plane is just the locus of the primitives.*

edge pattern loci are mutually symmetric spirals on the boundary of the pattern solid. We may also plot the projections of the locus of strong patterns. One of these projections we have already encountered. The locus is a twisted, closed space curve (figure 11).

A useful representation lets you appreciate the distribution of optimal patterns (figure 12). We plot curves of equal bar center and bar width in the $(\zeta_1, \zeta_2)$–diagram. The division of light and dark bars by the left and right edge patterns is also immediately apparent.

We are not yet done with the discussion of metamerism in $\mathcal{J}^2$: We have as yet developed no appreciation for the essential finite support of the operators. The aperture function basically only lets us "see" the interval $x \in (-1, +1)$. Everything outside that interval will be sampled by other $\mathcal{J}^2$'s, located at more advantageous positions. Outside the interval the operators are not oscillatory but just decay, thus they are not fit to capture any useful pattern structure.

We can demonstrate these problems by studying two very simple cases: A symmetric Gaußian light bar of variable width ($I(x) = \exp(-x^2/2\sigma^2)$), and an exponential ramp of varying slope ($\exp(\alpha x)$). Some remarkable cases are shown in figure 13 and figure 14. In figure 13 we illustrate a narrow and a broad Gaußian bar. The first one has an optimal pattern metamer that is also fairly narrow,

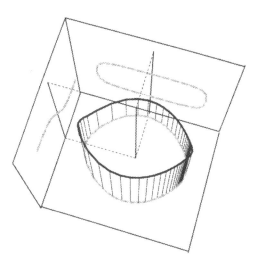

**Fig. 11.** *The strong patterns. The vertical line segment is the gray axis, its end points are the white and black points. The locus of strong patterns is a closed space curve that encircles the gray axis (it is also the locus of greatest orthogonal distance from this axis). The projections on the coordinate planes are also plotted. The projection on the base plane is the locus that also appears in the chromaticity diagram.*

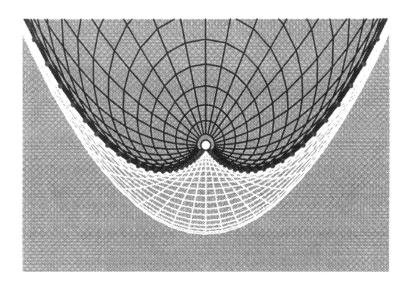

**Fig. 12.** *The optimal patterns parameterized by bar center and width. The curves that fan out from the point of uniform patterns are the loci of constant bar center location. The curves that run transversally to them are the loci of constant bar width. Notice that these are split into two families (depicted in different tone): the light and the dark bar patterns, separated by the loci of left and right edges.*

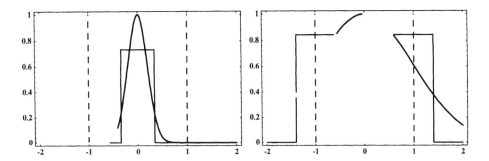

**Fig. 13.** Left: *A narrow Gaußian hill and an equivalent optimal pattern. The $\mathcal{J}^2$ cannot distinguish between these. The width and the height of the optimal pattern are close measures of the width of the actual pattern; Right: A broad Gaußian hill and its equivalent optimal pattern. The $\mathcal{J}^2$ cannot distinguish between these. Different from the previous case the transition of the equivalent optimal pattern occurs outside of the aperture of the operators. It would make sense then to classify this pattern as "uniform" (within the aperture).*

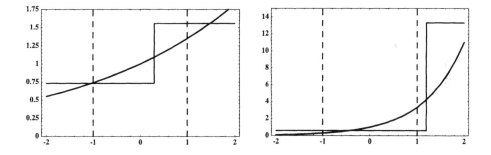

**Fig. 14.** Left: *An exponential ramp and its equivalent optimal pattern. The $\mathcal{J}^2$ cannot distinguish between these. Notice that the optimal pattern is indeed a close approximation to the actual pattern, at least within the aperture; Right: An exponential ramp and its equivalent optimal pattern. The $\mathcal{J}^2$ cannot distinguish between these. Different from the previous case the transition occurs outside of the aperture. It would make sense then to classify this pattern as "uniform" (within the aperture).*

the second one has an equivalent optimal pattern metamer that is a very broad bar. (The width and height of these optimal pattern metamers are shown in figure 15.)

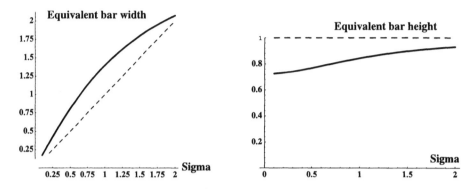

**Fig. 15.** Left: *The width of the equivalent optimal pattern for a Gaußian bar centered at the origin. The width indeed closely reflects the width of the Gaußian hill. (The dotted line indicates the perfect fit); Right: The ratio of the heights of the equivalent optimal pattern to that of the actual Gaußian hill pattern. The optimal pattern slightly underestimates the (maximum) height of the hill. (Dotted line). Clearly the optimal pattern approximates the actual pattern quite well.*

In case of the broad bar the transitions are *outside* of the aperture (that is to say the interval $[-1, +1]$) and it would make more sense to classify this image as *uniform* than as *a bar* purely on the basis of the observation. Likewise the shallow exponential ramp leads to an edge representation with transition near the origin, the steep one to an edge with transition far from the origin. The edge transition position is a monotonic function of the steepness of the exponential ramp. Again, it would make more sense to classify the steep ramp as *uniform* than as *an edge* purely on the basis of the observation. Clearly, we are in need of an extended discussion of metamerism in $\mathcal{J}^2$ that also takes the limited support of $\mathcal{J}^2$ into account.

One simple way to proceed is the following. Given an observation $(\varrho_0, \varrho_1, \varrho_2)$ we consider all possible interpretations, *i.e.*, the corresponding metamer. If this equivalence class contains a member that happens to be uniform over the interval $[-1, +1]$ we consider the observation to indicate a featureless image. In any case we try to pick the least committing interpretation in some reasonable sense. This will typically mean an interpretation with very small variation over the interval $[-1, +1]$. We need strong evidence to commit ourselves to a really articulate interpretation. The optimal patterns are thus excellent candidates. For any given observation we can find a 1–parameter family of optimal patterns that are valid interpretations. It is easy to construct this family: We construct the halfplane on the gray axis that contains our observation. Then we draw the lines through the observation and the white and black point. These lines meet the optimal pattern locus in two points. Thus we define a stretch on the optimal pattern locus: This is the required solution because the observation can be explained by the mixture of an optimal pattern on this stretch and a uniform gray image. Next we consider these possibilities and note whether there exist instances for which one or two of the transition locations fall(s) outside the stretch $[-1, +1]$. If both transitions may fall outside we declare the image to be locally uniform. If one transition may fall outside we declare it to be an edge. If both transitions fall inside the interval $[-1, 1]$ we declare the image to be a bar. In figure 16 we show the intersection of the plane $\varrho_1 = 0$ with the pattern solid. The shaded area indicates the loci of observations for which uniform interpretations are valid. As you see the size of this region is appreciable.

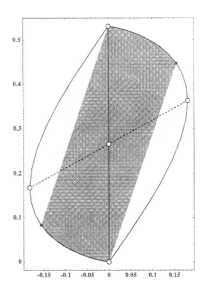

**Fig. 16.** *The intersection of the pattern solid with the plane $\varrho_1 = 0$. Indicated are the white point (top), black point (bottom) and medium gray point (center) on the gray axis (central vertical line). The oblique dashed line connects the two complementary strong patterns. The shaded area indicates all observations for which uniform image interpretations can be found. The areas outside the shaded area admit only of light bar (right hand side) or dark bar (left hand side) interpretations.*

It is fairly easy to find the segmentation of $(\zeta_1, \zeta_2)$–space into regions that admit of uniform, edge or bar interpretations (see figure 17). Thus we find that the observations obtained via the $^1\mathcal{D} \, \mathcal{J}^2$ are of one of the following categories:

**Invalid observations** some observations do not correspond to any interpretation with nonnegative pixel intensities. Such observations should probably be treated as "error messages";

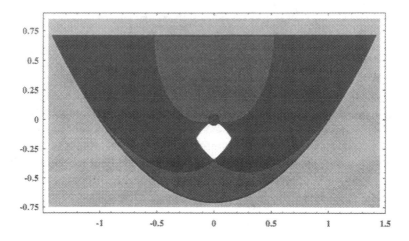

**Fig. 17.** *The segmentation of the $(\zeta_1, \zeta_2)$–plane into various regions. The area outside the convex region bounded by the parabola consists of points that can never occur as valid observations. The convex area can be subdivided into two disjunct areas for which edge interpretations are the simplest interpretation. One of these areas contains the right, the other the left edges. There are also two disjunct areas where bars are the simplest interpretation. One area contains light the other dark bars. The remaining area is the diamond shaped region that contains the white point as a boundary element. In this region a uniform interpretation suffices.*

**uniform patterns** some observations can be explained by patterns that are uniform within the region $[-1, +1]$. Outside this interval they are articulated, but other observations (centered at neighboring points) should take over there;

**edges** some observations cannot be explained by uniform patterns but can be explained by assuming the presence of a single transition from one constant value to another occurring in the interval $[-1, +1]$;

**bars** again, some valid observations cannot be explained by uniform patterns, nor by edge patterns. Such observations can be interpreted in terms of bars, *i.e.*, the occurrence of two transitions within the interval $[-1, +1]$ where transitions between two constant levels occur.

For a given observation there will still be some leeway left, *i.e.*, a limited 1–parameter variation of levels and transition locations. Thus one might be induced to augment the categorization scheme with a discretization of transition locations and/or levels.

The consequences of these considerations are important. For one thing, one should not think of the $\psi_1$ as "edge finders", nor of the $\psi_2$ as "bar detectors". The *meaning* of the observation $\varrho_1$ or $\varrho_2$ depends on the total activity of $\mathcal{J}^2$. This meaning can be fixed once the order of the observation has be decided upon. (For instance, in $\mathcal{J}^1$ one can *only* find uniform patterns and edges, but in $\mathcal{J}^2$ one may also find bars. Thus the "meaning" of $\varrho_1$ depends on the order of the jet.) For a given order the meaning of $\varrho_1$ still depends on the values of $\varrho_0$ and $\varrho_2$, *etc.* These simple facts are apparently not recognized in current state of the art image processing. We have shown how one may categorize local image structure by way of observations in $\mathcal{J}^2$. The jet $\mathcal{J}^2$ as a whole can recognize the categories "'bar", "edge", "uniform" or yield an "error message" and *only those*. In order to discriminate more (local) patterns one has to increase the order. Another way to discriminate more patterns is of course to consider multilocal observations.

# 4  $^2\mathcal{D}$ Images

The case of two dimensional images is not all that different. One difference is of course simply due to the dimension: In higher dimensions there is more freedom thus the possible generic singularities multiply.

Because the Gaußian kernel is separable in Cartesian coordinates there exists a close connection between the one and the two dimensional cases. If one has an image that changes only in one direction, the two dimensional case simply degenerates into the one dimensional case. Different from what one may perhaps expect this is actually an important case because *edges* and *lines* occur so frequently in practically interesting images.

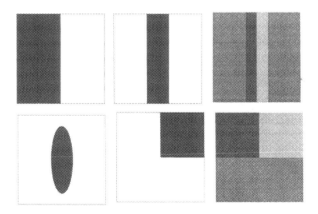

**Fig. 18.** *Examples of $^1\mathcal{D}$(!) images in the ($^2\mathcal{D}$!) plane.*

Truly $^2\mathcal{D}$ image features are more or less circular blobs, corners and such important entities as "T–junctions". In such cases one is again mainly interested in the effects of variation of a *single* parameter, in these cases the distance from the fiducial point is largely irrelevant. However, this case is quite different from that of a unidirectional variation of course. (Figure 18.) Much of computer vision is based virtually *completely* on effectively one–dimensional variations about special points ("edges", "bars", "junctions", "angles", "blobs").

The fact that these one–parameter variations are so often important leads to novel problems: How to *find* the direction of unidirectional variation (or, equivalently, the direction of invariancy) and how to find the fiducial "origins". Most often the answer will be relatively clear for only a sparse set of locations (because edges and lines don't fill areas, nor do singular points). In the former case one has to consider the additional *multilocal* problem of how to string the local directions together to extended edges or lines. I don't consider these later problems here because they arguably don't pertain to the front end proper. In order to "solve" the multilocal problem one will have to *commit* oneself to an *interpretation, i.e.*, one will select some evidence and ignore other evidence.

It is important to understand that such entities as "edges" are not so much properties of images or tasks as possible categories given by the order of representation of the observation. This is why the study of metamerism is so fundamental: It leads to a principled understanding of *what can be represented at all.*

## 4.1   Metamerism in $^2\mathcal{D}$ Images

One can easily transpose Schrödinger's method to higher dimensions and higher order representations. However, the number of categories will grow fast as either of these are increased. The case of dimension two, order two is still comparatively simple. Notice that the second order jet in dimension two is a *six* dimensional vector. One can construct a unique metameric image for any given image by considering additive combinations of sextuplets of point sources. Thus Schrödinger's argument works just as well for the lemma that optimal patterns have to be *binary* ones. The constraint on the *transitions* is now that it should be impossible to perturb it at six points in general position. If the transition curve is a general quadric one has exactly this situation: Since a quadric is fully determined by five points one can't find a sextuplet in general position. (In retrospect the pair of transition wavelengths in Schrödinger's original case are simply quadrics in one dimension.) One can indeed prove this hunch easily by writing down the condition $\det |\psi_j(x_k, y_k)| = 0$, where $j = 00, 10, 01, 20, 11, 02$, $k = 1, \ldots, 6$ and $(x_k, y_k)$ are points on the transition locus. This constraint is satisfied if $(x_6, y_6)$ satisfies a quadratic equation with the coefficients determined by the other coordinates: Thus the transition is indeed a general quadric. That it is indeed a *general* quadric is clear from the fact that one may specify the points $(x_1, y_1), \ldots, (x_5, y_5)$ arbitrarily.

Thus the optimal patterns are binary patterns with quadrics as boundaries. A general quadric in the plane is

$$1 + a_1 x + a_2 y + a_3 x^2 + a_4 xy + a_5 y^2 = 0, \tag{18}$$

and has five degrees of freedom. These degrees of freedom can be used to parameterize the boundary of the volume of possible jet activities. This volume is a convex subset of the space of *a priori* possible jet activities. Activities outside the volume cannot be induced by any image structure and if they nevertheless would occur can only be interpreted as "error messages".

**Fig. 19.** *The shape descriptor.*

Through a translation and a rotation of the coordinate system one can bring the quadric in the canonical form $(k_1 x^2 + k_2 y^2)/2 = 1$. Thus $(k_1, k_2)$ specify the *shape* of the quadric. Notice that one can still factor out the overall size and that an interchange of the axes leaves the shape invariant. If one introduces the parameter

$$\sigma = -\frac{2}{\pi} \arctan \frac{k_{max} + k_{min}}{k_{max} - k_{min}}, \quad k_{max} \geq k_{min}, \tag{19}$$

one has a convenient pure shape descriptor. (Figure 19.) The parameter $\sigma$ takes values on the segment $[-1, +1]$, where a sign change indicates a change of orientation (*e.g.*, white blob versus dark blob of the same shape). When $|\sigma| < 0.5$ one has hyperbolæ, for $|\sigma| > 0.5$ ellipses. For $|\sigma| = 1$ the quadrics are circles, whereas $|\sigma| = 0.5$ denotes the degenerated quadric (pair of parallel lines). This parameterization is useful because it gives us a handle on the topology of the image. When you consider maps from the image plane on the $\sigma$ domain you immediately see that generically you will have *regions* where the quadrics will be either elliptical or hyperbolical throughout, *curves* (that separate these regions) on which we find degenerated quadrics, and *isolated points* at which the quadrics are circular. In analogy with the classical differential geometry of surfaces I call the curves of degenerated quadrics "parabolic curves" and the points of circularity "umbilical points" or "umbilics".

Thus you can generically have light or dark blobs (elliptical boundaries) and light or dark bars (hyperbolical boundaries). If one restricts the attention to a circular disk of radius $\sqrt{4s}$ one finds

**Uniform images** either black or white,
**Edges** concave or convex according to some convention,
**Bars** black and white, curved and tapered and
**Blobs** either white or black on a black or white ground.

Thus one has obtained the conventional categories used to label simple cells in the mammalian primary visual cortex. You see that these categories are merely labels for certain canonical representatives of the metamers. A metamer is an *equivalence class* of local image structure, not any specific structure. Only if the "saturation" (to borrow a convenient term from color science) is very high does the metamer almost shrink to the canonical representative itself. The "meaning" of the activity of a jet is most uncertain for the low saturations. However, the uncertainty is due to not observable power at the higher orders, that is of higher resolution than one should consider in the first place: It is actually reasonable to disregard it. The fine detail should be studied at *another level resolution*. Thus one obtains both a *category* and a measure of *confidence*.

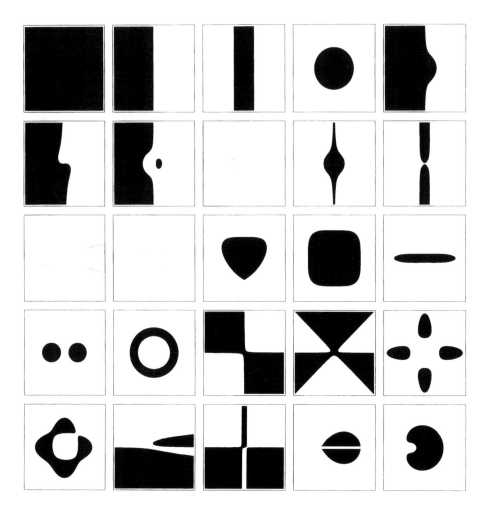

**Fig. 20.** *Some textons of $\mathcal{J}^4$.*

If you go to higher orders the mathematics becomes a little more unwieldy, but is not essentially different. For $\mathcal{J}^3$ the shapes are bounded by general cubics, thus the number of essentially different local image structures grows significantly: Newton counted seventy–two different cubics, to which he later added another six. For $\mathcal{J}^4$ the number is even larger. The classification of the quartics was first attempted by the abbé Bragelogne in the first half of the $18^{th}$ c., later refined by various authors, with decisive contributions by Euler, Cramer and Plücker, still later by Cayley and Zeuthen. Plücker counts 152 essentially different quartics. Since the order of representation in the human visual system is most likely four, this yields an estimate of the number of "textons" that the human observer might spontaneously ("pre–attentively") discriminate.

The $3^{rd}$ and $4^{th}$ order algebraic curves have been classified according to various criteria, some of them perhaps of little interest for the case of visual perception. One really needs a novel classification, based on *genericity* (many of the classical curves are singular cases) and up to affinities (many of the classical taxonomies are only up to projectivities). The major classes from a perceptual viewpoint are (articulated) blobs, bars and edges. The curves with a number of asymptotes exceeding three seem to hold little interest from a pragmatic point of view. (Figure 20.)

The distinction between *local* and *multilocal* properties is crucial. Strangely enough the distinction doesn't seem to have been made in image processing. This is probably the result of the fact that the notion of a coherent local representation of image structure (jets) is sadly lacking. It makes sense to distinguish sharply between *punctal properties* (the zeroth order), *local properties*, that are those properties that depend on derivatives at a point, that is to say the local jet, *multilocal properties* that depend on a jet, its neighbors and the connection, and *global properties* that depend on the image structure over arbitrary regions.

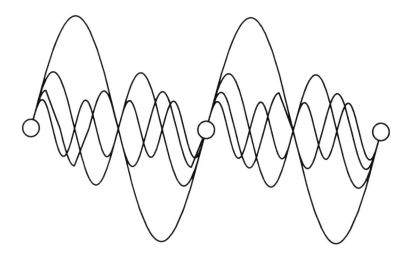

**Fig. 21.** *Some multilocal metameric images in* $\mathcal{J}^1$.

If one considers multilocal representations one has to take the sampling density into account. The sampling density that is required depends on the level of resolution and the order of representation. One can simply estimate the required sampling density from the Nyquist sampling theorem. This immediately suggests that there also will be *multilocal* metameric black images. For instance, suppose one has sampling locations in the one dimensional case at regular distances $\Delta$

(say) and that the order of representation is $n$. Then the functions

$$\sum_{k=n}^{\infty} \sin^{k+1} \frac{\pi x}{\Delta}, \tag{20}$$

are evidently metameric black patterns for the front end as a whole, for their value in the points $x = i$, $i = \ldots, -2, -1, 0, +1, +2, \ldots$ is zero and so are the values of the derivatives up to (and including) the $n^{th}$–order. In general one can construct metameric black patterns as deblurred $(n-1)^{th}$–order interpolating polynomials. Such patterns look like "noise" when you resolve them. (Figure 21.)

# References

1. Canny, J.F.: Finding edges and lines in images. Tech.Rep. **720** (1983) MIT, Cambridge Mass.
2. Hubel, D.H., Wiesel, T.,N.: Receptive fields of single neurons in the cat's striate cortex. J.Physiol. **148** (1959) 574–591
3. Koenderink, J.J., van Doorn, A.J.: Visual detection of spatial contrast: Influence of location in the visual field, target extent and illuminance level. Biol.Cybern. **30** (1978) 157–167
4. Koenderink, J.J., van Doorn, A.J.: The structure of images. Biol.Cybern. **50** (1984) 363–370
5. Koenderink, J.J.: The structure of the visual field. In: The physics of structure formation, Eds. W.Güttinger and G.Dangelmayr. (1986) Springer, Berlin, 68–77
6. Koenderink, J.J., van Doorn, A.J.: Representation of local geometry in the visual system. Biol.Cybern. **55** (1987) 367–375
7. Koenderink, J.J.: Color atlas theory. J.Opt.Soc.Am. **A4** (1987) 1314–1321
8. Koenderink, J.J., van Doorn, A.J.: Operational significance of receptive field assemblies. Biol.Cybern. **58** (1988) 163–171
9. Koenderink, J.J., van Doorn, A.J.: Receptive field families. Biol.Cybern. **63** (1990) 1–12
10. Koenderink, J.J.: The brain a geometry engine. Psychol.Res. **52** (1990) 122-127
11. Koenderink, J.J., van Doorn, A.J.: Receptive field taxonomy. In: Advanced neural computers, Ed. R.Eckmiller. (1990) Elsevier 295–301
12. Koenderink, J.J.: Mapping structures on networks. In: Artificial neural networks, Eds. T.Kohonen, K.Mäkisara, O.Simula and J.Kangas. (1991) Elsevier 93–98
13. Koenderink, J.J., van Doorn, A.J.: Generic neighborhood operators. IEEE Trans. PAMI **14** (1992) 597–605
14. Koenderink, J.J., van Doorn, A.J.: Receptive field assembly pattern specificity. J.Visual Comm. and Image Representation **3** (1992) 597–605
15. Koenderink, J.J., Kappers, A., van Doorn, A.J.: Local operations: The embodiment of geometry. In: Artificial and biological vision systems, Eds. G.A.Orban and H.–H.Nagel. (1992) Brussels: ESPRIT Basic Research Series 10–23
16. Ostwald, W.: Das absolute System der Farben. Z.Phys.Chem. **91** (1916) 132
17. Ostwald, W: Das absolute System der Farben. Z.Phys.Chem. **92** (1917) 222
18. Powell, J.L., Crasemann, B.: Quantum mechanics. (1961) Addison–Wesley, Reading Mass.
19. Ruskin, J.: The elements of drawing. (1857) Smith, Elder & Co., London

20. Schrödinger, E.: Grundlinien einer Theorie der Farbenmetrik im Tagessehen, I, II. Ann.Physik **63** (1920) 397 and 427
21. Schrödinger, E.: Theorie der Pigmente von größter Leuchtkraft. Ann.Physik **62** (1920) 603

# On Automatic Selection of Temporal Scales in Time-Causal Scale-Space

*Tony Lindeberg*

Computational Vision and Active Perception Laboratory (CVAP)
Department of Numerical Analysis and Computing Science
KTH, S-100 44 Stockholm, Sweden

**Abstract.** This paper outlines a general framework for automatic selection in temporal scale-space representations, and shows how the suggested theory applies to motion detection and motion estimation.

## 1 Introduction

A fundamental constraint on the design of a vision system originates from the fact that image structures are perceived as meaningful entities only over certain ranges of scale. In general situations, it is hardly ever possible to know in advance at what scales interesting structures can be expected to appear. For this reason, an image representation that explicitly incorporates the notion of scale is a crucially important tool when dealing with sensory data, such as images.

A multi-scale representation by itself, however, contains no explicit information about what image structures should be regarded as significant or what scales are appropriate for treating those. Early work addressing these problems for blob-like image structures was presented in (Lindeberg 1993a), leading to the notion of a scale-space primal sketch. Then, in (Lindeberg 1993b, 1996b) an extension to other aspects of image structures was presented by selecting scales for differential feature detectors (such as blobs, corners, edges and ridges) from maxima over scales of normalized differential entities.

The subject of this article is to address the problem of scale selection in the temporal domain, in order to deal with image data over time. Whereas, it is now rather generally accepted that some kind of "smoothing over time" is necessary when processing time-varying images, most current work on motion analysis is still carried out at a single temporal scale (see, e.g., (Barron *et al.* 1994; Beauchemin and Barron 1995)).

A main argument which will be advocated in this article, is that in analogy to earlier advances on spatial domains, the performance and robustness of algorithms operating over time can be improved substantially, if the spatio-temporal image data are considered at *several temporal scales simultaneously*, and if we incorporate *explicit mechanisms for automatic selection of temporal scales*.

To form the basis of a theory for temporal scale selection, we will start by showing how *time-causal normalized scale-space derivatives* can be defined for different types of time-causal scale-space concepts. Then, an adaptation of a previously proposed heuristic principle will presented, stating that in the absence of further information, important clues for spatio-temporal scale selection can be obtained from the scales at which (possibly non-linear) combinations of *normalized spatio-temporal derivatives assume maxima over scales*. Specifically, it will be shown how this approach applies to motion detection and velocity estimation.

## 2 Spatial and temporal scale-space: Overview

Traditionally, most work on scale-space representation has been concerned with the spatial domain, in which the values of the input signal are available in all coordinate directions. Given any $D$-dimensional signal $f: \mathbb{R}^D \to \mathbb{R}$, its (spatial) scale-space representation $L: \mathbb{R}^D \times \mathbb{R}_+ \to \mathbb{R}$ is defined by convolution

$$L(\cdot;\; s) = g(\cdot;\; s) * f \tag{1}$$

with the (rotationally symmetric) Gaussian kernel

$$g(x;\; s) = \frac{1}{(2\pi s)^{N/2}} e^{-x^T x/2s} \tag{2}$$

and scale-space derivatives are defined from this representation by $L_{x^\alpha}(\cdot;\; s) = \partial_{x^\alpha} L(\cdot;\; s)$ where $s \in \mathbb{R}_+$ is the scale parameter and $\alpha = (\alpha_1, \ldots, \alpha_D)$ represents the order of differentiation. As has been shown by several authors (Witkin 1983; Koenderink 1984; Yuille and Poggio 1986; Koenderink and van Doorn 1992; Florack 1993; Lindeberg 1994; Pauwels *et al.* 1995), the choice of the Gaussian kernel and its derivatives is basically a unique choice, given natural assumptions on a visual front-end (scale-space axioms).

This scale-space concept, however, cannot be directly applied to temporal data, since in a real-time situation it is essential that image operators do not extend into the future. One suggestion for how to deal with this problem was given by (Koenderink 1988), who proposed to transform the time axis so as to map the present moment to the unreachable infinity. In the transformed domain, he then applied the traditional scale-space concept given by (1) and (2). Based on a classification of scale-space kernels in the continuous and discrete domains, which guarantee non-creation of local extrema and respect the time direction as causal (Lindeberg 1990; Lindeberg and Fagerström 1996; Lindeberg 1997), three other types of temporal scale-space approaches can be distinguished:

*Continuous time and discrete scale parameter:* For continuous time, it turns out that all time-causal scale-space kernels can be decomposed into convolution with primitive *truncated exponential kernels*

$$h_{prim}(t;\; \mu) = \frac{1}{\mu} e^{-t/\mu} \qquad (t \geq 0) \tag{3}$$

having (possibly different) time constants $\mu$. For each such primitive filter, the mean is $\mu$ and the variance $\mu^2$. Hence, if we couple $k$ such filters in cascade, the equivalent convolution kernel will have a Laplace transform of the form

$$H_{composed}(s;\; \mu) = \int_{t=-\infty}^{\infty} \left( *_{i=1}^{k} h_{prim}(t;\; \mu_i) \right) e^{-st}\, dt = \prod_{i=1}^{k} \frac{1}{1 + \mu_i s}, \tag{4}$$

with mean (time delay) $\sum_{i=1}^{k} \mu_i$ and variance (effective integration time) $\sum_{i=1}^{k} \mu_i^2$.

*Discrete time with discrete scale parameter.* The discrete correspondence to the truncated exponential filters are *first-order geometric moving average filters* corresponding to the recurrence relation

$$f_{out}(t) - f_{out}(t-1) = \frac{1}{1+\mu}\left(f_{in}(t) - f_{out}(t-1)\right). \tag{5}$$

Such a primitive filter has mean $\mu$ and variance $\mu^2 + \mu$. Coupling $k$ such filters in cascade, gives a filter with generating function of the form

$$H_{composed}(z) = \sum_{n=-\infty}^{\infty} h_{composed}(n)\, z^n = \prod_{i=1}^{k} \frac{1}{1 - \mu_i\,(z-1)}, \tag{6}$$

with mean $\sum_{i=1}^{k} \mu_i$ and variance $\sum_{i=1}^{k}(\mu_i^2 + \mu_i)$. In the case of discrete time, also *time-shifted binomial kernels* satisfy temporal causality, and in this respect, discrete time allows for more degrees of freedom.

*Discrete time with continuous scale parameter.* The case of discrete time is special also in the sense that in this case, and only in this case, there is a non-trivial semi-group structure of scale-space kernels compatible with temporal causality. It corresponds to convolution with *Poisson kernels*

$$p(n;\ \lambda) = e^{-\lambda}\frac{\lambda^n}{n!} \tag{7}$$

which have mean $\lambda$, variance $\lambda$ and generating function $P(z;\ \lambda) = e^{\lambda(z-1)}$. Intuitively, this filter can be interpreted as the limit case of repeated convolution of geometric moving average filters (6) having time constants $\mu = \lambda/m$

$$\lim_{m\to\infty}\left(H_{geom}(z;\ \frac{\lambda}{m})\right)^m = \lim_{m\to\infty}\frac{1}{(1 - \frac{\lambda}{m}(z-1))^m} = P(z;\ \lambda). \tag{8}$$

For small values of the $\lambda$, these kernels are highly non-symmetric, whereas for large $\lambda$ they approach Gaussian kernels (having the same mean and variance).

This temporal scale-space concept can be regarded as the *canonical time-causal scale-space model*, since it is the *only* time-causal scale-space concept having a semi-group structure with a *continuous time-scale parameter* and guaranteeing non-creation of local extrema with increasing scales.

*Special properties of time-causal scale-spaces:* A fundamental difference between the temporal scale-space concepts and the spatial multi-scale representations is that the convolution kernels are non-symmetric. Each temporal channel is associated with an inherent time delay, reflecting the fact that there is no way to access real-world data at the very present moment. Any measurement requires a finite amount of energy, and hence integration over a certain time interval. This, implies computations over non-zero time-scales, and non-zero time delays.

# 3 Automatic scale selection: A general principle

The presentation so far provides a theoretical framework for *representing* image data at different spatial and temporal scales. When to use it in practice, basic problems concern how to determine what structures should be regarded as significant and what scales are appropriate for handling those.

A general principle for scale selection for feature detectors defined in a spatial scale-space representation has been proposed in (Lindeberg 1993b, 1994, 1996b). It is formulated in terms of the evolution properties over scales of image descriptors expressed in terms of $\gamma$-*normalized derivatives* defined by

$$\partial_\xi = s^{\gamma/2} \partial_x, \tag{9}$$

where $s \in \mathbb{R}_+$ denotes the scale parameter, $\gamma > 0$ is a free parameter and $\xi$ represents the $\gamma$-normalized coordinate of the (here, 1-D) variable $x$.

For an $r$th-order Gaussian derivative operator $g_{\xi^r}(\cdot;\ s)$ normalized in this way, it can be shown that the evolution over scales of its $L_p$-norm is given by

$$\|g_{\xi^r}(\cdot;\ s)\|_p = \sqrt{s}^{\,|r|\,(\gamma-1)+D(1/p-1)}\|g_{\xi^r}(\cdot;\ 1)\|_p. \tag{10}$$

Hence, this normalization corresponds to the $L_p$-norm of the equivalent normalized Gaussian derivative kernels $\partial_\xi^r g(x;\ s)$ being constant over scales, iff

$$p = \frac{1}{1 + \frac{|r|}{D}\,(1-\gamma)}. \tag{11}$$

The basic idea of the scale selection method is that in the absence of further evidence, *the scale levels at which some (possibly non-linear) combination of such normalized derivatives assume maxima over scales can be treated as reflecting characteristic lengths of corresponding structures in the data.* As support for this approach, the following evidence can be presented (Lindeberg 1996b):

- A general theoretical analysis showing that for large classes of differential invariants, local maxima over scales of such normalized differential entities will be preserved under rescalings of the input pattern.
- Theoretical analysis of model signals for which closed-form analysis is tractable.
- Simulation results for real-world and synthetic images.

The first works on this scale selection methodology (Lindeberg 1993b, 1994) were concerned with the case $\gamma = 1$, and it was shown that for this value of $\gamma$, the scale selection methodology commutes with size variations of the input pattern. More generally, from the transformation property of these $\gamma$-normalized derivatives under a rescaling of the input $f(x) = f'(sx)$ by a factor $s$

$$\partial_\xi^r L(x;\ s) = s^{r\,(1-\gamma)}\,\partial_{\xi'}^r L'(x';\ s'), \tag{12}$$

it is rather straightforward to show that homogeneous polynomial differential expressions scale according to a power law under size variations of the input, implying that local maxima of over scales (as well as in space) will be preserved.

Conversely, given the idea that scale selection should be performed in an analogous way as image features are computed on a spatial domain — from local spatial maxima of operator responses — one may then ask how responses from operators of different size should be normalized. Indeed, it can be shown (Lindeberg 1996b) that the $\gamma$-normalized derivative concept arises *by necessity*, given the following natural assumptions:

- local maxima over scales should be preserved under rescalings of any (non-trivial) image pattern,
- the only additional source of information that could be used for normalizing the operation is the scale parameter,
- at any scale, the spatial maxima should be preserved for feature detectors expressed as homogeneous differential expressions.

Hence, the $\gamma$-normalized derivative concept spans the class of reasonable normalizations for a scale selection procedure based on local maxima over scales.

## 4    Dense frequency estimation based on quasi quadrature

The scale selection methodology presented in previous section has mainly been applied to the detection of *sparse image features*, such as blobs, corners, edges and ridges. In many situations, however, we are also interested in the computation of *dense* image descriptors, such as texture descriptors and optic flow.

An obvious problem that arises if a scale selection mechanism is to be based on a linear combination of partial derivatives, such as the Laplacian operator, is that there could be large spatial variations in the operator response. In signal processing, a common methodology for reducing this so-called phase dependency is by using *quadrature filter pairs*, defined (from a Hilbert transform) in such a way that the Euclidean sum of the filter responses will be phase independent for any sine wave. The Hilbert transform of a Gaussian derivative kernel is, however, not within the Gaussian derivative family, and we are here interested in operators of small support which can be expressed within the scale-space framework.

Given the normalized derivative concept, there is a straightforward way of combining Gaussian derivatives into an entity that gives an approximately constant operator response at the scale given by the scale selection mechanism. At any scale $t$ in the scale-space representation $L$ of a one-dimensional signal $f$, define the following *quasi quadrature* entity in terms of normalized derivatives based on $\gamma = 1^1$ by

$$QL = L_\xi^2 + C\,L_{\xi\xi}^2 = s\,L_x^2 + C\,s^2\,L_{xx}^2, \tag{13}$$

where $C$ is a free parameter (to be determined). This approach bears close relationship to the idea by (Koenderink and van Doorn 1987) to regard derivatives

---

[1] Since the differential expression $QL$ is inhomogeneous, we must require $\gamma = 1$ for the scale selection procedure to commute with size variations in the input pattern.

of odd and even order as local sine and cosine functions. Specifically, for any sine wave $f(x) = \sin \omega_0 x$, we have

$$(QL)(x;\ s) = s\,\omega_0^2\,e^{-\omega_0^2 s}\left(1 + (C\,s\,\omega_0^2 - 1)\sin^2 \omega_0 x\right). \tag{14}$$

As can be seen, the spatial variations in $QL$ will be large when $s\,\omega_0^2$ is either much smaller or much larger than one, whereas the relative spatial oscillations decrease to zero when $s$ approaches $1/(C\,\omega_0^2)$.

To obtain an intuitive understanding of how the choice of $C$ affects local maxima of $QL$ over scales, let us differentiate (14):

$$s_{QL}(x) = \frac{1}{\omega_0^2}\left(1 + \frac{2C\,\sin^2(\omega_0 x)}{\cos^2(\omega_0 x) + \sqrt{\cos^4(\omega_0 x) + 4C^2\sin^4(\omega_0 x)}}\right). \tag{15}$$

Notably, the extreme values $s_{QL}|_{\omega_0 x = 0} = \frac{1}{\omega_0^2}$ and $s_{QL}|_{\omega_0 x = \frac{\pi}{2}} = \frac{2}{\omega_0^2}$ are independent of $C$, and graphs showing the spatial variation for a few values of $C$ are displayed in figure 1. Given the form of these curves, a natural symmetry requirement can be stated as

$$s_{QL}|_{\omega_0 x = \frac{\pi}{4}} = \frac{1}{2}\left(s_{QL}|_{\omega_0 x = 0} + s_{QL}|_{\omega_0 x = \frac{\pi}{2}}\right) \quad \Rightarrow \quad C = \frac{2}{3} \approx 0.6667. \tag{16}$$

In this respect, $C = \frac{2}{3}$ gives the most symmetric variation of selected scales w.r.t. the information contents in the first-order and second-order derivatives.

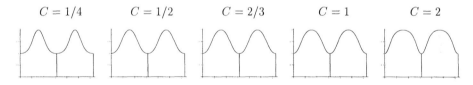

|  $C = 1/4$ |  $C = 1/2$ |  $C = 2/3$ |  $C = 1$ |  $C = 2$ |

**Fig. 1.** Spatial variation of the *selected scale levels* when maximizing the quasi quadrature entity (13) over scales for different values of the free parameter $C$ using a one-dimensional sine wave of unit frequency as input pattern. Observe that $C = 2/3$ (equation (16)) gives rise to the most symmetric variations in the selected scale values.

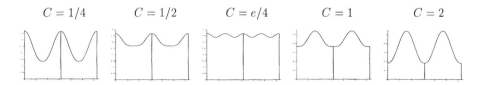

|  $C = 1/4$ |  $C = 1/2$ |  $C = e/4$ |  $C = 1$ |  $C = 2$ |

**Fig. 2.** Spatial variation of the *maximum value over scales* of the quasi quadrature entity (13) computed for different values of the free parameter $C$ for a one-dimensional sine wave of unit frequency. As can be seen, the smallest spatial variations in the amplitude of the maximum response are obtained for $C = e/4$ (equation (17)).

Another interesting factor to analyse is the variation in magnitude at the selected scales. Insertion of the scale values according to (15) into the quasi quadrature measure (13) gives spatial variations of as displayed in figure 2. To determine $C$, a simple minimum-ripple condition is to require that

$$QL|_{\substack{\omega_0 x = 0 \\ s = s_{Q,0}}} = QL|_{\substack{\omega_0 x = \frac{\pi}{2} \\ s = s_{Q,\frac{\pi}{2}}}} \quad \Rightarrow \quad C = \frac{e}{4} \approx 0.6796. \tag{17}$$

In other words, also a determination of $C$ based on small spatial variations in the magnitude measure computed at the selected scales gives rise to an approximately similar value of $C$ as the abovementioned symmetry requirement.

Moreover, note that $C = e/4$ corresponds to normalizing the first- and second-order Gaussian derivative kernels to having the same $L_1$-norm.

## 5  Normalized time-causal scale-space derivatives

The theory for automatic scale selection in section 3 applies to a scale-space on a continuous domain having a continuous scale parameter. Notably, however, the temporal scale-space concepts given by (4), (6) and (7) imply that either time or time-scale have to be discretized. Hence, a fundamental problem concerns how to normalize the derivative (or derivative approximation) responses over scales. Given the basic properties in section 3, natural constructions to consider are:

- multiply the derivative/difference operator by the variance of the smoothing kernel raised to the order of differentiation (in analogy with (9)),
- normalize the equivalent derivative/difference kernels to constant $L_p$-norm (or discrete $l_p$-norm) over scales (in analogy with (11)).

In this section, we shall describe properties of these approaches for the Poisson-type scale-space given by (7). A corresponding analysis for the two other temporal scale-space concepts is given in (Lindeberg 1996d).

To simplify the calculations, let us restrict ourselves to $\gamma = 1$, corresponding to normalization to constant $L_1/l_1$-norm for any order of differentiation.

*Normalized differences in the Poisson-type scale-space.* For the first-order backward difference of the Poisson kernel

$$(\delta_t p)(n;\ \lambda) = p(t;\ \lambda) - p(t-1;\ \lambda) = -\left(\frac{t}{\lambda} - 1\right) p(t;\ \lambda), \tag{18}$$

we can use the fact that $p(n;\ \lambda)$ assumes a local maximum at $n = [\lambda]$ to compute

$$\| (\delta_t p)(\cdot;\ \lambda) \|_1 = \sum_{n=-\infty}^{\infty} |p(n;\ \lambda) - p(n-1;\ \lambda)| = 2\, p([\lambda];\ \lambda) = \frac{2\, e^{-\lambda} \lambda^{[\lambda]}}{[\lambda]!}.$$

For small $\lambda$ ($\lambda < 1$), we have $\| (\delta_t p)(\cdot;\ \lambda) \|_1 = 2\, e^{-\lambda} = 2\,(1 - \lambda + \mathcal{O}(\lambda^2))$, whereas for large $\lambda$ Stirlings formula $n! = n^n e^{-n} \sqrt{2\pi n}\,(1 + \frac{1}{12n} + \mathcal{O}(\frac{1}{n^2}))$ gives

$$\| (\delta_t p)(\cdot;\ \lambda) \|_1 = \frac{2}{\sqrt{2\pi[\lambda]}} \left(1 - \frac{(\lambda - [\lambda])^2}{2\,[\lambda]} - \frac{1}{12\,[\lambda]} + \mathcal{O}\left(\frac{1}{[\lambda]^2}\right)\right). \tag{19}$$

Concerning the second-order differences, we can use the inflexion points at $n_{1,2} = \lambda + \frac{1}{2} \pm (\lambda + \frac{1}{4})^{1/2}$ to reduce the result to first-order differences (18)

$$\| (\delta_{tt}p)(\cdot; \ \lambda) \|_1 = 2 \sum_{n=n_1}^{n_2} |(\delta_{tt}p)(n; \ \lambda)| = 2 ((\delta_t p)([n_1]; \ \lambda) - (\delta_t p)([n_2]; \ \lambda)).$$

Unfortunately, it is hard to simplify this expression. For small $\lambda$, however, $\| \delta_{tt}p(\cdot; \ \lambda) \|_1 = -\delta_t p(1; \ \lambda) = (1 - \lambda)e^{-\lambda} = 1 - 2\lambda + \mathcal{O}(\lambda^2)$, and for large $\lambda$

$$\| (\delta_{tt}p)(\cdot; \ \lambda) \|_1 \approx \frac{4}{\sqrt{2\pi e}} \frac{1}{\lambda}. \tag{20}$$

As could be anticipated, variance-based normalization and normalization to constant $l_1$-norm approach each other[2] with increasing temporal scales (and decreasing effects of grid sampling). For small $\lambda$, on the other hand, where the sampling effects may be strong, the results differ significantly.

## 6    Selection of temporal scales: Intuitive ideas

To understand the consequences of selecting local maxima over scales of normalized temporal derivatives (derivative approximations), let us first consider the response properties for a (phase shifted) sine wave, which has been used as an illustrative example in the early developments of the scale selection methodology:

$$f(n) = \cos(\nu n + \varphi) \qquad (n \in \mathbb{Z}). \tag{21}$$

The Fourier transform of the Poisson kernel is $\mathcal{F}(p)(\omega; \ \lambda) = \sum_{n=-\infty}^{\infty} p(n; \ \lambda) e^{-in\omega} = P(e^{-i\omega}; \ \lambda) = e^{-\lambda(1-\cos\omega)} e^{i\lambda\sin\omega}$ and for the $r$:th order backward difference

$$\mathcal{F}(\delta_-^r)(\omega) = (1 - e^{-i\omega})^r = (1 - \cos\omega + i\sin\omega)^r = (2\sin(\tfrac{\omega}{2}))^r e^{ir(\pi-\omega)/2}.$$

Thus, the closed-form expression for the $r$:th order scale-space derivative of $f$ is

$$(\delta^{(r)}L)(n; \ \lambda) = e^{-\lambda(1-\cos\nu)} (2\sin(\tfrac{\nu}{2}))^r \cos(n\nu + \varphi + \lambda\sin\nu + r\tfrac{\pi-\nu}{2})$$

and the amplitude varies according to $\hat{L}_{t^r}(\lambda) = (2\sin(\tfrac{\nu}{2}))^r e^{-\lambda(1-\cos\nu)}$.

*Variance-based normalization.* If we normalize the discrete derivative approximation operator based on the variance of the equivalent convolution kernel, this corresponds to the *normalized backward difference operator*

$$\delta_\tau^r = \delta_{-,norm}^r = \lambda^{r/2} \delta_-^r, \tag{22}$$

where $\tau$ represents the temporal coordinate normalized with respect to temporal scale. Hence, the response will first increase and then decrease with scale,

$$\hat{L}_{\tau^r}(\lambda) = (2\sin(\tfrac{\nu}{2}))^r \lambda^{r/2} e^{-\lambda(1-\cos\nu)}, \tag{23}$$

---

[2] Recall that for the $L_1$-norms of the first- and second-order normalized Gaussian derivatives, we have $\| g_\xi(\cdot; \ s) \|_1 = (2/\pi)^{1/2}$ and $\| g_{\xi^2}(\cdot; \ s) \|_1 = (8/(\pi e))^{1/2}$.

and there is a unique maximum over scales at

$$\lambda_{\hat{L}_{\tau^r},max} = \frac{r}{4\sin^2(\frac{\nu}{2})}. \tag{24}$$

Thus, in a agreement with the results from the spatial domain (Lindeberg 1993b, 1994), *a local maximum over scales provides a qualitative measure of the approximate range of scales over which temporal variations occur*. If we insert the scale value (24) into (23), we see that the maximum normalized response

$$\hat{L}_{\tau^r}(\lambda_{\hat{L}_{\tau^r},max}) = \left(\frac{r}{e}\right)^{r/2} \tag{25}$$

is independent of $\nu$. Thus, *all frequencies are treated in a uniform manner*. Moreover, there are strong similarities between the results from this construction and their counterparts based on the Gaussian scale-space concept. For a sine wave with frequency $\nu$, there is a unique maximum over scales at $s = r/\nu^2$, and the maximum normalized response over scales is of the same form as (25).

This similarity is, in fact, not surprising. If we rewrite the Fourier transform of the Poisson kernel as $\mathcal{F}(p)(\omega;\ \lambda) = \exp(-\lambda\,(2\sin\frac{\omega}{2})^2/2)\exp(i\lambda\sin\omega)$, we see that the magnitude of $\mathcal{F}(p)(\omega;\ \lambda)$ is equal to the magnitude of the Fourier transform of the Gaussian kernel $\mathcal{F}(g)(\omega;\ s) = \exp(-s\omega^2/2)$ with $\omega$ replaced by $2\sin(\omega/2)$. Thus, disregarding the phase information, these kernels can be (formally) mapped to each other by a simple frequency warping.

*Normalization to constant $l_1$-norm.* To study the effect of normalization to constant $l_1$-norm, divide (22) by the $l_1$-norm of the first-order difference of the Poisson kernel (19). This gives a normalized response of the form

$$\hat{L}_{t^r,discrette}(\lambda) = \sin(\tfrac{\nu}{2})\,[\lambda]!\,\frac{e^{\lambda\,\cos\nu}}{\lambda^{[\lambda]}}. \tag{26}$$

Clearly, this function decreases with $\lambda$ when $\cos\nu < 1$, showing that a single frequency with $|\nu| \geq \pi/2$ cannot give rise to a local maximum with $\lambda > 0$. If we insert this frequency into the continuous expression (24), we obtain a rule of thumb saying that if we would like to compute a derivative of order $r$, the scale level should preferably not be lower than $\lambda_{min}(r) = r/2$.

When $[\lambda] = 0$, *i.e.*, when $\lambda \in [0,1[$, the normalized response increases with $\lambda$ if $|\nu| < \pi/2$ (and decreases otherwise). If $[\lambda] = 1$, *i.e.*, when $\lambda \in [1,2[$, there is a local maximum at $\lambda = 1/\cos\nu$ in this interval if $\cos\nu \in [\frac{1}{2},1[$. Similarly, if $[\lambda] = 2$, *i.e.*, when $\lambda \in [2,3[$, there is a local maximum at $\lambda = 2/(\cos\nu)$ in this interval if $\cos\nu \in [\frac{2}{3},1[$. This pattern shows how in the case of $l_1$-normalization, a single frequency may give rise to multiple responses over scales.

*Cascade-coupled first-order integrators.* If we couple $k$ truncated exponential filters having equal time constants $\mu_i = \mu$ in cascade, and define a variance-based *normalized derivative operator* by

$$\partial_\tau^r = (k\,\mu^2)^{r/2}\,\partial_t^r, \tag{27}$$

then the maximum over scales will be assumed in layer

$$k_{\hat{L}_{\tau^r},max} \approx \frac{r}{\log(1+\mu^2\nu^2)} = \frac{r}{\mu^2\nu^2}\left(1 + \mathcal{O}(\frac{1}{\mu^2\nu^2})\right) \qquad (28)$$

corresponding to variance $\lambda_{\hat{L}_{\tau^r},max} = k_{\hat{L}_{\tau^r},max}\,\mu^2$ of the equivalent convolution kernel. The maximum normalized response is

$$\hat{L}_{\tau^r}(\mu, k_{\hat{L}_{\tau^r},max}) = \left(\frac{r}{e}\right)^{r/2}\left(\frac{\mu^2\nu^2}{\log(1+\mu^2\nu^2)}\right)^{r/2}. \qquad (29)$$

*Cascade-coupled first-order recursive filters.* If we in an analogous way couple $k$ geometric moving average filters having equal time constants $\mu_i = \mu$ in cascade, and define a variance-based normalized difference operator by

$$\delta^r_{-,norm} = (k\,(\mu^2+\mu))^{r/2}\,\delta^r_-, \qquad (30)$$

the maximum over scales will be assumed in layer

$$k_{\hat{L}_{\tau^r},max} \approx \frac{r}{\log(1+4\,(\mu^2+\mu)\,\sin^2(\frac{\nu}{2}))} \qquad (31)$$

corresponding to variance $\lambda_{\hat{L}_{\tau^r},max} = k_{\hat{L}_{\tau^r},max}\,(\mu^2+\mu)$ of the equivalent convolution kernel. The maximum normalized amplitude over scales is

$$\hat{L}_{\tau^r}(\mu, k_{\hat{L}_{\tau^r},max}) = \left(\frac{r}{e}\right)^{r/2}\left(\frac{4\,(\mu^2+\mu)\,\sin^2(\frac{\nu}{2})}{\log(1+4\,(\mu^2+\mu)\,\sin^2(\frac{\nu}{2}))}\right)^{r/2}. \qquad (32)$$

Thus, local maxima over scales of these normalized derivative operators reflect similar properties as for the Poisson-type scale-space.

*Scale invariance properties.* Concerning the behaviour of the scale-selection method under size variations of the input pattern, we cannot, of course, aim at perfect scale invariance for the temporal scale-space concepts defined on discrete grids. For the temporal scale-space for a continuous domain (4), on the other hand, it can be shown that perfect scale invariance can be accomplished if we allow the time constants in the discrete set of filters to be variable.

## 7 Response properties for basic model patterns

To carry out closed-form analysis for signals having richer frequency contents, consider the behaviour in the Poisson-type scale-space of the following signals

$$f_{blob}(n) = p(n;\ \lambda_0), \qquad (33)$$

$$f_{edge}(n) = \Psi(n;\ \lambda_0) = \sum_{i=-\infty}^{n} p(i;\ \lambda_0). \qquad (34)$$

$f_{blob}$ can be interpreted as idealized models of a time pulse, while $f_{edge}$ models the edge of a new object that enters or leaves the visual field. The temporal extent of the pulse and the diffuseness of the edge are determined by $\lambda_0$.

$$L_{blob}(n;\ \lambda) \qquad (\delta_\tau L_{blob})(n;\ \lambda) \qquad (\delta_{\tau\tau} L_{blob})(n;\ \lambda)$$

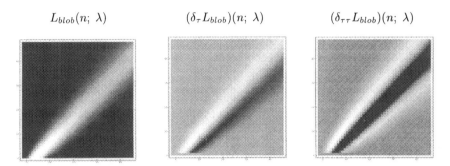

**Fig. 3.** The zero-, first- and second-order normalized responses in the Poisson-type scale-space representation of an *idealized blob* (time pulse) with a Poisson-shaped profile with temporal extent given by $\lambda_0 = 4$. Observe the characteristic increase in the time delay with increasing values of the scale parameter. Moreover, note that the scale at which the maximum over scales is assumed *increases with time and then decreases*. (Horizontal axis: time, vertical axis: temporal scale.)

$$L_{edge}(n;\ \lambda) \qquad (\delta_\tau L_{edge})(n;\ \lambda) \qquad (\delta_{\tau\tau} L_{edge})(n;\ \lambda)$$

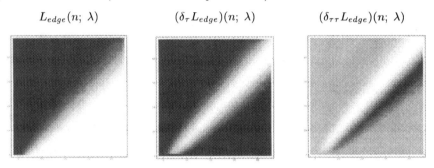

**Fig. 4.** The zero-, first- and second-order normalized responses in the Poisson-type scale-space representation of an *idealized edge* (time ramp) with intensity profile corresponding to an integrated Poisson kernel with diffuseness $\lambda_0 = 4$. Observe the characteristic increase in the time delay with increasing values of the scale parameter. Moreover, note that the scale at which the maximum over scales is assumed *increases monotonically with scale*. (Horizontal axis: time, vertical axis: time-scale.)

$$\sqrt{(\mathcal{Q}L_{blob})(n;\ \lambda)} \qquad \sqrt{(\mathcal{Q}L_{edge})(n;\ \lambda)}$$

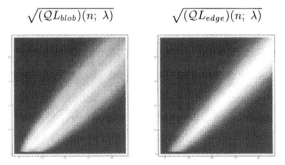

**Fig. 5.** The result of computing (the square root of) the *quasi quadrature measure* $\mathcal{Q}L = L_\tau^2 + C\, L_{\tau\tau}^2$ in the Poisson-type scale-space representations of the *idealized blob* signal in figure 3 and for the *idealized edge* signal in figure 4. Observe how this entity is less sensitive to the local phase information, while the qualitative properties of the scale selection are preserved. (Horizontal axis: time, vertical axis: time-scale.)

*Descriptors based on the Poisson-type scale-space.* From the semi-group property $p(\cdot;\ \lambda_1) * p(\cdot;\ \lambda_2) = p(\cdot;\ \lambda_1 + \lambda_2)$, it follows that the Poisson-type scale-space representations of these signals are given by

$$L_{blob}(n;\ \lambda) = p(n;\ \lambda + \lambda_0) \tag{35}$$

$$L_{edge}(n;\ \lambda) = \Psi(n;\ \lambda + \lambda_0), \tag{36}$$

and the first- and second-order normalized derivative approximations are

$$(\delta_\tau L_{blob})(n;\ \lambda) = \sqrt{\lambda}\,(\delta_t p)(n;\ \lambda + \lambda_0), \tag{37}$$

$$(\delta_{\tau\tau} L_{blob})(n;\ \lambda) = \lambda\,(\delta_{tt} p)(n;\ \lambda + \lambda_0), \tag{38}$$

$$(\delta_\tau L_{edge})(n;\ \lambda) = \sqrt{\lambda}\,p(n;\ \lambda + \lambda_0), \tag{39}$$

$$(\delta_{\tau\tau} L_{edge})(n;\ \lambda) = \lambda\,(\delta_t p)(n;\ \lambda + \lambda_0), \tag{40}$$

with $\delta_t p$ according to (18) and the second-order differences $\delta_{tt} p$ defined by

$$(\delta_{tt} p)(n;\ \lambda) = (\delta_i^2 p)(n;\ \lambda) = \left(\frac{n(n-1)}{\lambda^2} - \frac{2n}{\lambda} + 1\right) p(t;\ \lambda). \tag{41}$$

Figure 3–4 show $L_{blob}(n;\ \lambda)$ and $L_{edge}(n;\ \lambda)$ with their first- and second-order normalized differences as grey-level functions of $n$ and $\lambda$. Figure 5 shows corresponding results for a quadrature measure defined by

$$QL = L_\tau^2 + C\,L_{\tau\tau}^2 = \lambda(\delta_t L)^2 + C\,\lambda^2(\delta_{tt} L)^2, \tag{42}$$

where $L_\xi$ and $L_{\xi\xi}$ represent normalized discrete derivative approximations. Since for the non-symmetric backward difference operator, derivative approximation operators of different orders are associated with different time delays, the symmetric time-shifted difference operator $\delta_0 f(t) = (f(t) - f(t - 2))/2$ was used for computing the first-order derivative approximation. By this discretization, the first- and second-order difference operators will have the same time delay.

*Basic effects of the scale selection method.* For both signals, we see that there is a characteristic increase in the time delay when the scale parameter is increased. For small $t$ (compared to $\lambda_0$), the maximum over scales is assumed at fine scales, and this scale (as well as the maximum magnitude over scales) increase with time. These results illustrate one of the basic properties of temporal scale-space—for a new (isolated) object that enters the visual field, fine scales will be selected initially, since the object is only visible at the finest temporal scales. Then, with increasing time, this scale can be expected to increase, if we maximize a differential entity such as $Q$ over scales.

## 8 Spatio-temporal patterns and motion estimation

Let us now apply simultaneous selection of spatial scales and temporal scales to the computation of spatio-temporal derivatives and analyse its implications with respect to the analysis of motion data.

## 8.1 Motion of a one-dimensional sine wave

Consider first a one-dimensional sine wave moving with velocity $c$

$$f(x,t) = \cos(\nu(x - ct)),\tag{43}$$

and define the separable spatio-temporal scale-space representation of this signal as the tensor product of the spatial and temporal scale-space concepts. Moreover, to obtain compact closed-form expressions for the results (with the discretization aspects suppressed), let us first model the temporal scale-space representation as the convolution with Gaussian kernels having the same mean and variance

$$g(x; \lambda) = \frac{1}{\sqrt{2\pi\lambda}} e^{-(x-\lambda)^2/(2\lambda)}.\tag{44}$$

Whereas this family does not satisfy temporal causality, it constitutes a reasonable approximation of the behaviour at coarse temporal scales, and we obtain a (two-parameter) spatio-temporal scale-space representation $L$ of $f$ of the form

$$L(x,t;\ s,\lambda) = e^{-\nu^2 s/2} e^{-\nu^2 c^2 \lambda/2} \cos(\nu(x - ct + \lambda)).\tag{45}$$

If we independently select spatial and temporal scale levels from the maxima over scales of the amplitude of the first-order normalized spatial and temporal derivatives respectively, it follows that the selected spatial scale level will be $s = \frac{1}{\nu^2}$ and the temporal scale level $\lambda = 1/(c^2\nu^2)$. In other words, *the ratio between the selected spatial and temporal scales reflects the velocity*

$$c = \sqrt{\frac{s}{\lambda}}.\tag{46}$$

This behaviour bears close relationships to frequency-based motion approaches (Adelson and Bergen 1985; Heeger 1988; Fleet 1992). (A fundamental difference, however, is that there is no need for specifying frequencies manually, and the velocity is obtained fully automatically.) Moreover, from straightforward differentiation (assuming that it $L_x \neq 0$), it follows that the velocity estimate is

$$\hat{c} = -\frac{L_t}{L_x} = c = \sqrt{\frac{s}{\lambda}},\tag{47}$$

*i.e.*, for this ideal and noise free signal the resulting velocity estimate is independent of the scale parameters $s$ and $\lambda$.

To formulate an algorithm that allows for simultaneous determination of spatial and temporal scales adapted to the scale levels at which dominant variations occur in the signal, a straightforward approach is to extend the definition of the quadrature measure (13) to the spatio-temporal domain

$$\max_{s,\lambda}(\mathcal{Q}_{prod(x,t)}L)(x,t;\ s,\lambda) = \max_{s,\lambda}(L_\xi^2 + C\,L_{\xi\xi}^2)\,(L_\tau^2 + C\,L_{\tau\tau}^2).\tag{48}$$

and to maximize the entity over spatial as well as and temporal scales.

Figure 6 shows the result of computing this entity at different spatio-temporal scales for an arbitrary point $(x, t)$ in space-time and for a few combinations of $\nu$ and $c$. (Here, the scale parameters have been parameterized by the effective spatial scale $s_{\mathrm{eff}} = \log_2 s$ and the effective temporal scale $\lambda_{\mathrm{eff}} = \log_2 \lambda/2$.) Observe how we in this way obtain information about the dominant spatial and temporal frequencies around $(x, t)$. Specifically, the ratio between these scale levels serves as a direct estimate of the dominant motion of the first- and second-order image structures at this spatio-temporal scale.

Figure 7 shows corresponding results for two superimposed sine waves, of different spatial frequencies $\nu_i$, which move with different velocities $c_i$:

$$L(x, t; \ s, \lambda) = e^{-\nu_1^2 s/2} e^{-\nu_1^2 c_1^2 \lambda/2} \cos(\nu_1 (x - c_1 t + \lambda)) \tag{49}$$

$$+ \, e^{-\nu_2^2 s/2} e^{-\nu_2^2 c_2^2 \lambda/2} \cos(\nu_2 (x - c_2 t + \lambda)). \tag{50}$$

Note how multiple responses over scales are obtained, indicating the ability of this approach to capture multiple transparent motions.

$\nu = 1/4, \ c = 1$  $\nu = 1/16, \ c = 1/2$  $\nu = 1/64, \ c = 1/4$

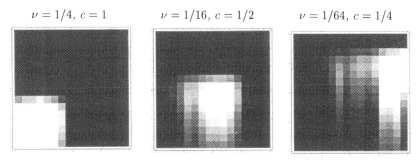

**Fig. 6.** $\lambda_{\mathrm{eff}}$-$s_{\mathrm{eff}}$-diagram showing the normalized spatio-temporal quasi quadrature measure $\mathcal{P}_{prod(x,t)} L$ as function of effective temporal scale $\lambda_{\mathrm{eff}} = \log_2 \lambda/2$ (horizontal axis) and effective spatial scale $s_{\mathrm{eff}} = \log_2 s/2$ (vertical axis). Observe how the dominant peak at $(\lambda, s) \approx 1/\nu^2 (1/c^2, 1)$, serves as an indicator of the dominant spatial frequency $\nu$ as well as the velocity $c$.

$\nu_1 = 1/64, \ c_1 = 1.$  $\nu_1 = 1/64, \ c_1 = 1/4.$
$\nu_2 = 1/4, \quad c_2 = 1/4.$  $\nu_2 = 1/16, \ c_2 = 1.$

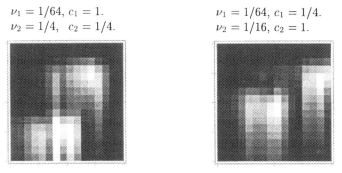

**Fig. 7.** Corresponding $\lambda_{\mathrm{eff}}$-$s_{\mathrm{eff}}$-diagram of $\mathcal{P}_{prod(x,t)} L$ for *two superimposed sine waves* of frequencies $\nu_1$ and $\nu_2$ which move with velocities $c_1$ and $c_2$, respectively. Notably, such transparent motion gives rise to multiple responses in spatio-temporal scale-space provided that the difference in frequency and velocity is sufficiently large.

## 8.2 Velocity estimation

A more traditional approach for computing velocity estimates is in terms of spatio-temporal derivatives or other filter outputs. Here, it will be illustrated how the proposed scale selection methodology can be applied to such problems.

Consider the motion constraint equation (Horn and Schunck 1981) in the case of a one-dimensional spatial domain, and differentiate this relation once:

$$\begin{cases} L_x + L_t = 0, \\ cL_{xx} + L_{xt} = 0. \end{cases} \tag{51}$$

In terms of normalized spatio-temporal derivatives, we can write:

$$\begin{cases} L_\xi + \sqrt{\frac{s}{\lambda}} L_\tau = 0, \\ cL_{\xi\xi} + \sqrt{\frac{s}{\lambda}} L_{\xi\tau} = 0. \end{cases} \tag{52}$$

Ideally, these equations should lead to the same solution. On real-world data, however, such consistency can hardly be expected. Therefore, let us solve them in a least-squares sense using weights as imposed by the normalized derivatives:

$$\min_c \ (cL_\xi + \sqrt{\frac{s}{\lambda}} L_\tau)^2 + C \, (cL_{\xi\xi} + \sqrt{\frac{s}{\lambda}} L_{\xi\tau})^2 = \min_c \ \alpha c^2 + 2\beta c + \gamma. \tag{53}$$

Differentiation with respect to $c$ shows that the velocity estimate is given by

$$c = -\frac{\beta}{\alpha} = -\sqrt{\frac{\sigma}{\lambda}} \left( \frac{L_\xi L_\tau + L_{\xi\xi} L_{\xi\tau}}{L_\xi^2 + L_{\xi\xi}^2} \right), \tag{54}$$

where the differential expression within parentheses gives the velocity estimates in units of current spatio-temporal scale. Insertion of this value into (53) gives a residual of the form $\rho = (\alpha\gamma - \beta^2)/\alpha$, and a *normalized residual*

$$\rho_{norm} = \frac{\rho}{\alpha c^2} = \frac{\alpha\gamma - \beta^2}{\alpha^2 c^2} = \frac{\alpha\gamma}{\beta^2} - 1. \tag{55}$$

In (Lindeberg 1996c), a scale selection methodology for stereo matching and flow estimation was presented based on the *region-based* flow estimation scheme by (Lukas and Kanade 1981; Bergen *et al.* 1992). The basic idea was to extend the least-squares methodology by (Lukas and Kanade 1981; Bergen *et al.* 1992) for computing the velocity estimates to *the minimization of a normalized residual over scales*. Given the least-squares formulation in (53), we propose to express a corresponding scale selection methodology for a *point-based* flow estimation scheme, by minimizing the normalized residual (55) over scales.

Figure 8 show three examples of computing flow estimates with automatic spatio-temporal scale selection in this way, from on a discrete separable temporal scale-space representation based on the time-causal temporal scale-space model in (Lindeberg and Fagerström 1996; Lindeberg 1997). To allow for the handling of two-dimensional image data, and to reduce the influence of the aperture problem, the least-squares formulation in (53) was extended in the following ways:

Uniform translation    Flow discontinuity    Expanding trees

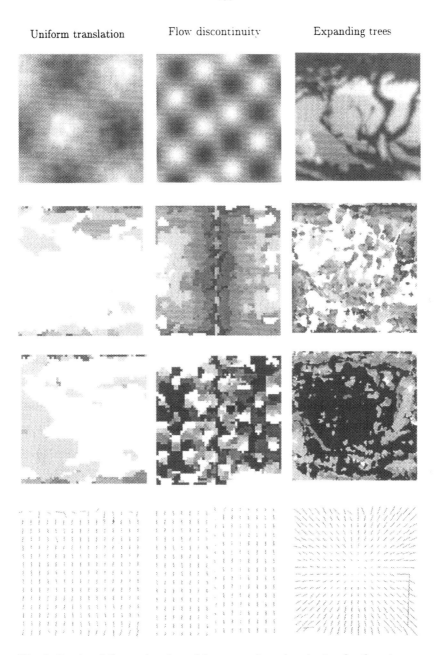

**Fig. 8.** Results of flow estimation with automatic scale selection for three image sequences: (left) noisy sine wave with uniform velocity, (middle) sine wave pattern where
the left half moves downwards and the right half moves upwards, (right) expanding
trees. The figures show from top to bottom: (to) a grey-level image from the sequence,
(top middle) selected spatial scales. (bottom middle) selected temporal scales. (bottom)
flow estimate. In these results, the following qualitative effects of the scale selection
methodology can be seen: (i) large image structures lead to the selection of coarser
scales than small image structures (compare left and right column), (ii) finer scales
will be selected when we approach a flow discontinuity (see middle column).

- from a one-dimensional to a two-dimensional spatial domain,
- from spatial derivatives up to order two to derivatives up to order four,
- the summation extended to nearest-neighbours (over a $3 \times 3$-neighbourhood) to enforce spatial consistency of local motions at the pixel level.

Finally, at each point, the global minimum of a corresponding normalized residual has been selected, and the flow estimate computed at that point. Whereas a more general scale selection methodology should also include a mechanism for explicit handling of multiple minima over scales as well as a mechanism for velocity adaptation (Lindeberg 1997), we can nevertheless notice that the following effects *arise as consequences* of the scale selection methodology:

- selection of larger scales with increasing size of image structures (compare the results in column 1 and column 2),
- selection of finer spatial scales near flow discontinuities (see column 2).

This algorithm was tested on the image sequences used by (Barron *et al.* 1994) and gave results which by visual inspection looked at least comparable to those reported in the evaluation. (One example is shown in third column.) A more detailed evaluation on calibrated reference data under variations of the size of image structures, the velocity and the noise level of superimposed Gaussian noise showed that the ratio between the error in the velocity estimates computed at the spatio-temporal scales given by this scale selection criterion and corresponding velocity estimates computed at the best spatio-temporal scales (as defined from comparisons with reference data) was typically within a factor of two.

## 9 Relations to previous work

The motion literature is large and it is impossible to give a fair review here. (See (Barron *et al.* 1994; Beauchemin and Barron 1995) for more extensive overviews.) As pointed out earlier, the velocity detector resulting from this scale selection methodology bears close relationships to frequency based motion approaches such as (Adelson and Bergen 1985; Heeger 1988; Fleet 1992). There is also a close relationship to Gabor-based approaches for estimating motion energy, such as the work by (Grzywacz and Yuille 1990). Compared to this type of motion detection, least-squares based motion approaches such as (Lukas and Kanade 1981; Bergen *et al.* 1992) and tensor filtering approaches such as (Bigün *et al.* 1991) allow motion estimates to be computed using a smaller set of spatio-temporal filters. A motion energy scheme, such as this one, on the other hand, is able to handle situations with multiple transparent motions, and is not restricted to computing the average motion direction.

(Jones and Malik 1992) performed stereo matching based on the responses of Gaussian derivative filters at different scales and of different orientations. Motion constraint equations involving derivatives of higher order have been studied by (Werkhoven and Koenderink 1990; Arnspang 1991). (Florack and Nielsen 1994) have analysed the effect of differentiating motion constraints equations to higher order and shown how this information relates to higher-order flow fields. The least squares formulation in (53) involving Gaussian derivatives of multiple orders in the motion constraint equation can be seen as a combination of these ideas with spatial integration approach in (Lukas and Kanade 1981; Bergen *et al.* 1992) for restricting parameterized motion models of low order by overdetermined systems of equations.

Minimizing a measure of uncertainty over scales bears relationship to the statistical methodology in (Kanade and Okutomi 1994) for adapting the window size for stereo correlation. The closest relation to previous works, however, is that it is shown how the general scale selection methodology proposed in (Lindeberg 1993b, 1994) can be applied to time-causal scale-space concepts over temporal domains, and be integrated with the region-based scale selection principle for stereo matching and flow estimation in (Lindeberg 1996c) to express mechanisms for motion detection and flow estimation based small-support point operations from the $N$-jet (Koenderink and van Doorn 1987).

# 10   Summary and discussion

The subject of scale selection is an essential complement to traditional scale-space representation concerning many computer vision problems. The main subject of this paper has been to outline a foundation for expressing mechanisms for automatic selection of temporal scales in multi-scale representations based on time-causal image operations. At a more technical level, the following contributions have been presented.

- It has described how the extension of the general scale selection methodology in (Lindeberg 1993b) to $\gamma$-normalized derivatives (Lindeberg 1996a) corresponds to *normalization in $L_p$-norm*, and how the $\gamma$-normalized derivative concept arises by *necessity* given natural commutative properties of a scale selection methodology under size variations (section 3).
- It has been shown how the concept of quasi quadrature can be defined to allow for *dense scale selection*, and the relative weighting of first- and second-order derivative information has been analysed (section 4).
- It has been shown how *time-causal normalized derivatives* can be defined for the three different types of time-causal temporal scale-space concepts that guarantee non-creation of local extrema with increasing scale (section 5–6).
- The effect of performing local maximization of normalized derivatives over scales has been analysed for different types of model patterns, and it has been shown how the resulting scales reflect characteristic length of corresponding (spatio-)temporal image structures as well as an analysis of how the inherent time delay in a temporal multi-scale representation affects the relations between image structures at different temporal scales (section 6–7)
- It has been shown how this scale selection methodology can be used for capturing motion energy at different spatio-temporal scales, and how velocity estimates can be obtained from the spatio-temporal scales selected by the scale selection procedure (section 8.1).

– It is shown how a velocity estimation scheme can be formulated by applying the motion constraint equation to the $N$-jet, and how the minimization of a normalized residual over scales in a least-squares formulation provides a complementary methodology for automatic scale selection (section 8.2).

In summary, these results show how the scale selection principles previously defined on spatial domains carry over to temporal data, and how they can be given well-founded formulations based on the strictly time-causal operations which are necessary to handle real-time image data.

## Acknowledgments

I would like to thank Daniel Fagerström for his kind help with the experiments underlying figure 8 as well as Kostas Daniliides for valuable discussions.

The support from the Swedish Research Council for Engineering Sciences, TFR, is gratefully acknowledged.

## References

[Adelson and Bergen, 1985] E.H. Adelson and J.R. Bergen. "Spatiotemporal energy models for the perception of motion". *JOSA*, A 2:284–299, 1985.

[Arnspang, 1991] J. Arnspang. *Motion Constraint Equations in Vision Calculus*. Doctoral dissertation. Doctoral dissertation, Dept. Med. Phys. Physics, Univ. Utrecht, NL-3508 Utrecht, Netherlands, 1991.

[Barron et al., 1994] J. J. Barron; D. J. Fleet, and S. S. Beachemin. "Performance of Optical Flow Techniques". *IJCV*, 12(1), 1994.

[Beauchemin and Barron, 1995] S.S. Beauchemin and J.L. Barron. "The computation of optical flow". *ACM Computing Surveys*, 27:433–467, 1995.

[Bergen et al., 1992] J. R. Bergen; P. Anandan; K. J. Hanna, and R. Hingorani. "Hierarchical Model-Based Motion Estimation". In *2nd ECCV*, 237–252, 1992.

[Bigün et al., 1991] J. Bigün; G. H. Granlund, and J. Wiklund. "Multidimensional orientation estimation with applications to texture analysis and optical flow". *IEEE-PAMI*, 13(8):775–790, 1991.

[Fleet and Langley, 1995] D. J. Fleet and K. Langley. "Recursive Filters for Optical Flow". *IEEE-PAMI*, 17(1):61–67, 1995.

[Fleet, 1992] D. Fleet. *Measurement of Image Velocity*. Kluwer, 1992.

[Florack and Nielsen, 1994] L. Florack and M. Nielsen. "The Intrinsic Structure of the Optic Flow Field". Technical Report ERCIM–07/94–R033, 1994.

[Florack, 1993] L. M. J. Florack. *The Syntactical Structure of Scalar Images*. PhD thesis. , Dept. Med. Physics, Utrecht Univ., NL-3508 Utrecht, Netherlands, 1993.

[Grzywacz and Yuille, 1990] N. Grzywacz and A.L. Yuille. "A model for the estimate of local image velocity by cells in the visual cortex". *Proc. Royal Soc. London*, B 239:129–161, 1990.

[Heeger, 1988] D. Heeger. "Optical flow using spatiotemporal filters". *IJCV*, 1:279–302, 1988.

[Horn and Schunck, 1981] B. K. P. Horn and B. G. Schunck. "Determining Optical Flow". *AI*, 17:185–204, 1981.

[Jones and Malik, 1992] D. G. Jones and J. Malik. "A computational framework for determining stereo correspondences from a set of linear spatial filters". In *2nd ECCV*, 395–410, 1992.

[Kanade and Okutomi, 1994] T. Kanade and M. Okutomi. "A Stereo Matching Algorithm with an Adaptive Window: Theory and Experiment". *IEEE-PAMI*, 16(9):920–932, 1994.

[Koenderink and van Doorn, 1987] J. J. Koenderink and A. J. van Doorn. "Representation of Local Geometry in the Visual System". *Biol. Cyb.*, 55:367–375, 1987.

[Koenderink and van Doorn, 1992] J. J. Koenderink and A. J. van Doorn. "Generic neighborhood operators". *IEEE-PAMI*, 14(6):597–605, 1992.

[Koenderink, 1984] J. J. Koenderink. "The structure of images". *Biol. Cyb.*, 50:363–370, 1984.

[Koenderink, 1988] J. J. Koenderink. "Scale-Time". *Biol. Cyb.*, 58:159–162, 1988.

[Lindeberg and Fagerström, 1996] T. Lindeberg and D. Fagerström. "Scale-Space with causal time direction". In *4th ECCV*, volume 1064, 229–240, Cambridge, UK, 1996.

[Lindeberg, 1990] T. Lindeberg. "Scale-Space for Discrete Signals". *IEEE-PAMI*, 12(3):234–254, 1990.

[Lindeberg, 1993a] T. Lindeberg. "Detecting salient blob-like image structures and their scales with a scale-space primal sketch: A method for focus-of-attention". *IJCV*, 11(3):283–318, 1993.

[Lindeberg, 1993b] T. Lindeberg. "On Scale Selection for Differential Operators". In *8th SCIA*, 857–866, 1993.

[Lindeberg, 1994] T. Lindeberg. *Scale-Space Theory in Computer Vision*. Kluwer, Netherlands, 1994.

[Lindeberg, 1996a] T. Lindeberg. "Edge detection and ridge detection with automatic scale selection". In *Proc. IEEE Comp. Soc. Conf. on Computer Vision and Pattern Recognition, 1996*, 465–470, San Francisco, California, June 1996.

[Lindeberg, 1996b] T. Lindeberg. "Feature Detection with Automatic Scale Selection". Technical Report ISRN KTH/NA/P--96/18--SE, KTH, Stockholm, Sweden, 1996. To appear in IJCV.

[Lindeberg, 1996c] T. Lindeberg. "A Scale Selection Principle for Estimating Image Deformations". Technical Report ISRN KTH/NA/P--96/16--SE, KTH, Stockholm, Sweden, 1996. To appear in IVC. Shortened version in 5th ICCV, 1995.

[Lindeberg, 1996d] T. Lindeberg. "On Automatic Selection of Temporal Scales in Time-Causal Scale-Space". in preparation, 1996.

[Lindeberg, 1997] T. Lindeberg. "Linear spatio-temporal scale-space". In *Proc. 1st Int. Conf. on Scale-Space Theory in Computer Vision*, July 1997. (To appear).

[Lukas and Kanade, 1981] B. D. Lukas and T. Kanade. "An iterative image registration technique with an application to stereo vision". In *Image Understanding Workshop*, 1981.

[Pauwels et al., 1995] E. J. Pauwels; P. Fiddelaers; T. Moons, and L. J. van Gool. "An extended class of scale-invariant and recursive scale-space filters". *IEEE-PAMI*, 17(7):691–701, 1995.

[Werkhoven and Koenderink, 1990] P. Werkhoven and J. J. Koenderink. "Extraction of motion parallax structure in the visual system". *Biol. Cyb.*, 1990.

[Witkin, 1983] A. P. Witkin. "Scale-space filtering". In *8th IJCAI*, 1019–1022, 1983.

[Yuille and Poggio, 1986] A. L. Yuille and T. A. Poggio. "Scaling Theorems for Zero-Crossings". *IEEE-PAMI*, 8:15–25, 1986.

*Technical reports on related subjects can be fetched from http://www.nada.kth.se/~tony.*

# Some Applications of Representations of Lie Algebras and Lie Groups

Reiner Lenz

Department Electrical Engineering,
Linköping University,
S-58183 Linköping,Sweden
reiner@isy.liu.se

**Abstract.** The paper gives a short overview over some basic facts from the representation theory of groups and algebras. Then we describe iterative algorithms to normalize coefficient vectors computed by expanding functions on the unit sphere into a series of spherical harmonics. Typical applications of the normalization procedure are the matching of different three-dimensional images, orientation estimations in low-level image processing or robotics. The algorithm illustrates general methods from the representation theory of Lie-groups and Lie-algebras which can be used to linearize highly-non-linear problems.It can therefore also be adapted to applications involving groups different from the group of three-dimensional rotations. The performance of the algorithm is illustrated with a few experiments involving random coefficient vectors.

## 1 Introduction

The representation theory of groups and algebras is an important part of mathematics with important applications in both pure and applied mathematics. It is also a corner stone of quantum mechanics [22] and recently it has been also received some attention in the signal processing community [21]. Its applications in image processing and pattern recognition have however been very limited. In this paper we will thus try to give a short summary of some of the basic ideas and a few results. Then we will describe an application where we use these tools in the design of an algorithm that computes the 3-D rotation which gives the best match between two 3-D patterns.

We will only consider the group SO(3) of 3-D rotations (and the closely related group SU(2) of unitary $2 \times 2$ matrices). Apart from the finite and the commutative groups these are the simplest groups and many of the general ideas can be illustrated using them as examples. Furthermore they are of considerable practical importance since 3-D rotations are related to orientation in 3-D space which is obviously of great importance in the generation and investigation of images.

# 2 Representations of algebra and groups

In this section we summarize briefly some basic facts about the representation theory of groups and algebras to give an intuitive understanding of the basic ideas. We consider mainly the 3-D rotation group which leads to the study of the spherical harmonics. Detailed information about Lie-theory and representations can be found in numerous textbooks on the subject[10,11,14,17,19,24,25].

We recall that a group is a set of elements together with a group operation. If the group multiplication and the operation of inversion are differentiable mappings then the group is called a Lie-group. If a subset of a group together with the multiplication inherited from the original group forms a group then it is called a subgroup. A subgroup of a Lie group which depends on one parameter only is called a one-parameter subgroup. Here we are mainly concerned with the group SO(3) of all three-dimensional rotations. The elements are $3 \times 3$ matrices which depend on three parameters. One way to parameterize this group is the Euler-angle parameterization which describes a rotation $R$ as the product of three rotations $R_z(\varphi)R_x(\theta)R_z(\psi)$ where $R_x(\alpha)$ is a rotation around the x-axis with angle $\alpha$ and $R_z(\beta)$ is a rotation around the z-axis with angle $\beta$.

It is also well-known that a three-dimensional rotation can be described by its rotation axis and its rotation angle. The description of the axis needs two parameters (the coordinates of a point on the sphere) and the third parameter is the rotation angle. If we keep the rotation axis fixed then we get a subgroup of SO(3) consisting of all rotations around this axis. This is a typical one-parameter group. For a one-parameter group with elements $R(t)$ and $R(0)$ the identity element we define the derivative: $X = \lim_{t \to 0} \frac{R(t) - R(0)}{t}$.

The objects obtained in this way form a vector space with dimension equal to the number of parameters of the original group. In the case of SO(3) this is a three-dimensional vector space. The operation of differentiation can be reversed by the exponential map defined as:

$$e^{tX} = \sum_{n=0}^{\infty} \frac{t^n}{n!} X^n \tag{1}$$

Besides the usual vector space operations of addition and scalar multiplication these vector spaces have another multiplication operation, the bracket, defined as: $[X, Y] = XY - YX$.

Vector spaces with the bracket multiplication are called Lie-algebras. For each Lie-group there is a Lie-algebra and for each element in the Lie-algebra we can construct an element in the Lie-group. The correspondence between Lie-groups and Lie-algebras is however not unique: different Lie-groups can generate the same Lie-algebra. Locally around $t = 0$ the exponential mapping is invertible. For the rotation group this construction leads to the exponential description $R = e^{(u_1 J_1 + u_2 J_2 + u_3 J_3)}$ with scalars $u_1, u_2$ and $u_3$ and the matrices:

$$J_1 = \begin{pmatrix} 0 & 0 & 0 \\ 0 & 0 & -1 \\ 0 & 1 & 0 \end{pmatrix} \quad J_2 = \begin{pmatrix} 0 & 0 & 1 \\ 0 & 0 & 0 \\ -1 & 0 & 0 \end{pmatrix} \quad J_3 = \begin{pmatrix} 0 & -1 & 0 \\ 1 & 0 & 0 \\ 0 & 0 & 0 \end{pmatrix} \tag{2}$$

which form a basis of the Lie-algebra $\mathfrak{so}(3)$ of $SO(3)$. When $f$ is a function of the rotation $R$ we will sometimes write $f(u)$ instead of $f(R)$. This description of the connection between Lie-groups and Lie-algebras is only intended as an intuitive introduction to the concepts involved. It is simplified and incomplete and the reader should consult the literature for a correct description of the subtleties involved.

Next we introduce the concept of representations of groups and algebras. An m-dimensional matrix representation $T$ of a group $G$ is a map from the group to the space of $m \times m$ matrices which preserves the group operation, i.e.: $T : R \mapsto T(R)$ with $T(R_1 R_2) = T(R_1)T(R_2)$.

If $T_1(R)$ and $T_2(R)$ are two representations of the same group and if there is one matrix $B$ such that: $B^{-1}T_1(R)B = T_2(R)$ for all group elements $R$ then the two representations are equivalent. A representation $T_1$ is called reducible if we can find a matrix $B$ such that all matrices $T_2(R)$ have the block struc-

ture $\begin{pmatrix} * & * \\ 0 & * \end{pmatrix}$ where the $*$ denote some non-zero blocks. A representation is called irreducible if it is not reducible.

For a one-parameter group with elements $R(t)$ and an m-dimensional representation $T$ of the group $T(R(t))$ is a one parameter group of matrices in the representation space. Computing the derivative

$$\widehat{T} = \lim_{t \to 0} \frac{T(R(t)) - T(R(0))}{t}$$

gives a mapping from the Lie-algebra to the set of $m \times m$ matrices. This mapping (which we sometimes denote by $\widehat{T}$ and sometimes by $T$) preserves the Lie-algebra operations, i.e. $\widehat{T}([X,Y]) = \left[\widehat{T}(X), \widehat{T}(Y)\right]$ and it therefore defines a representation of the Lie-algebra. Again we can locally connect the two objects by the exponential mapping.

The irreducible representations are the basic building blocks from which all representations can be build. For the $SO(3)$ group these representations are connected to the spherical harmonics as follows: Assume that $o$ is a function on the unit sphere and $R$ is a three-dimensional rotation. It is well-known [24] that the spherical harmonics $\{Y_l^m, l = 0, \ldots, \infty, m = -l, \ldots l\}$ form a complete orthonormal system for functions defined on the unit-sphere. (In polar coordinates $(\theta, \varphi)$ they are given by $Y_l^m(\theta, \varphi) = P_l^m(\cos\theta)e^{im\varphi}$ where $P_l^m$ is the associated Legendre function). There are thus coefficients $c_{lm}$ such that: $o = \sum_{l=0}^{\infty} \sum_{m=-l}^{l} c_{lm}Y_l^m$. For a fixed value of $l$ we collect the coefficients in the $(2l+1)$-dimensional vector $c_l$ and the spherical harmonics in $Y_l$ and write the expansion of $o$ as: $o = \sum_{l=0}^{\infty} c_l'Y_l$ where $c_l'$ is the transpose of the vector $c_l$.

If $x$ is a point on the unit sphere and $Rx$ its image under the rotation $R$ then we can define the rotated function $o^R$ by $o^R(x) = o(R^{-1}x)$. Expanding the rotated function $o^R$ and comparing its coefficient vectors $c_l(R)$ with $c_l$ gives: $c_l(R) = T_l(R)c_l$, where $T_l(R)$ are $(2l+1)$-dimensional matrices which satisfy the transformation equation: $T_l(R_1 R_2) = T_l(R_1)T_l(R_2)$. They define thus a representation of $SO(3)$ in the space of spherical harmonics of degree $l$.

Since rotations can be characterized by Euler angles it is sufficient to calculate the representation matrices for rotations around the x- and the z-axis. For rotations around the z-axis the matrices $T_l(R)$ are diagonal matrices but for a rotation around the x-axis their entries are polynomials in cosines and sines of the rotation angle (the computation of these matrices is described in [3]).

## 3 Representation theory and iterative matching procedures

We now apply the theory derived so far to the simple image processing problem of finding the rotation $R$ which produces the best match between two given functions $o_1$ and $o_2$ on the sphere. This is a special case of the general matching problem which can be described as follows: Given are two functions $o_1, o_2$ and a set of transformations $\{R_i : i \in I\}$, find the index $j \in I$ such that $R_j o_1$ is as similar to $o_2$ as possible. In the most general case with arbitrary transformations $R_i$ there is only one solution to the problem: to compare $R_i o_1$ and $o_2$ for all $i$. An important special case in which the problem is easier to solve is to assume that the transformations form a group. One example is the group of permutations in which the functions $o_i$ are vectors and two vectors are defined as equal if they contain the same elements in a different order. Here we do not consider such finite groups but concentrate on Lie-groups and especially on the group $SO(3)$. The problem is then to find out if the two patterns $o_1$ and $o_2$ are similar up to a change in three-dimensional orientation. We also restrict the functions $o_k$ to be linear combinations of spherical harmonics of a fixed degree. This choice is motivated by the properties of the spherical harmonics that every function on the unit sphere can be developed in a series of spherical harmonics and that the set of linear combinations of spherical harmonics of a fixed degree is closed under the action of 3-D rotations. Even this restricted case has a number of interesting applications in low-level filtering or image registration: [5,4,6,7,9,12,13,15–18,20,26]. Special cases of this problem were investigated in [5,4,6,7] which mainly concentrated on second order surface harmonics.

The matching problem in terms of representation theory is as follows: Given are two functions $o_1$ and $o_2$ which are linear combinations of spherical harmonics of degree $l$. They are thus completely characterized by two coefficient vectors $c_l^{(1)}$ and $c_l^{(2)}$. For those two vectors the rotation $R$ which minimizes the distance between $T_l(R)c_l^{(1)}$ and $c_l^{(2)}$ is sought.

This is a difficult non-linear problem due to the complicated form of the representation matrices $T_l(R)$. For $l = 1$ the solution can be easily found. For the case $l = 2$ it can be shown that the solution can (in principle) be solved by a closed expression of the vector elements. This requires however the solution of a polynomial equation of degree three. For larger values of $l$ it is impossible to find closed form solutions. It is therefore necessary to find iterative techniques to solve these equations. Such methods (based on Euler angles) were for the case $l = 2$ developed in [6], direct methods were investigated in [4]. Here we show

how Lie-theory can be used to linearize the problem and to find fast iterative algorithms.

From a representation of a Lie group a representation of its Lie algebra can be computed by differentiation. Differentiation of the representation matrices belonging to the three one-parameter subgroups corresponding to the rotations around the three coordinate axis gives three matrices $D_1^{(l)}, D_2^{(l)}$ and $D_3^{(l)}$ which correspond to the matrices $J_k$ introduced in (2). They are:

$$D_1^{(l)} = \frac{1}{2}\begin{pmatrix} 0 & d_l & & & & \\ -d_l & 0 & d_{l-1} & & & \\ 0 & -d_{l-1} & 0 & d_{l-2} & & \\ & & & & & d_{-l+1} \\ & & & & -d_{-l+1} & 0 \end{pmatrix}$$

$$D_2^{(l)} = \frac{1}{2i}\begin{pmatrix} 0 & d_l & & & & \\ d_l & 0 & d_{l-1} & & & \\ 0 & d_{l-1} & 0 & d_{l-2} & & \\ & & & & & d_{-l+1} \\ & & & & d_{-l+1} & 0 \end{pmatrix}$$

$$D_3^{(l)} = \frac{1}{i}\begin{pmatrix} l & & & \\ 0 & l-1 & & \\ & & -l+1 & 0 \\ & & & -l \end{pmatrix} \tag{3}$$

with $d_k = \sqrt{l(l+1) - k(k-1)}, k = l, l-1, \ldots, -l+1$. We will usually assume that the index $l$ is fixed and therefore write $D_k$ instead of $D_k^{(l)}$.

A general element of the representation of the Lie-algebra is thus given by: $D = u_1 D_1 + u_2 D_2 + u_3 D_3$ and each such element can be exponentiated to give a representation matrix of the Lie-group: $e^D = e^{u_1 D_1 + u_2 D_2 + u_3 D_3}$. Using this exponential form of the representation matrices gives the following approximation of the transformation properties of the coefficient vectors:

$$T_l(R)c_l = T_l(u)c_l = T_l(e^{u_1 J_1 + u_2 J_2 + u_3 J_3})c_l = e^D c_l$$
$$= e^{u_1 D_1 + u_2 D_2 + u_3 D_3}c_l \approx c_l + (u_1 D_1 + u_2 D_2 + u_3 D_3)c_l \tag{4}$$

This approximation replaces the non-linear expression $T_l(R)c_l$ by the simpler, linear combination

$$c_l + (u_1 D_1 + u_2 D_2 + u_3 D_3)c_l$$

and the general idea behind the following algorithms is to find a solution vector $u = (u_1, u_2, u_3)$ by using this linear approximation. Then these values are inserted in the exponential and a new approximation is obtained.

# 4 The algorithm

The set $\{o^R : R \in SO(3)\}$ where $o$ is a function and $o^R$ is the rotated function is known as the orbit of $o$ under $SO(3)$. The matching problem is to find the relation between two elements of the orbit. A related problem is to find in each orbit a simple representative $o_0$. The representative can be used as a description of the whole class of functions. In the matching problem the existence of such a reference function can be used to get simpler matching algorithms by splitting the complete matching process into two parts: first both functions $o_1$ and $o_2$ are matched to the reference function $o_0$ obtaining transformations $R_1$ and $R_2$ respectively. Then the transformation matching $o_1$ to $o_2$ is given by $R_2^{-1}R_1$. Using such a strategy is often preferable since the special properties of the known element $o_0$ can be used. In low-level image processing the definition of a representative is often necessary since only the function $o$ to be analyzed is available.

For linear combinations of spherical harmonics of order two we have the following characterization of the orbits (see [4])

**Theorem 1.** *For each real function $o$ of the form $o = c_2'Y_2$ with $\|c_2\| = 1$ there is a rotation $R$ such that the coefficient vector of the transformed function $o^R$ has the form $(\xi_2, \quad 0, \quad \xi_0, \quad 0, \quad \xi_2)$ with $\xi_0 \geq 0.5$ and $\xi_2 \geq 0$.*

From the theorem follows that the main problem is to find a rotation such that the new vector has entries: $c_2^1 = Im(c_2^2) = 0$. For arbitrary values of $l$ we say that a vector is reduced if

$$c_l^1 = Im(c_l^2) = 0. \tag{5}$$

From the diagonal form of the transformation matrices for z-axis rotations it follows that a vector with $c_l^1 = 0$ can be easily reduced by applying a z-axis rotation.

In the rest of this section we assume that the value of $l$ is fixed and we will mostly ignore it, thus: $c^k = c_l^k$. We will also use $\xi$ and $\eta$ for the real and imaginary parts of the coefficients: $c^k = \xi_k + i\eta_k$. If necessary the effect of applying a rotation $R$ will be indicated as: $c_l(u) = c_l(R) = T_l(R)c_l$ and $c^k(R) = \xi_k(R) + i\eta_k(R)$. It is a well-known property of the spherical harmonics that for real functions $o$ the coefficients satisfy the relation: $c_l^{-k} = (-1)^k \overline{c_l^k}$ where $\bar{c}$ is the conjugate complex of $c$. For the real and imaginary parts this gives: $\xi_{-k} = (-1)^k\xi_k$ and $\eta_{-k} = (-1)^{k+1}\eta_k$.

Usually a closed form solution of the equation $c^1(R) = 0$ does not exist and we will therefore describe an algorithm that computes a series of rotations $R_i$ such that: $c^1 \to c^1(R_1) \to c^1(R_2R_1) \to \cdots \to 0$. The parameters of these rotations will be computed by the linear approximation introduced in the previous section. Thus we approximate the equation $c(u) = T(u)c$ by the linear expression (see (4)): $c(u) \approx c + (u_1D_1 + u_2D_2 + u_3D_3)c$. The entry $c^1(u)$ is given by the sum of $c^1$ and the product of the row number $l + 2$ of the matrix $(u_1D_1 + u_2D_2 + u_3D_3)$ with the vector $c$. From the form of the $D_k$ matrices (equation (3)) follows that row $l + 2$ of $(u_1 D_1 + u_2 D_2 + u_3 D_3)$ is equal to

$$\left[ \ldots, 0, 0, 0, -\frac{1}{2} u_1 d_1 - \frac{1}{2} i u_2 d_1, i u_3, \frac{1}{2} u_1 d_2 - \frac{1}{2} i u_2 d_2, 0, \ldots \right]. \quad (6)$$

Inserting the real and imaginary parts of the components of $c$ gives for the entry $(l + 2)$ of $c + (u_1 D_1 + u_2 D_2 + u_3 D_3)c$ the expression:

$$\left( \frac{1}{2} d_2 \, \xi_2 + \frac{1}{2} i \, d_2 \, \eta_2 - \frac{1}{2} \xi_0 \, d_1 \right) u_1 +$$

$$\left( -\frac{1}{2} i \, d_2 \, \xi_2 - \frac{1}{2} i \, \xi_0 \, d_1 + \frac{1}{2} d_2 \, \eta_2 \right) u_2 +$$

$$(i \, \xi_1 - \eta_1) \, u_3 + \xi_1 + i \, \eta_1. \quad (7)$$

Separating the real and imaginary parts of the equation $c^1(u) = 0$ leads to the matrix equation:

$$A \cdot u = -b \quad (8)$$

with:

$$A = \begin{bmatrix} -\dfrac{1}{2} \xi_0 \, d_1 + \dfrac{1}{2} d_2 \, \xi_2 & \dfrac{1}{2} d_2 \, \eta_2 & -\eta_1 \\[2mm] \dfrac{1}{2} d_2 \, \eta_2 & -\dfrac{1}{2} d_2 \, \xi_2 - \dfrac{1}{2} \xi_0 \, d_1 & \xi_1 \end{bmatrix} \quad (9)$$

and

$$b' = [\xi_1 \; \eta_1]. \quad (10)$$

This equation can be solved for the unknown vector $u$ if the matrix $A$ has rank 2.

In the numerical implementation we do not use the most general form of the solution but we consider the following three cases separately:

1. The matrix $A_{12}$ (consisting of columns one and two of $A$) has full rank. Then there is a solution of the form $u = (u_1, u_2, 0)$.
2. Considering the submatrix $A_{13}$ consisting of columns one and three gives a solution: $(u_1, 0, u_3)$.
3. And deleting the first column leads to $A_{23}$ and a solution $(0, u_2, u_3)$.:

No solution can be obtained if all $2 \times 2$ submatrices of $A$ are singular. In this case the equation system $\{\det A_{12} = \det A_{13} = \det A_{23} = 0\}$ has six solutions. For two of them the vector is already reduced. The remaining four cases require special relations between the components of the vector. These relations can be easily broken by applying a rotation $R$. This was never a problem in our experiments.

Summarizing we have now the following iterative algorithm to find a rotation $R$ such that $c^1(R) = 0$ :

1. From the computed series coefficients $c^k, k = -l \ldots l$ the matrix $A$ and the vector $b$ in equations (9 and 10) are computed
2. The linearized problem is solved resulting in a solution $u$.
3. The new vector $c(u) = T(u)c$ is computed. If the entry $c^1(u)$ is sufficiently small the iteration is stopped, otherwise $c(u)$ becomes the input vector for the new iteration.

# 5 Implementation and experiments

The important advantage of the Lie approach is the reduction of a highly non-linear problem to a problem that can be solved by standard linear algebra methods. This makes it possible to do most of the symbolic calculations involved in a computer algebra system. In our implementation we used the MAPLE system which computes from a given value of $l$ (the degree of the spherical harmonics involved) the symbolic solutions $u$. It also translates these expressions automatically to the MATLAB code which forms the main part of the numerical implementation of the algorithm. In the numerical implementation we compute the three solutions $u = (u_1, 0, u_3), (u_1, u_2, 0), (0, u_2, u_3)$ (if they exist) and select the solution which gives the smallest value of $|c^1|$. If the new vector was no improvement then a line search (in which the length of the parameter vector $u$ is reduced) is tried first. If this leads to no improvement after a predetermined number of steps (usually 5 to 10) then the effect of three rotations ($\pi/2$ rotations around the x- and the y-axis and around the space diagonal) is tested and the resulting vector with the smallest value of $|c^1|$ is selected as input to the next iteration.

In the tests we first generated 1000 data vectors of a given degree. The components $c^k (k \leq 0)$ were initialized with equally distributed random numbers. From them the components $c^k$ with $k > 0$ were computed using the relations $c_l^{-k} = (-1)^k \overline{c_l^k}$ which are valid if the underlying function is real valued. Finally the vectors are normalized to unit length.

In the first three experiments we used data vectors of length 5, 7 and 9 corresponding to second, third and fourth order spherical harmonics. The threshold for the length of $c^1$ was always 0.01 in these experiments. In the last experiment second order spherical harmonics were used but the threshold was now 0.001. The mean number and the maximum number of iterations for these experiments are summarized in Table 5 and Tables 2 to 5 show how many iterations were needed for the different data vectors in these experiments.

| Experiment (order/threshold) | (2/0.01) | (3/0.01) | (4/0.01) | (2/0.001) |
|---|---|---|---|---|
| Mean number of iterations | 1.76 | 2.603 | 2.602 | 2.16 |
| Max. number of iterations | 5 | 31 | 28 | 9 |

**Table 1.** Mean and maximum number of iterations needed for convergence

| Loop | 1 | 2 | 3 | 4 | 5 |
|---|---|---|---|---|---|
| No. of vectors | 409 | 457 | 106 | 21 | 7 |

**Table 2.** Stopping times for second order spherical harmonics, threshold 0.01

| Loop | 1 | 2 | 3 | 4 | 5 | 6-10 | 11-15 | 16-20 | 21-25 | 26-30 | 31-35 |
|------|---|---|---|---|---|------|-------|-------|-------|-------|-------|
| No. of vectors | 434 | 354 | 73 | 28 | 23 | 44 | 25 | 12 | 5 | 1 | 1 |

**Table 3.** Stopping times for third order spherical harmonics, threshold 0.01

| Loop | 1 | 2 | 3 | 4 | 5 | 6-10 | 11-15 | 16-20 | 21-25 | 26-30 |
|------|---|---|---|---|---|------|-------|-------|-------|-------|
| No. of vectors | 445 | 318 | 80 | 43 | 25 | 50 | 25 | 7 | 4 | 3 |

**Table 4.** Stopping times for fourth order spherical harmonics, threshold 0.01

| Loop | 1 | 2 | 3 | 4 | 5 | 6-10 |
|------|---|---|---|---|---|------|
| No. of vectors | 126 | 660 | 160 | 43 | 8 | 3 |

**Table 5.** Stopping times for second order spherical harmonics, threshold 0.001

## 6  Further comments and conclusions

In the experiments we only reduced the incoming vectors to a simple form by applying a class of transformations. In many cases (for example in the low-level image processing detection tasks) it is also of interest to know which rotation brought the input vector into its final form. This rotation will then describe the orientation of the input pattern with respect to the output pattern. Such information is often useful when the results are used in higher level processing since the orientation of neighboring points can be compared and evaluated. The corresponding rotation matrices can be easily computed since the algebras spanned by the matrices $D_k$ and $J_k$ (defined in equations (3) and (2)) are the same from an algebraical point of view. The rotation generated by the parameter vector $(u_1, u_2, u_3)$ is $R(u) = e^{u_1 J_1 + u_2 J_2 + u_3 J_3}$. This matrix is computed in each iteration and the product of these matrices gives the rotation which transforms the input vector to the estimated output vector. Other parameterizations such as Euler angles or rotation axis and rotation angle can be computed from the final rotation matrix (for more information on parameterizations of the rotation group see [23]).

From an implementation point of view an expensive step in the algorithm is the computation of the matrix exponential $e^{u_1 D_1 + u_2 D_2 + u_3 D_3}$ (and perhaps $e^{u_1 J_1 + u_2 J_2 + u_3 J_3}$) defined in (1). This infinite sum of matrix products is usually computed by diagonalizing the matrix $u_1 D_1 + u_2 D_2 + u_3 D_3$. In this special case the computations can however be simplified by using the properties of the matrices $D_k$ and $J_k$. One important relation is the connection between the matrices $D_1$ and $D_2$ and the raising and lowering operators $L_+$ and $L_-$ defined as: $L_- = D_1 + i \cdot D_2$ and $L_+ = D_1 - i \cdot D_2$. These are matrices that have nonzero entries only in the entries above and below the diagonal. The matrices $L_-^{2l+1}$ and $L_+^{2l+1}$ are therefore zero.

With the technique described above it is also possible to linearize all three equations (5) simultaneously. This leads to a linear equation of the form $B \cdot u = d$ similar to equation (8). But now $B$ is a square matrix of size $3 \times 3$. In the cases where $B$ has full rank this leads directly to a unique solution for the parameter vector $u$. A common problem with this approach is that the matrix $B$ is often ill-conditioned which makes it necessary to use a search procedure to find a good new iteration vector. Furthermore it is computationally more expensive. We experimented with this solution but found it to be inferior to the method described above.

The implementation of the basic algorithm described above is ad hoc and only intended to serve as an illustration. More effective implementations should incorporate more advanced numerical techniques [8] which should improve the performance of the method considerably. In such an implementation it is also possible to incorporate higher order approximations of $T(R)$ which are easily computable.

In the previous discussion we kept the filter vector (consisting of the surface harmonics) fixed and computed its effect on the original and the transformed input function. Exchanging the role of the filters and the input functions leads directly to the theory of steerable filter systems. All results described above are therefore applicable in the analysis of steerable filter systems.

Finally we want to point out a connection between representation theory and the theory of special functions which has found very little attention in the signal processing literature. In the description of the basic facts of representation theory we introduced the irreducible representations of a group as the smallest building blocks from which all other representations can be constructed. If $T$ is such a representation and $R$ is an element of the group under consideration then $T(R)$ is a matrix with elements $t_{kl}(R)$. These functions $t_{kl}(R)$ are known as the matrix elements of the representation. It is a well-known fact that for a large class of groups the collection of all matrix elements is a complete set of functions defined on the group. All functions on such a group are therefore series in the matrix elements. Furthermore most of the known special functions can be "explained" as matrix elements of group representations and relations between them originate often in the representation properties of the matrices. As an example how this can be used in signal processing we summarize here briefly the work of Atakishiyev and Wolf [1,2] on the Wigner distribution for finite systems. Their basic observation is that the Lie algebras for $SU(2)$ and $SO(3)$ are identical and that their representations are given by the exponentials of the matrices $D_i^{(l)}$. These exponentials are unitary matrices and the matrix $e^{t \cdot D_3^{(l)}}$ is diagonal. It's eigenvectors are thus the unit vectors and the usual component-wise description of a vector is thus the decomposition of the vector in a sum of eigenvectors of the representation matrix $e^{t \cdot D_3^{(l)}}$. Considering the representation matrix $e^{t \cdot D_1^{(l)}}$ instead leads to a decomposition of the finite signal vector into a sum of Kravchuk functions since the Kravchuk functions are the matrix elements of the irreducible representations of the group of rotations around the x-axis considered as a function of the index (instead of the rotation angle). These functions are furthermore the basis of functions defined on a finite set when the binomial distribution is used as a weight function. When the number of sampling points is increased the weight function approaches the Gaussian and the Kravchuk expansion can thus be seen as an approximation of Gabor filtering for finite signals.

Summarizing, we demonstrated how the theory of Lie-groups and Lie-algebras can be used to construct efficient iterative numerical methods to compute the orientation parameters from the coefficients of a series expansion in spherical harmonics. The basic structure of the algorithm is the same for all Lie-groups groups which depend on a finite number of parameters and can therefore be easily adopted to problems which involve other groups of this type. We also mentioned the connection between the representation theory of groups and special functions which can be used to develop new "Fourier-like" methods for signal processing.

**Acknowledgments** Part of this work was done while R. Lenz was at the Mechanical Engineering Laboratory, Tsukuba, Japan.

# References

1. N. M. Atakishiyev, S. M . Chumakov, and K. B. Wolf. Wigner distribution functions for finite systems. Technical report, IIMAS, UNAM, Mexico, 1997.
2. N. M. Atakishiyev and K. B. Wolf. Fractional Fourier Kravchuk transform. Technical report, IIMAS, UNAM, Mexico, 1996.
3. L. C. Biedenharn. *Angular momentum in quantum physics*, volume 8 of *Encyclopedia of mathematics and its applications*. Addison-Wesley Publishing Company, Reading, Massachusetts, 1984.
4. Gilles Burel and Hugues Henocq. Determination of the orientation of 3d objects using spherical harmonics. *CVGIP-Graphical Models and Image Processing*, 57(5):400–408, 1995.
5. Gilles Burel and Hugues Henocq. Three-dimensional invariants and their application to object recognition. *Signal Processing*, 45:1–22, 1995.
6. Per-Erik Danielsson. Orientation and shape from second derivatives in 3d volume data. Technical Report LiTH-ISY-R-1696, Dept. EE, Linköping University, S-58183 Linköping, October 1994.
7. Per-Erik Danielsson. Analysis of 3d volume data using 2nd derivatives. In *Proceedings DICTA-95, Brisbane*, pages 14–19, 1995.
8. J.E. Dennis and R.B. Schnabel. *Numerical Methods for Unconstrained Optimization and Nonlinear Equations*. Prentice-Hall, 1983.
9. W. T. Freeman and E. H. Adelson. The design and use of steerable filters for image analysis. *IEEE Transactions on Pattern Analysis and Machine*, 13(9):891–906, September 1991.
10. I. M. Gelfand, R. A. Minlos, and Z. Y. Shapiro. *Representations of the rotation and Lorentz groups and their applications*. Pergamon Press, 1963.
11. M. Hamermesh. *Group Theory and Its Applications to Physical Problems*. Addison-Wesley, 1962.

12. H. Härtl. *Darstellungstheorie in der Bildverarbeitung mit Schwerpunkt in der Bewegungsanalyse.* PhD thesis, Universität Karlsruhe, 1991.

13. Y. Hel-Or and P. C. Teo. A common framework for steerability, motion estimation and invariant feature detection. Technical report, Dept. CS. Stanford, 1996. STAN-CS-TN-96-28.

14. Ken-Ichi Kanatani. *Group Theoretical Methods in Image Understanding.* Springer Verlag, 1990.

15. R. Lenz. Optimal filters for the detection of linear patterns in 2-D and higher dimensional images. *Pattern Recognition,* 20(2):163–172, 1987.

16. Reiner Lenz. Group-invariant pattern recognition. *Pattern Recognition,* 23(1/2):199–218, 1990.

17. Reiner Lenz. *Group Theoretical Methods in Image Processing.* Lecture Notes in Computer Science (Vol. 413). Springer Verlag, Heidelberg, Berlin, New York, 1990.

18. M. Michaelis and Gerald Sommer. A Lie group approach to steerable filters. *Pattern Recognition Letters,* 16:1165–1174, 1995.

19. M. A. Naimark and A. I. Stern. *Theory of Group Representations.* Springer Verlag, New York, Heidelberg, Berlin, 1982.

20. Joseph Segman, Jacob Rubinstein, and Yehoshua Zeevi. The canonical coordinates method for pattern deformation: Theoretical and computational considerations. *IEEE Transactions on Pattern Analysis and Machine,* 14(12):1171–1183, December 1992.

21. Ramachandra Ganesh Shenoy. *Group Representations and Optimal Recovery in Signal Modeling.* PhD thesis, Cornell University, 1991.

22. S. Sternberg. *Group Theory and Physics.* Cambridge University Press, Cambridge, England, first paperback edition edition, 1995.

23. J. Stuelpnagel. On the parametrization of the three-dimensional rotation group. *SIAM Review,* 6(4):422–430, 1964.

24. N.Ja. Vilenkin and A.U. Klimyk. *Representation of Lie groups and special functions.* Mathematics and its applications : 72. Kluwer Academic, 1991-1993.

25. D. P. Zelobenko. *Compact Lie Groups and their Representations.* American Mathematical Society, Providence, Rhode Island, 1973.

26. Steven W. Zucker and Robert A. Hummel. A three-dimensional edge operator. *IEEE Transactions on Pattern Analysis and Machine,* 3(3):324–331, 1981.

# A Systematic Approach to Geometry-Based Grouping and Non-accidentalness

Luc Van Gool

Katholieke Universiteit Leuven, ESAT-MI2
Kardinaal Mercierlaan 94, B-3001 Leuven, BELGIUM

**Abstract.** Geometric regularities have often been used for grouping. Nonetheless, their foundations have typically been rather *ad hoc* - with "regular" or "non-accidental" features being listed according to intuition or based on application-specific considerations. This paper describes a more systematic line of thought towards such visual grouping. Based on an earlier observation that fixed structures in images are directly related to object regularities and grouping specific invariants, fixed structures are propounded as a theoretical glue. Moreover, grouping strategies with less than combinatorial complexity are difficult to develop. The propounded approach is also intended to keep grouping complexity under control. To that end, it combines the use of invariants with a Cascaded Hough Transform to efficiently extract candidate fixed structures.

## 1  Introduction

Grouping is the process of combining bits and pieces of visual information into perceptually salient structures. Grouping is important as a stepping stone from low-level image features to scene interpretation. According to the principle of non-accidentalness, out of a plentitude of image cues grouping forms configurations that hardly could have arisen by pure chance [3, 6]. Here we focus on grouping (planar) edges.

The work reported here has the following goals:

**To develop a more systematic approach:** In the literature, grouping rules are often selected on rather intuitive grounds, e.g. the catalogues of Gestalt laws [21] or non-accidentalness configurations [4]. Here, a more systematic classification of grouping configurations is put forward, which simulteneously underpins their detection. This is based on classificaitons of subgroups of the planar projectivities.

**Taking full perspective effects into account:** Grouping has often been restricted to cases where the projection can be simplified to (pseudo-)orthographic projection. Features such as parallel joins between mirror symmetric points survive in skewed symmetry in such case. Here, the perspective nature of projection will be taken into account completely, and far fewer features survive. Obviously, this makes grouping more difficult.

**To increase grouping efficiency:** Grouping algorithms are typically plagued by one or several steps of combinatorial complexity. This is not surprising as grouping has to do with combining parts into larger entities. Here, the combined use of invariants and the Hough transform should allow to do away with much of such combinatorics.

Of course, also this framework is restricted in a number of ways. First, it takes a geometry-driven approach to grouping, i.e. grouping is based on shape rather than intensity or texture. Second, current results focus on grouping *planar* shapes. Nevertheless, these planar shapes can be part of groupings of a 3D nature (examples will be given) and the same geometrical constraints following from this framework are found when grouping curved surfaces such as surfaces of revolution [22].

The structure of the paper is as follows. Section 2 discusses the kind of subgroups that will be considered. In particular, the concept of fixed structures as a guiding principle in this analysis is discussed. There the paper also recapitulates some issues of general, projective invariants, to then press on with their specialisation towards the subgroups in the subsequent sections (section 3 for fixed points and lines, section 4 for fixed sets of points, and section 5 for fixed conics). Section 6 introduces the Cascaded Hough transform to aid in the detection of the fixed structures. The results are then brought together in a strategy for geometry-based grouping based on the fixed structures and the invariants of the subgroups that they define. Section 7 gives an example. Section 7 concludes the paper and comments on possible future work.

# 2 Identifying subgroups for grouping

## 2.1 Fixed structures and subgroups

Consider two planar shapes in 3D space. Suppose that there exists a 3D projective transformation that maps one to the other. This is the basic grouping configuration studied here. A special case is when the two planes coincide, as with the two halfs of a mirror symmetry.

The existence of the projectivity in 3D implies that in an image of such configuration, the projections of the shapes are related by a projectivity in 2D. Furthermore, if the 3D projectivity maps certain structures onto themselves, i.e. keeps these structures fixed in three-dimensional space, then *a fortiori* their images will remain fixed under the 2D projectivity in the image. Trivial as this observation may be, it is important to keep in mind that not too many features survive the projection onto the image. Taking mirror symmetry as an example, symmetric points have the same distance to the axis, the same curvature, etc. in 3D space, but *not* in the image [2] Yet, the projectivity that maps the symmetric halfs onto each other in the image, still has a symmetry axis, i.e. a straight line all points of which are mapped onto themselves. Also pairs of symmetric points still form fixed pairs in the sense that one point is mapped to the other. The *fixed structures* survive in the image and hence provide a solid basis for the non-accidentalness paradigm.

Moreover, projectivities that keep the same structures fixed, e.g. a specific line or point, form subgroups of the projectivities [19]. Thus, if groupings are organised according to the fixed structures of the corresponding projectivities, these projectivities belong to specific subgroups, for which specific invariants can be extracted. This is the crux of the matter. The subgroups defined by the fixed structures yield invariants. These allow to match the parts of the grouping without a combinatoric search, e.g. using hashing as for recognition [9, 1]. These invariants are also grouping-specific, i.e. geared towards a specific type of configuration such as a symmetry, rather than being general projective invariants. This adds to the efficiency of the search and makes it possible to match curve segments that would be too small for effective projective matching.

Finally, going via fixed structures cuts down on the number of cases to be considered, by conveying equivalence of cases where this is intuitively not necessarily clear. As an example, the detection of mirror and point symmetries in perspective views can be proved to be one and the same problem from a mathematical point of view, precisely because they have the same kind of fixed structures in the image [20].

## 2.2 Selection of fixed structures

When searching for structures that could remain fixed under projectivities, it stands to reason to first concentrate on the simplest kind of structures that remain qualitatively invariant. Examples are points, lines and conics, because points are mapped to points, lines to lines, and conics to conics. There are other such structures, like curves with constant projective curvature, but these are considered too intricate to be of practical use here. Most of the analysis will be carried out in the real plane.

A further distinction can be made between cases where the remaining structures – points, lines, and conics – are fixed individually or as a set. For example, under a rotational symmetry (also when viewed obliquely), only the center of rotation is a point that is fixed individually, but other points belong to sets of points that remain fixed as a set. In addition to this distinction, it is also useful to consider combinations of fixed structures, like transformations that keep 2 points and a line fixed. In particular, it comes out that complete pencils of fixed structures are a particularly relevant case, as will be seen later.

All in all, the number of cases to be considered seems to become quite high. Nevertheless, not just any combination of fixed structures is possible. Although a complete classification of consistent combinations has not been developed yet, the following cases can be given already:

1. combinations of fixed points and lines
2. fixed sets of points
3. combinations of fixed conics

The focus will be on the first case.

As to the related, grouping-specific invariants, previous work on semi-differential invariants is extended. These are invariants for the description of curves, that combine point coordinates with their derivatives [16]. In order to keep this paper more or less self-contained, the sequel of this section gives a short overview of how semi-differential invariants can be derived for the general case of plane projectivities.

It is not difficult to show that for a general projectivity with matrix $P = (p_{ij})$ acting on points $\mathbf{x}_k = (x_k, y_k)^T$ as

$$x'_k = \frac{p_{11}x_k + p_{12}y_k + p_{13}}{p_{31}x_k + p_{32}y_k + p_{33}} \quad , \quad y'_k = \frac{p_{21}x_k + p_{22}y_k + p_{23}}{p_{31}x_k + p_{32}y_k + p_{33}}$$

one has

$$\left| \mathbf{x}'_1 - \mathbf{x}'_2 \;\; \mathbf{x}'_1 - \mathbf{x}'_3 \right| = \frac{|P|}{N_1 N_2 N_3} \left| \mathbf{x}_1 - \mathbf{x}_2 \;\; \mathbf{x}_1 - \mathbf{x}_3 \right| \;, \tag{1}$$

where vertical bars indicate determinants and

$$N_i = p_{31}x_i + p_{32}y_i + p_{33} \;.$$

Indicating the order of coordinate derivatives with respect to a projectively invariant parameter between parentheses, also

$$\left| \mathbf{x}'_1 - \mathbf{x}'_2 \;\; \mathbf{x}'^{(1)}_1 \right| = \frac{|P|}{N_1^2 N_2} \left| \mathbf{x}_1 - \mathbf{x}_2 \;\; \mathbf{x}^{(1)}_1 \right| \tag{2}$$

and

$$\left| \mathbf{x}'^{(1)}_1 \;\; \mathbf{x}'^{(2)}_1 \right| = \frac{|P|}{N_1^3} \left| \mathbf{x}^{(1)}_1 \;\; \mathbf{x}^{(2)}_1 \right| \;. \tag{3}$$

At a discontinuity like a vertex, $\mathbf{x}_1$ say, where a left (l) and right (r) derivative can be distinguished, one has

$$\left| \mathbf{x}'^{(1:\ell)}_1 \;\; \mathbf{x}'^{(1:r)}_1 \right| = \frac{|P|}{N_1^3} \left| \mathbf{x}^{(1:\ell)}_1 \;\; \mathbf{x}^{(1:r)}_1 \right| \;. \tag{4}$$

One could consider (4) as the counterpart of (3) for discontinuities.

The expressions (1), (2), (3), and (4) can be considered building blocks for the generation of projective invariants. A possible strategy is to take products of these building blocks raised to appropriate powers to eliminate all the factors that they produce under projective transformations [16].

In general an invariant parameter will not be available and also invariance under reparameterisation has to be realised. Fortunately, the same building blocks can be used. If the left hand sides are calculated on the basis of a parmeter $t'$ and the right hand sides use $t$, then building blocks (3) and (4) change with $\left(\frac{dt}{dt'}\right)^3$ and building block (2) with $\left(\frac{dt}{dt'}\right)$. Again these factors should cancel [16].

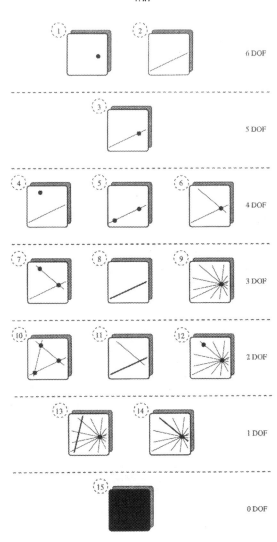

**Fig. 1.** *Classification of fixed structure subgroups for fixed points and lines.*

# 3   Combinations of fixed points and lines

## 3.1   A classification of subgroups

The possible combinations of fixed points and fixed lines that projective transformations can share are shown in fig. 1. Every square corresponds to a different type of subgroup, with a qualitatively different combination of fixed structures. A point in such square indicates a specific (but arbitrary) fixed point; the same for a line. Note that sometimes a fixed point lies on a fixed line. In the cases indicated with numbers 8, 11, 13, and 14 a thicker line is drawn. This is to mean that every point on such line is a fixed point and hence thick lines represent lines of fixed points. In cases number 9, 12, 13, and 14 a bunch of concurrent lines has been drawn. These are supposed to represent pencils of fixed lines, where all lines through a point – the so-called vertex – remain fixed. The vertex is a fixed point. Such pencils are the projective duals of lines of fixed points. The black square at the bottom is the trivial case (case 15), where all points are fixed points and the remaining subgroup only contains the identity.

Fig. 1 in effect is more than an enumeration of subgroup types. Going down the classification, additional fixed structures are added, thereby gradually decreasing the dimensionality of the subgroups. The dimension of the corresponding subgroups is indicated on the right. A more detailed discussion of the subgroups and their invariants is given elsewhere [19]. Note that the classes of fixed points and lines for *individual* projectivities as they are given by Springer [14] correspond to only 7 out of the 15 classes for the subgroups. These two classifications must not be confused.

Six of the subgroup types of fig. 1 are of special interest: these are the cases 8, 9, 11, 12, 13, and 14 which all contain a line of fixed points, a pencil of fixed lines, or both. These cases are of special interest because both the line of fixed points and the pencil of fixed lines fix 5 degrees of freedom (d.o.f.) whereas only two parameters need to be specified to fully chararacterise them: the two parameters to specify the line or the vertex. This gain in d.o.f.'s yields invariants that require strictly less information than needed for the general projective invariants. One might argue that this also applies to the other cases in fig. 1, but having a fixed point would e.g. lead to invariants based on the fixed point and 4 *additional* points, still requiring a total of 5 points. A similar observation can be made for all the other cases without a line of fixed points or a pencil of fixed lines.

Next it is shown how the existence of lines of fixed points or pencils of fixed lines yields invariants specific for the corresponding subgroups. Compared to the building blocks of section 2.2 for the general projective case, these fixed structures yield additional building blocks and factors that are easier to eliminate.

## 3.2 A pencil of fixed lines

If there is a pencil of fixed lines, then every point is known to stay on the line of the pencil on which it lies. Denoting the pencil vertex with $\mathbf{x}_v = (x_v, y_v)^T$, one therefore knows that there exists a factor $k_i$ such that

$$(x_i' - x_v) = k_i(x_i - x_v) \; ,$$
$$(y_i' - y_v) = k_i(y_i - y_v) \; .$$

Hence $(x_i - x_v)$ and $(y_i - y_v)$ are additional building blocks. Such a factor $k_i$ exists for every point $\mathbf{x}_i$. It immediately follows that

$$\frac{(y_i - y_v)}{(x_i - x_v)}$$

is an invariant, requiring only two points, one of which is the vertex.

In order to derive additional invariants (combinations with the different building blocks of section 2.2), it is important to know more about the factor $k_i$. Consider

$$|\mathbf{x}_1' - \mathbf{x}_v \ \ \mathbf{x}_2' - \mathbf{x}_v| = \frac{|P|}{N_1 N_2 N_v} |\mathbf{x}_1 - \mathbf{x}_v \ \ \mathbf{x}_2 - \mathbf{x}_v| \ ,$$

then it follows that it can also be written as

$$|\mathbf{x}_1' - \mathbf{x}_v \ \ \mathbf{x}_2' - \mathbf{x}_v| = k_1 k_2 |\mathbf{x}_1 - \mathbf{x}_v \ \ \mathbf{x}_2 - \mathbf{x}_v|$$

and therefore

$$k_1 k_2 = \frac{|P|}{N_1 N_2 N_v} \ .$$

From the fact that this latter equality holds for any choice of the points $\mathbf{x}_1$ and $\mathbf{x}_2$, it follows that

$$k_i = \pm \sqrt{\text{abs}\left(\frac{|P|}{N_v}\right) \frac{1}{N_i}} \ .$$

We conclude that $(x_i - x_v)$ and $(y_i - y_v)$ come a additional building blocks with the pencil of fixed lines, easing the construction of invariants. An example invariant parameter is then

$$\int \frac{|\mathbf{x} - \mathbf{x}_v \ \ \mathbf{x}^{(1)}|}{(x - x_v)^2} dt \ .$$

## 3.3 A line of fixed points

If there is a line of fixed points – in the sequel referred to as *the axis* – then any point $\mathbf{x}_{ai}$ on it is fixed. Hence,

$$|\mathbf{x} - \mathbf{x}_{a1} \ \ \mathbf{x} - \mathbf{x}_{a3}| \ = l|\mathbf{x} - \mathbf{x}_{a1} \ \ \mathbf{x} - \mathbf{x}_{a2}|$$
$$|\mathbf{x}' - \mathbf{x}_{a1} \ \ \mathbf{x}' - \mathbf{x}_{a3}| = l|\mathbf{x}' - \mathbf{x}_{a1} \ \ \mathbf{x}' - \mathbf{x}_{a2}|$$

with

$$l = \frac{||\mathbf{x}_{a1} - \mathbf{x}_{a3}||}{||\mathbf{x}_{a1} - \mathbf{x}_{a2}||} \ .$$

It follows that

$$\frac{|P|}{N N_{a1} N_{a3}} = \frac{|\mathbf{x}' - \mathbf{x}_{a1} \ \ \mathbf{x}' - \mathbf{x}_{a3}|}{|\mathbf{x} - \mathbf{x}_{a1} \ \ \mathbf{x} - \mathbf{x}_{a3}|} = \frac{|\mathbf{x}' - \mathbf{x}_{a1} \ \ \mathbf{x}' - \mathbf{x}_{a2}|}{|\mathbf{x} - \mathbf{x}_{a1} \ \ \mathbf{x} - \mathbf{x}_{a2}|} = \frac{|P|}{N N_{a1} N_{a2}}$$

and thus $N_{a1} = N_{a2} = N_{a3} = N_a$ where $N_a$ is one and the same value for all the points on the axis.

It then immediately follows that e.g.

$$\frac{|\mathbf{x}_{a1} - \mathbf{x}_1 \ \ \mathbf{x}_{a1} - \mathbf{x}_2|}{|\mathbf{x}_{a2} - \mathbf{x}_1 \ \ \mathbf{x}_{a2} - \mathbf{x}_2|}$$

is an invariant, which requires knowledge about the axis and only two additional points, hence a total of 6 parameters (the two points on the axis can be chosen arbitrarily). A geometrical interpretation is that the lines $\langle \mathbf{x}_1, \mathbf{x}_2 \rangle$ and $\langle \mathbf{x}_1', \mathbf{x}_2' \rangle$ interesect the axis in the same point.

## 3.4 A pencil and an axis

If both a pencil of fixed lines and a line of fixed points exist, then the previous results can be combined. If one considers $(x_a - x_v)$ where both the point on the axis $\mathbf{x}_a$ and the pencil vertex $\mathbf{x}_v$ are fixed points now, this expression is a trivial invariant, i.e.

$$k_a = \pm\sqrt{\text{abs}\left(\frac{|P|}{N_v}\right)\frac{1}{N_a}} = 1$$

and therefore $N_a = \pm\sqrt{\text{abs}\,(|P|/N_v)}$, or, equivalently, $N_v = |P|/N_a^2$.

Cases with such combination of an axis and a pencil come out to be of particular, practical importance. Such *planar homologies* seem to pop up virtually everywhere in vision.

For examples where planar homologies emerge, consider fig. 2: the relation in the image between a planar shape and its shadow or the top and bottom plane of an extruded surface. Another example where planar homologies pop

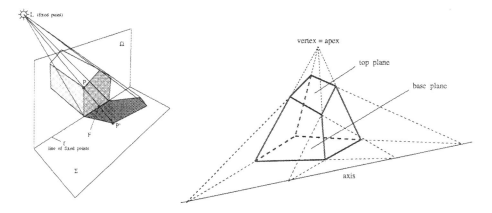

**Fig. 2. Left:** *Geometry of the object-shadow configuration.* **Right:** *Example of an extruded surface. Such shapes are formed by cutting a general cone by two planes (top and base plane).*

up is in determining the epipolar geometry of a pair of cameras. As has been noted before [13], the knowledge of the projectivities $P_1$ and $P_2$ for two planes between the two views suffices. What matters are the composed transformations $P_2^{-1}P_1$ and $P_1^{-1}P_2$. As a matter of fact, these correspond to planar homologies. The fixed points off the axis (the vertices) correspond to the epipoles. The lines of fixed points are the intersections of the planes as seen in each of the stereo views. Connecting the epipoles with corresponding points on the lines of fixed points yields pairs of corresponding epipolar lines.

All transformations that share the same axis and vertex form a one-parameter subgroup. Note that this reduction from the 8-d.o.f. group of general projectivities to a 1-d.o.f. subgroup of planar homologies only requires the specification of a line (the axis) and a point (the vertex), i.e. 4 parameters. The remaining degree of freedom can be expressed as a fixed cross ratio for corresponding points, as illustrated in fig. 3. In terms of the classification of fig. 1, the planar homologies

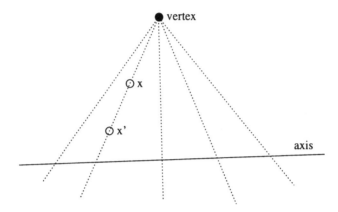

**Fig. 3.** *A planar homology is a plane projective transformation which has a line of fixed points, called the* axis, *and a distinct fixed point, not on the line, called the* vertex *of the homology. Corresponding point pairs* $x \leftrightarrow x'$ *and the vertex are collinear. The cross-ratio defined by the vertex, the corresponding points* $x, x'$ *and the intersection of their join with the axis is the* same *for all corresponding point pairs.*

correspond to case 13. A more detailed account on how to exploit the special characteristics of planar homologies is given in [17].

# 4   Fixed sets of points

A set of points may, rather than being fixed individually, map onto eachother. The set is fixed, not its points. Such cases are important, because they correspond to discrete symmetries. Mirror symmetry is an example where every point belongs to a fixed pair of symmetric points. Ornamental symmetries include all cyclic and dihedral symmetry groups of different orders. Cyclic symmetry of order $n$ is synonymous to $n$-fold rotational symmetry. Dihedral symmetry groups add mirror symmetries. As a matter of fact, there is an isomorphism between the skewed symmetries as observed in the image and the 'ornamental symmetry group' of the shape. Vice versa, the existence of fixed sets of points typically are a strong indication for the presence of skewed ornamental symmetries, and in some cases it even gives a guarantee (e.g. if there is a fixed triple [12]).

As in the case of a line of fixed points or a pencil of fixed lines, the presence of fixed discrete sets of points yields specialised invariants. And again, these are based on further constraints on the factors of the building blocks in section 2.2.

Consider a fixed $n$-tuple of points, $x, x', x'', \ldots, x^{[n-1]}$. Consider what happens to $|x - x' \ \ x - x''|$. Applying the transformation $n$ times brings all the points back to their original positions. Hence, following the factor brought about by such building blocks according to eq. 1

$$\frac{|P|^n}{(N N' N'' \dots N^{[n-1]})^3} = 1$$

and therefore

$$N N' N'' \dots N^{[n-1]} = |P|^{n/3} . \tag{5}$$

A degenerate case of a point cycle is the $n$-fold repetition of the rotation center. It follows from eq. 5 that for this point – $\mathbf{x}_c$ say – $N_c^n = |P|^{n/3}$, i.e. $N_c = \pm |P|^{1/3}$. This holds irrespective of the angle of rotation.

As an example, if one is looking obliquely at a 3-fold rotational symmetry,

$$\frac{|\mathbf{x}_1 - \mathbf{x}_2 \ \ \mathbf{x}_1' - \mathbf{x}_2| \ |\mathbf{x}_1 - \mathbf{x}_3 \ \ \mathbf{x}_1'' - \mathbf{x}_3|}{|\mathbf{x}_1 - \mathbf{x}_2 \ \ \mathbf{x}_1 - \mathbf{x}_3|} .$$

is an invariant under the transformation that corresponds to the $120°$ rotation as seen in the image. Although this invariant uses a total of 5 points as a general point-based projective invariant would, it is both simpler and more selective. This expression is not invariant under general projectivities. Note that – as usual – this symmetry-specific invariant contains information on the fixed structures of the symmetry,' i.c. the fixed triple $\mathbf{x}_1, \mathbf{x}_1', \mathbf{x}_1''$.

Skewed mirror symmetry deserves some special attention, both because of its special status in the grouping literature and because of its rich collection of fixed structures. In fact, it can be considered the combination of fixed pairs of points with a planar homology, with its line of fixed points (axis) and pencil of fixed lines. In the terminology of projectivities, it is referred at as a harmonic homology. Note that oblique point symmetries yield exactly the same combination of fixed structures [20] (still assuming perspective projection). Hence, all invariants derived for one case are directly applicable to the other. This is an example of how cases that seem quite different at first, admit identical geometrical analysis, as brought to the fore by them sharing qualitatively the same set of fixed structures. Compared to other planar homologies, the extra fixed sets of points also yield the additional constraint that $N_v = |P|^{1/3}$ for the pencil vertex. This follows from $N_v = |P|/N_a^2$ for planar homologies (with $N_a$ for the axis points) and $N_a^2 = |P|^{2/3}$ because points on the axis form degenerate fixed pairs.

In the case of *affinely* skewed symmetry which is valid if the model for the viewing conditions is simplified to pseudo-orthographic projection, Ponce [8] has shown that in a pair of symmetric points

$$\frac{\kappa}{\kappa'} = \frac{\sin^3 \theta}{\sin^3 \theta'}$$

where $\kappa$ and $\kappa'$ are the curvatures of the symmetric contours in these points and the meaning of $\theta$ and $\theta'$ is explained in fig. 4. From the relation (5) with $n = 2$

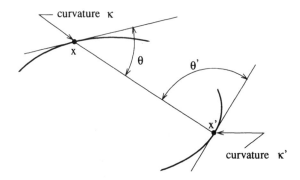

**Fig. 4.** *variables used in Ponce's symmetry invariant.*

it follows that this relation also holds under *perspective* skewing. It suffices to rewrite the Ponce relation as

$$\frac{|\mathbf{x}^{(1)} \ \mathbf{x}^{(2)}|}{|\mathbf{x} - \mathbf{x}' \ \mathbf{x}^{(1)}|^3} = \frac{|\mathbf{x}'^{(1)} \ \mathbf{x}'^{(2)}|}{|\mathbf{x}' - \mathbf{x} \ \mathbf{x}'^{(1)}|^3}$$

and to check the cancelling of building block factors. Hence, this ratio is an invariant of the harmonic homology. It was tested on the fly flapper outline of fig. 5. Fig. 6 overlays the values for corresponding points along the ordinate,

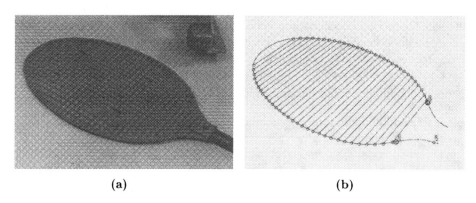

| (a) | (b) |

**Fig. 5.** (a) *Picture of a mirror symmetric fly flapper taken from an oblique viewpoint,* (b) *Corresponding points on the flapper's spline fit. Note the perspective distortion.*

with the abscissa indicating subsequent pairs, starting from the indicated pair on fig. 5(b). The indicated pair was identified as the pair of inflections. For the identification of the pairs a parameterisation was used that is invariant under the harmonic homology. As can be seen, the computed ratio is very similar for corresponding points indeed.

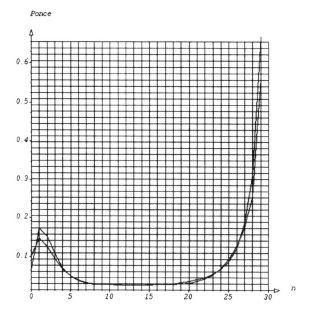

**Fig. 6.** *The Ponce relation for the different corresponding points.*

## 5  Combinations of fixed conics

Projectivities map conics to conics and they therefore are obvious fixed structure candidates. From a practical perspective, the important case of pure rotation is *the* rationale for considering conics.

In order to render to problem more amenable to mathematical analysis, the projective reference frame of fig. 7 will be used. It is based on a reference triangle,

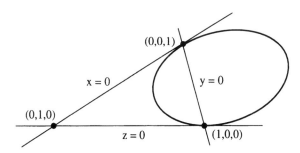

**Fig. 7.** *Conic and the reference triangle used as reference projective frame.*

formed by choosing a point and the two tangent lines with the conic through that point. In this reference frame the conic has the canonical equation

$$y^2 = xz \qquad (6)$$

in homogeneous point coordinates $(x, y, z)^T$.

As Semple & Kneebone [12] have shown, there are two families of projective transformations that keep a conic fixed. One family has two (complex) fixed points on the conic, the other one (i.e. two coincident fixed points that are real). The set of projectivities of the first category that keep the conic (6) fixed (actually a projective subgroup) is given by the projective transformation matrices

$$P_{1\alpha} = \begin{pmatrix} \alpha^2 & 0 & 0 \\ 0 & \alpha & 0 \\ 0 & 0 & 1 \end{pmatrix} \qquad (7)$$

and those of the second category (also forming a subgroup) by

$$P_{2\alpha} = \begin{pmatrix} 1 & 2\alpha & \alpha^2 \\ 0 & 1 & \alpha \\ 0 & 0 & 1 \end{pmatrix} \qquad (8)$$

where $\alpha$ is a complex number. Note that the first group includes a harmonic homology (involution) by putting $\alpha = -1$.

It is interesting to ask which other conics (if any) will also remain fixed under the foregoing projectivities. Given the original conic as a fixed structure, it is not difficult to check that the first subgroup will keep a complete pencil of conics fixed, where the members of the pencil are given by

$$C_{1\lambda} = y^2 + \lambda(xz) = 0 \qquad (9)$$

when expressed in complex, projective coordinates $(x, y, z)^T$ and with $\lambda$ ar variable parameter. The second subgroup keeps the pencil

$$C_{2\lambda} = (y^2 - xz) + \lambda z^2 = 0 \qquad (10)$$

fixed, but will not be discussed further here (focus is on the first case because it includes the important case of rotations about a point).

We may conclude that keeping a conic fixed implies keeping a whole pencil of conics fixed. All transformations conjugate with the tranformations $P_{\alpha 1}$ and $P_{\alpha 2}$ by a *complex* projectivity $P_t$ (i.e. transformations $P_t P_{1\alpha} P_t^{-1}$ and $P_t P_{1\alpha} P_t^{-1}$) will keep the pencils of conics $P_t^{-T} C_{1\lambda} P_t^{-1}$ and $P_t^{-T} C_{2\lambda} P_t^{-1}$ fixed, resp.

Springer [14] gives a classification of pencils of conics, ordered according to the nature of the degenerate conics in the pencil. The analysis is based on finding the roots of the cubic equation in $\mu$

$$|C_\lambda + \mu C_{\lambda'}| = 0$$

where $\lambda$ and $\lambda'$ represent two different conics of the pencil. According to the multiplicity $m$ of the roots and the corresponding ranks $r$ of the above matrix,

5 classes are distinguished. It can be shown that the above pencils belong to Springer's classes 3 and 5, resp., which we will coin $S_3^C$ and $S_5^C$. $S_3^C$ has two roots, one with $m = 2$ and $r = 1$ and one with $m = 1$ and $r = 2$. As to the degenerate conics (Springer calls them composite), there are three of them. For $S_3^C$ the situation is depicted in fig. 8. $S_3^C$ has two distinct degenerate conics, two

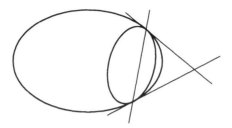

**Fig. 8.** *Degenerate conics and two non-degenerate conics for $S_3^C$.*

lines and a line counted twice. The conics are tangent to the former where the latter intersects them. These two points correspond to the fixed points on the conics.

In contrast to the point/line classifications of section 3 and 4, the foregoing analysis pertains to the complex plane. Computer vision deals with a real image and fixed structure is particularly relevant if it can be observed there. This has consequences for class $S_3^C$, since a further distinction has to be made between cases where the degenerate conic consisting of two lines has two lines in the real plane, or two lines with complex conjugate line coordinates. This distinction also corresponds to the dividing line between having two (real) fixed points on the conics or none. Hence, we add a second canonical representation for pencils of conics without fixed points, derived as

$$\begin{pmatrix} 2c & 0 & 0 \\ 0 & d & 0 \\ 0 & 0 & 2c \end{pmatrix} = \begin{pmatrix} 1 & 0 & 1 \\ 0 & 1 & 0 \\ i & 0 & -i \end{pmatrix} \begin{pmatrix} 0 & 0 & c \\ 0 & d & 0 \\ c & 0 & 0 \end{pmatrix} \begin{pmatrix} 1 & 0 & i \\ 0 & 1 & 0 \\ 1 & 0 & -i \end{pmatrix} ,$$

where the corresponding transformation on the points maps the intersections $(1, 0, 0)$, $(0, 1, 0)$, and $(0, 0, 1)$ of the lines composing the degenerate conics to $(1, 0, -i)$, $(0, 1, 0)$, and $(1, 0, i)$, resp. Only one of these points remains in the real plane. The corresponding pencil of fixed conics can be written

$$C_{3\lambda} = (x + iz)(x - iz) + \lambda y^2 = (x^2 + z^2) + \lambda y^2 = 0$$

and the transformation group that keeps the conics fixed is given by

$$P_{3\alpha} = \begin{pmatrix} \alpha^2 + 1 & 0 & (\alpha^2 - 1)i \\ 0 & 2\alpha & 0 \\ -(\alpha^2 - 1)i & 0 & \alpha^2 + 1 \end{pmatrix} .$$

Putting $\alpha = e^{i\theta}$ yields a rotation over the angle $\theta$. Obliquely viewed rotations $(P_t P_{3\alpha} P_t^{-1})$ yield the degenerate structures that we expect: a fixed rotation center as the transformed version of the real intersection $(0, 1, 0)$ of the two complex lines and the transformed horizon line $y = 0$. The fixed conics can be used to form invariants, possibly in combination with points and lines (examples of such invariants can be found in [7]).

# 6    A strategy for grouping

Invariants are useful tools for grouping because they allow systems to find matches while avoiding combinatorial search. Using general projective invariants isn't necessarily the optimal approach. This may be because such invariants need a minimum of contour information for their extraction, e.g. a "bitangent segment" [1], and as grouping is about matching parts these may lack such rich local structure even more easily. A second problem may be that far too many matches might result if the problem would be like finding one out of many copies of the same object, e.g. the razor example used by Lowe [5]. To clarify the latter point further, consider a bin full of identical, mirror symmetric shapes. All half shapes would match under projective invariants. This may result in hundreds of possible matches which then have to be checked further whether they really represent a mirror symmetry. In such cases symmetry-specific invariants can increase efficiency considerably, as they selectively pick out half shapes that are in symmetric positions.

Whatever the reason for using grouping-specific invariants – out of necessity or efficiency considerations – they require the explicit knowledge of the fixed structures they contain. So we need a strategy to find fixed structure candidates. An important point to make is that however complicated the shapes to be grouped might be, e.g. the halfs of a complicated ornament, the fixed structures themselves always are equally simple. The search for fixed structures remains limited to fixed points, lines, and conics. Fortunately, finding such simple parametric shapes has been studied for a long time in computer vision.

A well-known technique in computer vision to detect simple parametric shapes is the Hough transform. Here we will focus on the simplest part: detecting the fixed points and lines. These, however, also represent the majority of cases. In a companion paper [15] it is described how an iterative application of the Hough transform can yield different types of fixed structures along the way. This will be referred to as the Cascaded Hough Transform or CHT for short. To fix ideas, the straight lines coming out of the first step might eg already contain the horizon, which is a fixed line for quite a few types of grouping, or some fixed lines of a pencil. The second stage may yield the pencil's vertex, vanishing points, or a center of rotation. The third step again could give out the horizon, a symmetry axis, etc.

Although no hard guarantees can be given that the fixed structures of all groupings present in the image will be found by the CHT, this is not necessarily required. Some groupings will be formed efficiently based on general projective invariants. The fixed structures of other groupings could be found by considering the complete set of fixed structures of groupings found already. That set may contain fixed structures not found via the CHT. It happens very regularly that different groupings share some of their fixed structures. Anyway, those fixed structures that are found with the CHT will be found without combinatorial search. Hence, the non-combinatorial Hough yields fixed structures that can activate grouping-specific invariants, that allow to perform non-combinatorial matching between segments that actually belong to the corresponding groupings.

The grouping strategy can be sketched as in fig. 9. It is important to em-

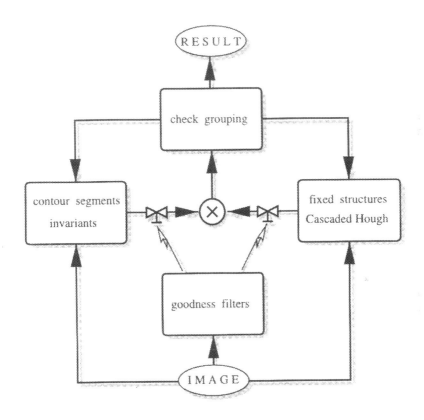

**Fig. 9.** *Proposed grouping strategy (see text).*

phasise that this work is ongoing. The strategy proceeds along two simultaneous tracks. On the one hand, invariants are exploited at the earliest opportunity. Initially, projective invariants are calculated for segments spanned by bitangent lines. These are matched and the matches might yield some groupings already. If there aren't many, these groupings can be analysed by considering the projective transformations that bring the segments in registrations. The fixed structures of these transformations can be reused for other groupings. In parallel, the cascaded

Hough scheme yields candidate fixed structures. Starting with the strongest candidates (getting the most support from image features), the invariants of the corresponding subgroups are used, mainly to those contour segments that have contributed in the extraction of the fixed structures. The strategy is to use as few fixed structures in combination as possible. Assuming a structure to be fixed introduces the risk that it is wrong. Hence, if an assumption is made that several structures are fixed simultaneously, the change of errors increases. Strong assumptions are only made as a last resort. The 'goodness filters' introduce the possibility to let background knowledge play a role and to rank hypotheses for the order in which they should be tried. Fixed structures can be given higher preference if fewer of them are combined, if they are more outspoken in the CHT, if they group longer edges, or if they don't yield too many possible matches.

Note that combinatorial procedures are avoided at both ends. The invariant based matching proceeds through efficient hashing, and finding fixed structures is based on a cascade of Hough transforms. Also note that far from rejecting grouping based on general projective invariants, that is exactly what the system might try first. As we have found, however, there are quite a few cases where such strategy does not suffice. That is where the fixed structures come in.

## 7   A grouping example

This section gives an example of how fixed structures allow to perform simple tests. Consider the scene of fig. 10. The goal could be to detect the mirror symmetry in the top key, e.g. as a cue for finding it in the first place. For the shape

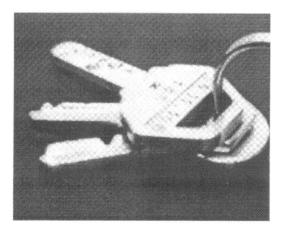

**Fig. 10.** *Scene with keys.*

of the symmetric key, this would not be that easy, because the structure of each half is too simple to admit the traditional projective invariants or at least would such invariants be quite unreliable.

Using the Cascaded Hough Transform, candidate vanishing points were search-ed. These in turn are good candidates to be the vertices of pencils of fixed lines as their are found with (perspectively) skewed mirror symmetries. Fig. 11 shows the lines that contribute to the detection of three vanishing points. The lines of

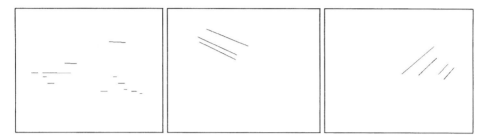

**Fig. 11.** *Line segments that contribute to the vanishing points detected by the CHT.*

the figure on the right yield the correct vertex. The lines in the middle figure yield a point on the symmetry axis. The vanishing point corresponding to the left figure is unrelated to the symmetry of the key.

Calling the vertex $\mathbf{x}_v$ and the point on the axis $\mathbf{x}_a$, it is illustrated how simple invariants can be used to detect the symmetry, given these two points and their respective roles... Fig. 12 shows the edges that were used for the matching. The skewed symmetry-invariant description that was used for the matching is the following. The invariant

$$\frac{\left|\mathbf{x} - \mathbf{x}_v \quad \mathbf{x}^{(1)}\right|}{\left|\mathbf{x} - \mathbf{x}_a \quad \mathbf{x}^{(1)}\right|}$$

was calculated as a function of

$$\frac{(x - x_v)}{(y - y_v)} \ .$$

The resulting descriptions are compared in fig. 13. On the left, the descriptions of both halfs of the symmetric key are superimposed. The match is reasonably good. At the right, the descriptions for the bottom edge of the symmetric key is compared against the bottom edge of the second key. Now the matching fails completely.

## 8  Conclusions and future research

An attempt was made to get a more systematic handle on geometric grouping and the use of invariants towards that purpose in particular. Fixed structures under planar projectivies were the key element. The ability to stick to invariance even in cases where general projective invariants can no longer be extracted allows to avoid combinatorics in the matching. Finding the fixed structures is crucial, however, and the Cascaded Hough Transform was proposed as a non-combinatorial approach for that part. The actual grouping then proceeds as the interplay between hypothesising fixed structures – but as few and least far-fetched as possible – and invariant-based matching. This is the subject of ongoing research, and much work remains to be done to arrive at a fully automatic grouping algorithm.

Several key issues remain. First, the three types of fixed structures were discussed completely independently. Yet, the coexistence of the three types should be investigated and a joint classification made. As an example, following a theorem of Desargues, conics of a pencil intersect any line not through the base points in points that are in involution. Hence, if such a line would be assumed fixed, the intersections are automatically consistent with two possibilities. The first is that the line is a line of fixed points. The second is that the homography on the line is an involution and therefore has two fixed points and an infinity of fixed point pairs. Another example is the fixed line that emerges from the line of Pascal for a fixed set of 6 points lying on a conic.

Another issue is the definition of the "goodness filters" of fig. 9, which determine in what order different possible groupings are to be investigated, and which edges will be tried in combination with which fixed structures. In combining edges and fixed structures still lies hidden a danger of introducing combinatorics. A way of reducing that risk is to first consider those edges that have contributed to the extraction of the fixed structures, or at least edges that lie close to edges that have contributed.

The research will also be directed more strongly towards 3D patterns. There are e.g. strong relations between results presented here for the line of fixed points case and the butterfly and caging invariants presented in the literature [10, 11].

**Acknowledgements:** The work reported in this paper has been supported by the Flemish Fund for Scientific Research (FWO). The implementations of the CHT and the invariants have been produced by Marc Proesmans and Tinne Tuytelaars.

# References

1. S. Carlsson, R. Mohr, T. Moons, L. Morin, C. Rothwell, M. Van Diest, L. Van Gool, F. Veillon, and A. Zisserman, Semi-local projective invariants for the recognition of smooth plane curves, to appear in Int. Journal of Computer Vision

2. R. Glachet, J. Lapreste, M. Dhome, Locating and modelling a flat symmetric object from a single projective image, *Computer Vision, Graphics, and Image Processing: Image Understanding*, Vol.57, pp.219-226, 1993

3. T. Kanade, Recovery of the 3-dimensional shape of an object from a single view, Artificial Intelligence, Vol.17, pp.75-116, 1981

4. D. Lowe, Perceptual Organization and Visual Recognition Stanford University technical report STAN-CS-84-1020, 1984

5. D. Lowe, The Viewpopint Consistency Constraint, International Journal of Computer Vision, pp.57-72, Vol.1, 1987

6.  D. Lowe, *Perceptual organisation and visual recognition*, Kluwer Academic Publishers, 1985

7.  J. Mundy and A. Zisserman (eds.), *Geometric invariance in computer vision*, MIT Press, Cambridge, 1992

8.  J. Ponce, On characterizing ribbons and finding skewed symmetries, Proc. Int. Conf. on Robotics and Automation, pp. 49-54, 1989

9.  C. Rothwell, Recognition using perspective invariance, PhD Thesis, Univ. of Oxford, 1993

10. C. Rothwell, D. Forsyth, A. Zisserman, and J. Mundy, Extracting projective structure from single perspective views of 3D points sets, Proc. 3rd Int. Conf. Computer Vision, Belin, pp. 573-582, 1993

11. C. Rothwell and J. Stern, Understanding the shape properties of trihedral polyhedra, Eur. Conf. Computer Vision, pp. 175-185, Cambridge, UK, 1996

12. J. Semple and G. Kneebone, *Algebraic projective geometry*, Clarendon, 1979

13. D. Sinclair, H. Christensen, and C. Rothwell, Using the relation between a plane projectivity and the fundamental matrix, Britsh Machine Vision Conference,

14. C. Springer, *Geometry and analysis of projective spaces*, Freeman, 1964

15. T. Tuytelaars, M. Proesmans, and L. Van Gool, The Cascaded Hough Transform as support for grouping and finding vanishing points and lines, published in these proceedings.

16. L. Van Gool, T. Moons, E. Pauwels, and A. Oosterlinck, Semi-differential invariants, in *Applications of invariance in vision*, J. Mundy and A. Zisserman (eds.), pp. 157-192, MIT Press, Boston, 1992

17. L. Van Gool, M. Proesmans, and A. Zisserman, Grouping and invariants using planar homologies, Proc. Europe-China Workshop on Geometrical Modeling and Invariants for Computer Vision, pp. 182-189, 1995

18. L. Van Gool, T. Moons, D. Ungureanu, and A. Oosterlinck, The characterisation and detection of skewed symmetry, *Computer Vision and Image Understanding* (previously CVGIP), Vol. 61, No. 1, pp. 138-150, 1995

19. L. Van Gool, T. Moons, and M. Proesmans, Groups for grouping: a strategy for the exploitation of geometrical constraints, Proc. 6th Int. Conf. on Computer Analysis of Images and Patterns, Prague, pp. 1-8, sept. 1995

20. L. Van Gool, T. Moons, and M. Proesmans, Mirror and point symmetry under perspective skewing, IEEE Conf. Computer Vision and Pattern Recognition, pp. 285-292, San Francisco, june 1996

21. M. Wertheimer, Laws of organization in perceptual forms, in *A source-book of Gestalt Psychology*, ed. D. Ellis, Harcourt, Brace and Co., pp.71-88, 1938

22. A. Zisserman, J. Mundy, D. Forsyth, and J. Liu, Class-based grouping in perspective images, Proc. Int. Conf. Computer Vision, pp.183-188, 1995

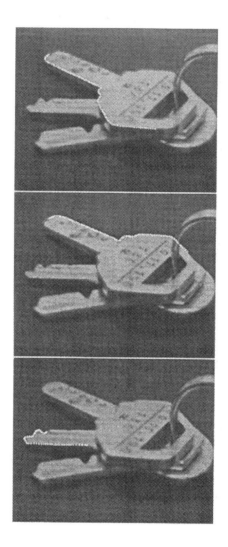

**Fig. 12.** *Edges used for the matching experiment.*

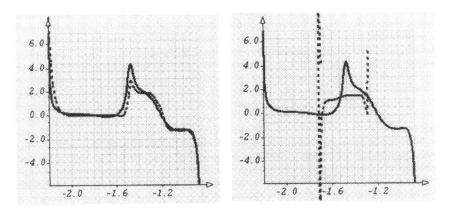

**Fig. 13.** *Invariant descriptions for the edges shown in fig. 12.*

# Multi–Dimensional Signal Processing Using an Algebraically Extended Signal Representation*

Thomas Bülow and Gerald Sommer

Institute of Computer Science
Christian–Albrechts–University of Kiel,
Preußerstr. 1–9, 24105 Kiel, Germany
{tbl,gs}@informatik.uni-kiel.de

**Abstract.** Many concepts that are used in multi–dimensional signal processing are derived from one–dimensional signal processing. As a consequence, they are only suited to multi–dimensional signals which are intrinsically one–dimensional. We claim that this restriction is due to the restricted algebraic frame used in signal processing, especially to the use of the complex numbers in the frequency domain. We propose a generalization of the two–dimensional Fourier transform which yields a quaternionic signal representation. We call this transform quaternionic Fourier transform (QFT). Based on the QFT, we generalize the conceptions of the analytic signal, Gabor filters, instantaneous and local phase to two dimensions in a novel way which is intrinsically two–dimensional. Experimental results are presented.

## 1 Introduction

Realizations of autonomous technical systems which are designed on principles of the perception–action–cycle (PAC) are supposed to act in the real world which can be described as taking place in a four–dimensional Euclidean space–time. Therefore, a PAC–system has to be able to percept events and to organize processes in such a world.

Focusing on the perceptional part of a PAC system we face some serious shortcomings in low–level processing of multi–dimensional signals. These shortcomings have been recognized for a long time but by now are not solved satisfactorily. In the authors opinion the root of the problems seems to lie in the restricted algebraical embedding of multi–dimensional signal–processing which has not yet been recognized. The algebraical embedding of signal processing is meant to be the choice of algebra in which a signal is represented in the frequency domain. Usually this role is played by the algebra of complex numbers but we will show that it is useful to apply algebras of higher dimensions here. Most of the valuable tools that have been developed in one–dimensional signal processing are nowadays used in multi–dimensional signal processing but in a way which leaves them intrinsically one–dimensional.

* This work was supported by the Studienstiftung des deutschen Volkes (Th.B.) and by the DFG (So-320-2-1) (G.S.).

One example for this is the concept of local phase. The local phase of a one–dimensional signal can be estimated by applying a quadrature filter — e.g. a complex Gabor filter — and evaluating the argument of the resulting complex filter response. This concept is usually generalized to two dimensions by defining the two–dimensional Gabor filters as the Gaussian windowed basis function of the Fourier transform for some frequency $u$. Again, we get as the local phase the argument of the complex filter response and we get different values for different orientations of the Gabor filter. Granlund [7] defines an $(n+1)$–dimensional phase vector for $n$–dimensional signals, consisting of the real phase and the directional vector of the chosen orientation.

In one dimension the local phase yields information about the local structure of the signal. In two dimensions the variety of possible local structures is much higher than in one dimension and so we cannot hope to characterize the local image structure using only one real number. Looking for a concept which yields a higher–dimensional value for the local phase we find that the main restriction of the phase dimension lies in the fact that the responses of the Gabor filters are complex–valued. Thus, we will study filters with responses which are elements of a higher–dimensional algebra than the complex numbers. We will show that a generalization to quaternion–valued filters is possible in two dimensions. A short review on quaternions will be given in the following section.

One–dimensional Gabor filters are based on the Fourier transform. Therefore we will extend the two–dimensional Fourier transform in such a way that it yields a quaternion–valued representation in the frequency domain. We call this transform quaternionic Fourier transform (QFT). We will demonstrate the shift theorem in the case of the QFT, analyze the symmetry properties of the QFT and show its relation to the Fourier transform and to the Hartley transform.

In order to define the instantaneous phase of a two–dimensional signal we will introduce the quaternionic analytic signal of a two–dimensional signal via the QFT. Finally we introduce the quaternionic Gabor filters based on the QFT and the two–dimensional local phase and demonstrate some experimental results.

## 2 Quaternions

As motivated in the introduction we need as the range of a generalized two–dimensional Fourier transform an algebra whose dimension is higher than the dimension of the algebra of complex numbers. In the following we will use the four–dimensional $\mathbb{R}$–algebra

$$\mathbb{H} = \{q = a + bi + cj + dk \mid a, b, c, d \in \mathbb{R}\} \quad , \tag{1}$$

where $i$, $j$ and $k$ obey the following multiplication rules:

$$i^2 = j^2 = -1, \quad k = ij = -ji \implies k^2 = -1 \quad . \tag{2}$$

The algebra IH was invented in 1843 by Hamilton[2] [8] who called it the *algebra of quaternions*. There is a whole lot of literature on quaternions (see e.g. [9, 10]). For the sake of brevity we will only introduce the properties which will be needed in the course of this article.

For a quaternion $q = a + bi + cj + dk$ the component $a$ is called the *scalar part* of $q$, whereas $bi + cj + dk$ is called the *vector part* of $q$. A quaternion consisting only of a vector part is called a *pure quaternion*. Like in the algebra of complex numbers we can define the operation of conjugation for quaternions. The conjugate of a quaternion $q = a + bi + cj + dk$, denoted by $q^*$, is defined by changing the sign of the vector part of $q$:

$$q^* = a - bi - cj - dk \quad . \tag{3}$$

The operation of conjugation is a vector space involution. The *magnitude* of $q$ is defined as

$$|q| = \sqrt{qq^*} = \sqrt{a^2 + b^2 + c^2 + d^2} \quad . \tag{4}$$

By $\epsilon$, $\alpha$, $\beta$ and $\gamma$ we denote the four algebra involutions of IH. They are given by

$$
\begin{aligned}
\epsilon : q &\mapsto q, & \epsilon(q) &= a + bi + cj + dk, \\
\alpha : q &\mapsto -iqi, & \alpha(q) &= a + bi - cj - dk, \\
\beta : q &\mapsto -jqj, & \beta(q) &= a - bi + cj - dk, \\
\gamma : q &\mapsto -kqk, & \gamma(q) &= a - bi - cj + dk.
\end{aligned}
$$

In analogy to a Hermitian function $f : \mathbb{R} \to \mathbb{C}$ with $f(x) = f^*(-x)$ for every $x \in \mathbb{R}$ we introduce the notion of a *quaternionic Hermitian function* for a function $f : \mathbb{R}^2 \to$ IH which obeys the rules

$$f(-x, y) = \beta(f(x, y)) \quad \text{and} \quad f(x, -y) = \alpha(f(x, y)) \quad , \tag{5}$$

for each $(x, y) \in \mathbb{R}^2$. For a quaternionic Hermitian function also the relation

$$f(-x, -y) = \gamma(f(x, y)) \tag{6}$$

holds true.

We will need the exponential function $exp :$ IH $\to$ IH of a quaternion $q$ which is defined via the series

$$\exp(q) = \sum_{k=0}^{\infty} \frac{q^k}{k!}, \quad q \in \text{IH} \quad . \tag{7}$$

It can be shown that this sum converges for every quaternion $q$. Let us write the quaternion $q$ in the form $q = s + v$, where $s$ and $v$ denote the scalar part and

---

[2] Blaschke [3] states that they were already known to Euler — who used them to describe rotations in $\mathbb{R}^3$ — in 1748.

the vector part of $q$, respectively. We can then evaluate $exp(q)$ in the following way:

$$\exp(q) = \exp(s + v) = \exp(s)\left(\cos(|v|) + \frac{v}{|v|}\sin(|v|)\right) \quad . \tag{8}$$

In the last step we used the fact that the Euler formula

$$e^{i\phi} = \cos(\phi) + i\sin(\phi) \tag{9}$$

is not only valid if $i$ is the imaginary unit of the complex numbers but also in the form

$$e^{r\psi} = \cos(\psi) + r\sin(\psi) \quad ,$$

where $r$ is an arbitrary pure unit quaternion[3].

## 3  Quaternionic Fourier Transform

Here we want to give a review of the recently introduced quaternionic Fourier transform (QFT) [5]. The QFT is a transform for two–dimensional signals which on the first glance seems to be only a slight modification of the well–known two–dimensional Fourier transform. For detailed information on Fourier transform see e.g. [4]. The two–dimensional Fourier transform is given by

$$F(u) = \int\limits_{-\infty}^{\infty} \int\limits_{-\infty}^{\infty} e^{-i2\pi ux} f(x) e^{-i2\pi vy} d^2x \quad , \tag{10}$$

whereas the quaternionic Fourier transform is defined as

$$F^q(u) = \int\limits_{-\infty}^{\infty} \int\limits_{-\infty}^{\infty} e^{-i2\pi ux} f(x) e^{-j2\pi vy} d^2x \quad , \tag{11}$$

with the only difference of using two different imaginary units in the exponential functions. Here $x$ denotes the vector $(x, y)$ in the image plane and $u$ denotes the two–dimensional frequency vector $(u, v)$. The units $i$ and $j$ are supposed to be two of the imaginary units of the quaternion algebra defined above. This leads to a significant difference between the two–dimensional Fourier transform and the QFT. Let $\mathcal{F}$ and $\mathcal{F}_q$ denote the operators of the Fourier transform and the QFT, respectively. For a two–dimensional signal $f : \mathbb{R}^2 \to \mathbb{R}$ the Fourier transform $F = \mathcal{F}\{f\}$ maps the image signal to a complex–valued representation whereas the QFT $F^q = \mathcal{F}_q\{f\}$ maps the image signal onto a quaternion–valued spectral representation.

---

[3] For the proof of the Euler formula we only need the definition of the exponential function, the cosine and the sine function as series and the algebraic properties of $i$, i.e. $i^2 = -1$. Hence, to proof the Euler formula for quaternions we must only show that $r^2 = -1$ which is straightforward.

Also Chernov [6] used quaternions in the Fourier transform. However, his aim was to find fast algorithms for the evaluation of the two–dimensional discrete Fourier transform, while we are interested in constructing a more complex phase representation.

The QFT of a real signal is a quaternionic Hermitian function as defined in the previous section. The inverse of the QFT is given by

$$
f(x) = \int\limits_{-\infty}^{\infty} \int\limits_{-\infty}^{\infty} e^{i2\pi u x} F^q(u) e^{j2\pi v y} d^2 u \quad . \tag{12}
$$

In order to translate some of the properties of the Fourier transform to the QFT, we will consider the shift–theorem here. Let $F^q(u)$ and $F_T^q(u)$ be the QFT's of a real signal $f(x)$ and the translated signal $f_T(x) = f(x - d)$. It follows easily that

$$
F_T^q(u) = e^{-i2\pi u d_1} F^q(u) e^{-j2\pi v d_2} \quad . \tag{13}
$$

In the following we will write this down in matrix–notation representing the quaternion $F^q(u) = F_0^q(u) + iF_1^q(u) + jF_2^q(u) + kF_3^q(u)$ as the vector $\mathbf{F}^q(u) = (F_0^q(u), F_1^q(u), F_2^q(u), F_3^q(u))^T$. Eq. (13) then reads $\mathbf{F}_T^q(u) = T(\phi, \theta)\mathbf{F}^q(u)$, with

$$
T(\phi, \theta) = \begin{pmatrix} \cos(\phi)\cos(\theta) & -\sin(\phi)\cos(\theta) & -\cos(\phi)\sin(\theta) & -\sin(\phi)\sin(\theta) \\ \sin(\phi)\cos(\theta) & \cos(\phi)\cos(\theta) & -\sin(\phi)\sin(\theta) & \cos(\phi)\sin(\theta) \\ \cos(\phi)\sin(\theta) & \sin(\phi)\sin(\theta) & \cos(\phi)\cos(\theta) & -\sin(\phi)\cos(\theta) \\ \sin(\phi)\sin(\theta) & -\cos(\phi)\sin(\theta) & \sin(\phi)\cos(\theta) & \cos(\phi)\cos(\theta) \end{pmatrix} \quad . \tag{14}
$$

Here $\phi$ and $\theta$ denote $-2\pi d_1 u$ and $-2\pi d_2 v$, respectively. Now we can show explicitly how the translation vector $d$ can be recovered from a pair $\mathbf{F}^q(u)$ and $\mathbf{F}_T^q(u)$ for a single value of $u$. It is straightforward to show that

$$
\mathbf{F}_T^q(u) = \begin{pmatrix} F_0^q(u) & -F_1^q(u) & -F_2^q(u) & F_3^q(u) \\ F_1^q(u) & F_0^q(u) & -F_3^q(u) & -F_2^q(u) \\ F_2^q(u) & -F_3^q(u) & F_0^q(u) & -F_1^q(u) \\ F_3^q(u) & F_2^q(u) & F_1^q(u) & F_0^q(u) \end{pmatrix} \begin{pmatrix} \cos\phi\cos\theta \\ \sin\phi\cos\theta \\ \cos\phi\sin\theta \\ \sin\phi\sin\theta \end{pmatrix} =: F(u) \begin{pmatrix} \alpha \\ \beta \\ \gamma \\ \delta \end{pmatrix}
$$

If $F(u)$ is invertible we get $\alpha, \beta, \gamma$ and $\delta$ directly from

$$
F^{-1}(u)\mathbf{F}_T^q(u) = (\alpha, \beta, \gamma, \delta)^T \quad . \tag{15}
$$

It is possible to recover $(\phi, \theta)$ within the interval $[0, 2\pi[ \times [0, \pi[$ from $(\alpha, \beta, \gamma, \delta)$ by a function $arg : \mathbb{H}\backslash\{0\} \mapsto \mathbb{R}^2$, $arg(\alpha, \beta, \gamma, \delta) = (\phi, \theta)$. Because of the bulkiness of the definition of $arg$ we will give it in the appendix A. From $\phi$ and $\theta$ we get $d_1$ and $d_2$ by

$$
d_1 = \frac{\phi}{2\pi u}, \quad d_2 = \frac{\theta}{2\pi v} \quad . \tag{16}
$$

Thus, we recovered the translation vector $d$ from the QFT's of $f$ and $f_T$ for one specific value of $u$.

# 4   Symmetries of the QFT

The concept of symmetry of a signal is well known in one dimension. Globally as well as locally signals can be split into an even and an odd component. While the global symmetry is not an inherent signal property but depends on the choice of the origin, the local symmetry describes the local structure of the signal; it is even for a peak–like structure and odd for step–like structure in the signal.

In this section we will show how the QFT deals with signals of different combinations of even and odd symmetries. It is a well known fact that the Fourier transform of the even part of a real one–dimensional signal is real and even. The Fourier transform of the odd part of the same signal is imaginary and odd. As said above this splitting depends on the choice of the origin. It is independent of scaling, though.

A two–dimensional signal can be split into even and odd parts along the $x$–axis and along the $y$–axis as well. So, every real two–dimensional signal can be written in the form $f = f_{ee} + f_{oe} + f_{eo} + f_{oo}$, where $f_{ee}$ denotes the part of $f$ which is even with respect to $x$ and $y$, $f_{oe}$ denotes the part which is odd with respect to $x$ and even with respect to $y$ and so on. In this case the splitting is not only dependent of the choice of the origin but also of the orientation of the image. Because the two–dimensional Fourier transform has only two components — one real and one imaginary component — we are not able to immediately recognize the four components of different symmetry.

However, the quaternionic Fourier transform has symmetry–splitting properties that are analogous to the properties of the one–dimensional Fourier transform: The transform of the $f_{ee}$–part of a real two–dimensional signal is real, the $f_{oe}$–part is transformed into a $i$–imaginary part, $f_{eo}$ into the $j$–imaginary and $f_{oo}$ into the $k$–imaginary part. The symmetry of the signal is preserved by the quaternionic Fourier transform. We can see this easily by looking at the quaternionic Fourier transform as two sequentially performed one–dimensional Fourier transforms: First we perform a one–dimensional Fourier transform on $f(x)$ with respect to $x$ keeping $y$ fixed and call the result $\tilde{f}$:

$$\tilde{f}(u, y) = \int\limits_{-\infty}^{\infty} e^{-i2\pi ux} f(x, y) dx \quad . \tag{17}$$

In a second step we perform a Fourier transform on $\tilde{f}$ with respect to $y$ keeping $u$ fixed:

$$F^q(\boldsymbol{u}) = \int\limits_{-\infty}^{\infty} \tilde{f}(u, y) e^{-j2\pi vy} dy \quad . \tag{18}$$

Actually, this two step procedure is the way we implemented the QFT on the computer. Hence, the implementation is similar to the one of the two–dimensional Fourier transform. The difference is that while calculating the two–dimensional Fourier transform we add up some components which we keep separately when calculating the QFT. An overview over the symmetry properties of the QFT is given in table 1.

| | $f = f_{ee}$ | | | | $f = f_{oe}$ | | | | $f = f_{eo}$ | | | | $f = f_{oo}$ | | | |
|---|---|---|---|---|---|---|---|---|---|---|---|---|---|---|---|---|
| $f$ | r | i | j | k | r | i | j | k | r | i | j | k | r | i | j | k |
| $\tilde{f}$ | r | i | j | k | i | r | k | j | r | i | j | k | i | r | k | j |
| $F$ | r | i | j | k | i | r | k | j | j | k | r | i | k | j | i | r |

**Table 1.** Symmetry properties of the QFT. $r$ stands for the real part, $i$ for the $i$–imaginary and so on.

In order to clarify the position of the QFT among the existing transforms, we relate the QFT to the Fourier transform and to the Hartley transform (see e.g. [4]):

The Hartley transform of a one–dimensional signal $f$ is defined by

$$H(u) = \int\limits_{-\infty}^{\infty} f(x)\{\cos(2\pi u x) + \sin(2\pi u x)\}dx \quad . \tag{19}$$

It is related to the Fourier transform of $f$ by

$$F(u) = H_e(u) - iH_o(u) \quad , \tag{20}$$

where $H_e$ and $H_o$ denote the even and odd part of $H$ respectively. So the Fourier transform separates the parts of different symmetry — which are mixed in the Hartley transform — by putting them into different components.

The two–dimensional Hartley transform is given by

$$H(\boldsymbol{u}) = \int\limits_{-\infty}^{\infty} \int\limits_{-\infty}^{\infty} f(\boldsymbol{x})\{\cos(2\pi \boldsymbol{u} \cdot \boldsymbol{x}) + \sin(2\pi \boldsymbol{u} \cdot \boldsymbol{x})\}d^2\boldsymbol{x}$$

$$= \int\limits_{-\infty}^{\infty} \int\limits_{-\infty}^{\infty} f(\boldsymbol{x}) \{\cos(2\pi u x)\cos(2\pi v y) - \sin(2\pi u x)\sin(2\pi v y)$$

$$+ \cos(2\pi u x)\sin(2\pi v y) + \sin(2\pi u x)\cos(2\pi v y)\} \, d^2\boldsymbol{x}$$

$$= H_{ee}(\boldsymbol{u}) + H_{oo}(\boldsymbol{u}) + H_{eo}(\boldsymbol{u}) + H_{oe}(\boldsymbol{u}) \quad . \tag{21}$$

Again it is possible to get the Fourier transform of $f$ from the Hartley transform:

$$F(\boldsymbol{u}) = \big(H_{ee}(\boldsymbol{u}) + H_{oo}(\boldsymbol{u})\big) - i\big(H_{eo}(\boldsymbol{u}) + H_{oe}(\boldsymbol{u})\big) \quad . \tag{22}$$

Hence, also in this case the Fourier transform separates parts of different symmetry, which are mixed in the Hartley transform, but this separation is only halfway. The complete separation is only given by the quaternionic Fourier transform:

$$F^q(\boldsymbol{u}) = H_{ee}(\boldsymbol{u}) - kH_{oo}(\boldsymbol{u}) - jH_{eo}(\boldsymbol{u}) - iH_{oe}(\boldsymbol{u}) \quad . \tag{23}$$

It follows from this that the two–dimensional Fourier transform stands between the QFT and the two–dimensional Hartley transform in the sense that we can derive the QFT from the two–dimensional Fourier transform in a similar way as we derive the Fourier transform from the Hartley transform.

As shown in section 2 we can represent the QFT of a signal also in polar representation. Thus, we can write $F^q(\boldsymbol{u})$ in the form

$$F^q(\boldsymbol{u}) = |F^q(\boldsymbol{u})| \exp(s(\boldsymbol{u})\psi(\boldsymbol{u})) \tag{24}$$

with

$$s(\boldsymbol{u}) = i\cos(\phi(\boldsymbol{u}))\sin(\theta(\boldsymbol{u})) + j\sin(\phi(\boldsymbol{u}))\sin(\theta(\boldsymbol{u})) + k\cos(\theta(\boldsymbol{u})) \tag{25}$$

The three angles $\psi$, $\phi$ and $\theta$ could be regarded as the phase of a two–dimensional signal. Nevertheless, the phase $(\psi, \phi, \theta)$ for special values of the angles is an element of the three–dimensional hypersphere $S^3$ which makes an interpretation of the values complicated. Another approach to a multidimensional phase concept will be represented in the next section.

## 5  The Analytic Signal

The analytic signal plays an important role in one–dimensional signal processing. One of the main reasons for this fact is, that it is possible to read the instantaneous amplitude and the instantaneous phase from a signal $f$ at a certain position $x$ simply by taking the magnitude and the phase of the analytic signal $f_A$ at the position $x$, where $f_A(x)$ is a complex number. The analytic signal $f_A$ of a real signal $f$ is defined as $f_A = f - i\mathcal{H}\{f\}$ where $\mathcal{H}\{f\}$ is the Hilbert transform of $f$. It can be derived from $f$ by taking the Fourier transform $F$ of $f$, suppressing the negative frequencies and multiplying the positive frequencies by two. Applying this procedure, we do not lose any information about $f$.

One way to extend the concept of the analytic signal to two dimensions is to split the frequency plane into two half planes with respect to a direction $e = (\cos(\theta), \sin(\theta))$. A frequency $\boldsymbol{u} = (u, v)$ with $e \cdot \boldsymbol{u} > 0$ is called positive while a frequency $\boldsymbol{u} = (u, v)$ with $e \cdot \boldsymbol{u} < 0$ is called negative. With this definition the one–dimensional construction rule for the analytic signal can be applied to two–dimensional signals [7]. Using this construction, the conception of the analytic signal remains a one–dimensional one, though.

We will present here another extension of the analytic signal conception using the QFT: For the one–dimensional analytic signal it is important that the Fourier transform of a real signal is a Hermitian function, i.e. that the equation

$$F(-u) = F^*(u) \tag{26}$$

holds, where $F^*$ is the complex conjugate function of $F$. Therefore, if we want to examine what the notion of the analytic signal means in the conception of QFT we have to remember the notion of a quaternionic Hermitian function which was introduced in section 2.

In section 4 we found out that the QFT of a real image $f$ obeys some symmetry rules, e.g. that the real part of the QFT is even with respect to both arguments of $F^q$. We can restate these properties in the form

$$F(-u, v) = \beta(F(\boldsymbol{u})) \tag{27}$$

$$F(u, -v) = \alpha(F(\boldsymbol{u})) \tag{28}$$

$$F(-\boldsymbol{u}) = \gamma(F(\boldsymbol{u})) \quad , \tag{29}$$

where $\alpha$, $\beta$ and $\gamma$ are the nontrivial involutions of $\mathbb{H}$ defined in section 2. Writing $F(\boldsymbol{u}) = F_0(\boldsymbol{u}) + iF_1(\boldsymbol{u}) + jF_2(\boldsymbol{u}) + kF_3(\boldsymbol{u})$ we can restate (27) as

$$F(-u, v) = \beta(F(\boldsymbol{u})) = -jF(\boldsymbol{u})j$$
$$\implies \quad F_0(-u, v) = F_0(\boldsymbol{u}), \qquad F_1(-u, v) = -F_1(\boldsymbol{u})$$
$$F_2(-u, v) = F_2(\boldsymbol{u}), \qquad F_3(-u, v) = -F_3(\boldsymbol{u}) \quad ,$$

which means that the $i$–imaginary and the $k$–imaginary part of $F$ are odd with respect to the first argument whereas the real and the $j$–imaginary part are even with respect to the first argument. Analogously we can find from (28) that with respect to the second argument the real and the $i$–imaginary component are even while the $j$–imaginary and the $k$–imaginary part are odd. These are the symmetry properties of the QFT we found in the previous section.

Hence, it follows that the QFT of a real signal is a quaternionic Hermitian function. It is easy to see that a quaternionic Hermitian function contains redundant information in three quadrants of its domain and, therefore, can be reconstructed from the values $f(x, y)$ for $x \geq 0$ and $y \geq 0$. In order to reconstruct the function $f$ from these values we need only to apply the equations (27), (28) and (29).

For this reason it seams reasonable to define the quaternionic analytic signal of a real 2–dimensional signal in the following way: We suppress all frequencies $\boldsymbol{u}$ in the quaternionic frequency domain for which either $u$ or $v$ or both of them are negative. The values at the double positive frequencies are multiplied by four. By the inverse QFT we transform the result into the spatial domain again and get the quaternionic analytic signal which, of course, is quaternion valued. The three imaginary components of the quaternionic analytic signal can be seen as the *quaternionic Hilbert transform* of $f$.

**Definition**: The quaternionic analytic signal of a two–dimensional signal $f$ is given by

$$f_A(x, y) = \mathcal{F}_q^{-1}\{Z^q(u, v)\} \quad , \tag{30}$$

where $Z^q$ is defined as

$$Z^q(u, v) = \begin{cases} 4F^q(u, v) & \text{if } u \geq 0 \text{ and } v \geq 0 \\ 0 & \text{else.} \end{cases} \tag{31}$$

We will prove that the real part of the quaternionic analytic signal of a real signal is equal to the signal itself.

**Proof**: In the following we will use the fact that for each quaternion $q$ the relations $Re(q) = Re(\alpha(q)) = Re(\beta(q)) = Re(\gamma(q))$ hold. Following our definition the quaternionic analytic signal $z^q$ of $f$ is given by

$$z^q(x,y) = 4 \int_0^\infty \int_0^\infty e^{j2\pi vy} F^q(u,v) e^{i2\pi ux} du\, dv \quad . \tag{32}$$

Regarding only the real part of $z^q$ and omitting the factor four we find

$$Re\left( \int_0^\infty \int_0^\infty e^{j2\pi vy} F^q(u,v) e^{i2\pi ux} du\, dv \right)$$

$$= Re\left( \int_0^\infty \int_0^\infty \alpha(e^{j2\pi vy} F^q(u,v) e^{i2\pi ux}) du\, dv \right)$$

$$= Re\left( \int_{-\infty}^0 \int_0^\infty e^{j2\pi vy} F^q(u,v) e^{i2\pi ux} du\, dv \right). \tag{33}$$

Analogously using the involutions $\beta$ and $\gamma$ instead of $\alpha$ we get

$$Re\left( \int_0^\infty \int_0^\infty e^{j2\pi vy} F^q(u,v) e^{i2\pi ux} du\, dv \right)$$

$$= Re\left( \int_0^\infty \int_{-\infty}^0 e^{j2\pi vy} F^q(u,v) e^{i2\pi ux} du\, dv \right) \tag{34}$$

$$= Re\left( \int_{-\infty}^0 \int_{-\infty}^0 e^{j2\pi vy} F^q(u,v) e^{i2\pi ux} du\, dv \right) \quad , \tag{35}$$

respectively. Substituting (33) and (35) in (32) completes the proof:

$$Re(z^q(x,y)) = 4Re\left( \int_0^\infty \int_0^\infty e^{j2\pi vy} F^q(u,v) e^{i2\pi ux} \right) du\, dv$$

$$= Re\left( \int_{-\infty}^\infty \int_{-\infty}^\infty e^{j2\pi vy} F^q(u,v) e^{i2\pi ux} \right) du\, dv = f(x,y) \tag{36}$$

$\square$

Like in the one–dimensional case also here we can use the (quaternionic) analytic signal to define the instantaneous phase of a signal. We will demonstrate this in the following section.

## 6 Two–dimensional Phase

In one dimension the analytic signal of the cosine function $f(x) = \cos(x)$ is $f_A(x) = e^{ix}$. Hence, for each $x \in \mathbb{R}$ we can get the instantaneous phase of $f$ by evaluating the argument of $f_A$ at the position $x$. For the cosine function we simply get $\arg(f_A(x)) = x$. We can generalize this concept (see [7]) to all functions $f : \mathbb{R} \to \mathbb{R}$ for which the analytic signal $f_A$ exists. We call $\arg(f_A(x))$ the *instantaneous phase* of $f$ at $x$.

We want to generalize this concept to two dimensions. In order to start with the same motivation as in the one–dimensional case we consider the function $f(x,y) = \cos(x)\cos(y)$ first. We will show that the quaternionic analytic signal of $f$ is $f_A(x,y) = e^{ix}e^{jy}$.

**Proof:** In the one–dimensional case we know that

$$f(x) = \cos(x) \Rightarrow f_A(x) = e^{ix} \quad ,$$

which follows from:

$$e^{ix} = \int\limits_{0}^{\infty} \int\limits_{-\infty}^{\infty} \left( e^{i2\pi ux} e^{-i2\pi ux'} \cos(x') \right) dx' du \quad . \tag{37}$$

Therefore, using (37), we obtain

$$f_A(x,y) = \int\limits_{0}^{\infty}\int\limits_{0}^{\infty}\int\limits_{-\infty}^{\infty}\int\limits_{-\infty}^{\infty} e^{i2\pi ux} e^{-i2\pi ux'} \cos(x') \cos(y') e^{-j2\pi vy'} e^{j2\pi vy} dx'\, dy'\, du\, dv$$

$$= e^{ix} \int\limits_{0}^{\infty}\int\limits_{-\infty}^{\infty} \cos(y') e^{-j2\pi vy'} e^{j2\pi vy} dy'\, dv = e^{ix} e^{jy} \quad . \tag{38}$$

$\square$

Since we are looking for a concept of two–dimensional phase it is now straightforward to define the phase of $f(x,y) = \cos(x)\cos(y)$ at position $(x,y)$. In section 3 we already mentioned the $arg$–function that maps the quaternions without zero to $\mathrm{I\!R}^2$ in such a way that $arg(|q|e^{ix}e^{jy}) = (x,y)$ for $(x,y) \in [0,2\pi[ \times [0,\pi[$. The function $arg$ is defined in the appendix A.

In one dimension the phase is defined within the interval $[0,2\pi[$. In order to clarify why the two–dimensional phase is only defined within $[0,2\pi[ \times [0,\pi[$, we show in figure 1 how the function $f(x,y) = \cos(x)\cos(y)$ is made up of patches of the size $[0,2\pi[ \times [0,\pi[$.

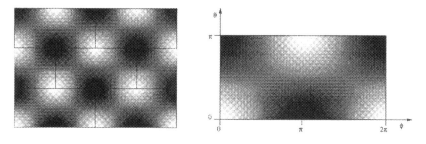

**Fig. 1.** The function $f(x,y) = \cos(x)\cos(y)$ with $(x,y) \in [0,4\pi[\times[0,3\pi[$ (left) and $(x,y) \in [0,2\pi[\times[0,\pi[$ (right).

According to the definition of the two–dimensional argument function we can define the instantaneous phase of a two–dimensional signal $f$ at $(x,y)$ as

$$instantaneous\ phase\ of\ f(x,y) = \arg(f_A(x,y)) \quad . \tag{39}$$

As Granlund [7] states for the one–dimensional case also we have to say that the instantaneous phase in general will not describe the local behavior of $f$. For this reason we will introduce the concept of *local phase* here.

In one dimension the local phase concept is well known. The local phase can be estimated using a quadrature filter, e.g. a Gabor filter with a central frequency $u_0$ which is defined by

$$g_{u_0}(x) = e^{-\pi x^2/\sigma^2} e^{i2\pi u_0 x} \quad . \tag{40}$$

The Gabor filter consists of a real part which is even and an odd imaginary part. Convolving the signal with the Gabor filter leads to a complex filter response. The argument of the response at position $x$ is then called the local phase of the signal at $x$.

Of course the local phase of a signal is dependent of the central frequency of the Gabor filter. In order to demonstrate the local phase concept we borrow a figure from Granlund's book ([7], p. 262) which shows in which way the local phase corresponds to the local form of the signal (figure 4a).

There are several attempts to use the local phase for multi–dimensional signals. One possibility is to extend a Gabor filter to two dimensions in the following way[4]:

$$g_{\boldsymbol{u}_0}(x, y) = e^{-\pi \boldsymbol{x}^2/\sigma^2} e^{i2\pi \boldsymbol{u}_0 \cdot \boldsymbol{x}} \tag{41}$$

An example of such a Gabor filter is shown in figure 2. For arbitrary $\boldsymbol{u}_0$ we can

**Fig. 2.** A two–dimensional complex Gabor filter with an even real part (left) and an odd imaginary part (right).

obtain $g_{\boldsymbol{u}_0}(x, y)$ by rotating the Gabor filter

$$g(x, y) = e^{-\pi \boldsymbol{x}^2/\sigma^2} e^{i2\pi(u_0 x + 0 y)} \tag{42}$$

by some angle $\theta$ about the origin. The local phase can then be defined along the direction $e = (\cos(\theta), \sin(\theta))$ by evaluating the argument of the filter response of $g_{\boldsymbol{u}_0}(x, y)$. Thus, we find that this generalized filter is in principle the same as a one–dimensional Gabor filter. Therefore, we will define the notion of a quaternionic Gabor filter. As the real part of this Gabor filter we take the function $f(x, y) = \cos(2\pi u_0 x) \cos(2\pi v_0 y)$ windowed with a Gaussian function:

$$g_{ee}^q(x, y) = e^{-\pi(x^2+y^2)/\sigma^2} \cos(2\pi u_0 x) \cos(2\pi v_0 y) \quad . \tag{43}$$

---

[4] We restrict ourselves to to the usage of isotropic Gaussian windows here. It is also possible to use different values of $\sigma$ for the directions $x$ and $y$.

A one–dimensional filter that is an analytic function itself, is called a quadrature filter. We want to apply this notion also in the two–dimensional case and call a filter which is a quaternionic analytic signal a quaternionic quadrature filter. One–dimensional Gabor filters are quadrature filters, so we should require this also in the two–dimensional case. By taking the analytic signal of $g_{ee}$ we get

$$g^q(x, y) = g^q_{eeA}(x, y)$$
$$= e^{-\pi(x^2+y^2)/\sigma^2} \left(\cos(2\pi u_0 x) \cos(2\pi v_0 y) + i \sin(2\pi u_0 x) \cos(2\pi v_0 y)\right.$$
$$\left. + j \cos(2\pi u_0 x) \sin(2\pi v_0 y) + k \sin(2\pi u_0 x) \sin(2\pi v_0 y)\right) \quad . \tag{44}$$

In the following we will call such a filter a two–dimensional quaternionic Gabor filter. It is depicted for $u_0 = v_0$ in figure 3.

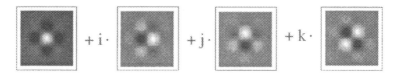

**Fig. 3.** A quaternionic Gabor filter with $u_0 = v_0$.

In one dimension the local phase gives information about the local symmetry or form of the signal, especially whether there is a peak or a step in the signal at the considered position. Using the quaternionic Gabor filters and evaluating the local signal phase by the two–dimensional $arg$–function we get the analogous information for an image signal, which is more complicated and contains more possible symmetries as the one–dimensional phase. In analogy to figure 4a we show the relation between the two–dimensional phase and the local signal structure in figure 4b and 4c.

As mentioned earlier we can evaluate the two–dimensional phase in a region $[0, 2\pi[ \times [0, \pi[$ which can be thought of as a half torus. The circles in figures 4b and 4c result from cutting through the torus for different values of $\theta$.

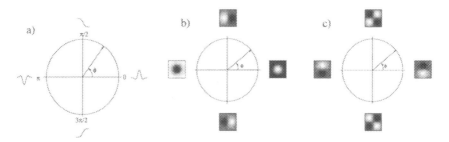

**Fig. 4.** Relation between the local phase and the local signal strcture: **a.** the one–dimensional case (see [7]), **b.** the two–dimensional case with $\theta = 0$, **c.** the two–dimensional case with $\theta = \pi/2$.

# 7 Experimental Results

Some experiments have been made which show how the local phase can be estimated from the answer of a quaternionic Gabor filter. We estimate the local phase of the function $f(\phi, \theta) = \cos(\phi)\cos(\theta)$ along some path through its domain in the following way. The signal function $f(\phi, \theta)$ is convolved with the quaternionic Gabor filter shown in figure 3. The filter response at each position in the $(\phi, \theta)$-plane is given by a quaternion. Along the line $s$ shown in figure 5a the quaternionic argument function which is defined in the appendix A is applied to the quaternion–valued filter response. We denote the estimated local phase by $(\hat{\phi}, \hat{\theta})$ and compare it to the instantaneous phase that can be evaluated analytically for $f(\phi, \theta) = \cos(\phi)\cos(\theta)$ as $(\phi, \theta)$ for $(\phi, \theta) \in [0, 2\pi[ \times [0, \pi[$.

The central frequency of the used Gabor filters is four times higher than the frequency of the signal $f$. In Fig. 5b and 5c the estimated values $\hat{\phi}$ and $\hat{\theta}$ are compared to $\phi$ and $\theta$, respectively. The straight lines are the values of the instantaneous phase $(\phi, \theta)$ while the slightly curved lines represent the estimated local phase $(\hat{\phi}, \hat{\theta})$.

The arguments $\phi$ and $\theta$ of the Gaborian's answers are nearly linear and give a good approximation to the instantaneous phase of the signal.

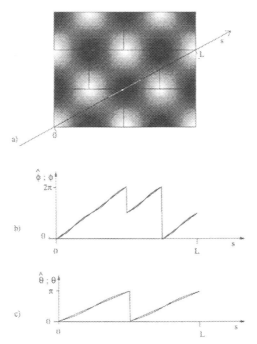

**Fig. 5.** The function $f(\phi, \theta) = \cos(\phi)\cos(\theta)$ with the path along which the local phase $(\phi, \theta)$ is estimated, **b)** Variation of $\phi$ and $\hat{\phi}$ along the depicted path, **c)** Variation of $\theta$ and $\hat{\theta}$ along the depicted path.

# 8 Conclusion

In this article we presented the quaternionic Fourier transform (QFT), an integral transform for two–dimensional signals which is based on the Fourier transform but provides a quaternion–valued representation of the signal in the frequency domain.

Based on the QFT we generalized the concepts of the analytic signal, of Gabor filters and the local phase to two dimensions in a novel way.

This generalization could be of interest especially in PAC systems for the following reason. There are recent attempts to embed the different tasks of a PAC system into one mathematical system using Clifford algebras [11]. Clifford algebras in the form of Geometric algebras have already been applied to neural computation [1] and to computer vision [2]. Since quaternions are a special Clifford algebra, it should be possible to integrate the QFT approach into a Geometric algebra PAC system.

# A  The *arg*–function

**Definition:** For every quaternion $q$ which can be given in the form $q = |q| e^{i\phi} e^{j\theta}$ the angles $\phi$ and $\theta$ within a range $(\phi, \theta) \in [0, 2\pi[ \times [0, \pi[$ are called the *argument* of $q$. We define the function $arg : \mathbb{H} \backslash \{0\} \mapsto \mathbb{R}^2$ that recovers for quaternions $q$ of the mentioned form the argument of $q$. Let $q = a + bi + cj + dk, q \neq 0$.

$$\arg(q) = (\phi, \theta) \quad , \tag{45}$$

with

$$\phi = \begin{cases} \pi - d' \frac{\pi}{2} & \text{for } a = b = c = 0 \\[2mm] \pi - \text{sign}(b)\,\text{sign}(bd)\,\frac{\pi}{2} & \text{for } a = c = 0, b \neq 0 \\[2mm] \begin{array}{l} \text{sign}(c)\,\arcsin(d') + \text{step}(-c)\,\pi \\ +2\pi\,\text{step}(c)\,\text{step}(-d) \end{array} & \text{for } a = 0, c \neq 0 \\[4mm] \begin{array}{l} \arctan(b/a) + \pi\,\text{step}(-a) \\ +2\pi\,\text{step}(a)\,\text{step}(-ab) \end{array} & \text{for } a \neq 0, c = d = 0 \\[4mm] \begin{array}{l} \arctan(b/a) + \pi\,\text{step}(-c) \\ +2\pi\,\text{step}(-d)\,\text{step}(c) \end{array} & \text{for } a \neq 0 \wedge (c \neq 0 \vee d \neq 0) \end{cases} \tag{46}$$

and

$$\theta = \begin{cases} \frac{\pi}{2} & \text{for } a = b = c = 0 \\[2mm] \arcsin(\text{sign}(b)\,d') + \pi\,\text{step}(-bd) & \text{for } a = c = 0, b \neq 0 \\[2mm] \frac{\pi}{2} & \text{for } a = 0, c \neq 0 \\[2mm] 0 & \text{for } a \neq 0, c = d = 0 \\[2mm] \arctan(c/a) + \pi\,\text{step}(-c/a) & \text{for } a \neq 0 \wedge (c \neq 0 \vee d \neq 0) \end{cases} \quad , \tag{47}$$

with $d' = d/|q|$. Here we used the *step-* and the *sign-*function which are defined by

$$\text{step}(x) = \begin{cases} 0 & \text{for} & x \leq 0 \\ 1 & \text{for} & x > 0 \end{cases} \tag{48}$$

and

$$\text{sign}(x) = \begin{cases} -1 & \text{for} & x < 0 \\ 1 & \text{for} & x \geq 0 \end{cases}. \tag{49}$$

**Acknowledgment**

We would like to thank Dr. Kostas Daniilidis for his interest in this work and for valuable discussions on the subject of this article.

# References

1. E. BAYRO–CORROCHANO, S. BUCHHOLZ & G. SOMMER, *A new self-organizing neural network using geometric algebra*, in: Proc. ICPR '96, vol.: D, 555–559, Vienna, 1996

2. E. BAYRO-CORROCHANO, J. LASENBY & G. SOMMER, *Geometric Algebra: A framework for computing point and line correspondences and projective structure using n uncalibrated cameras*, in: Proc. ICPR '96, vol.: A, 334–338, Vienna, 1996

3. W. BLASCHKE, *Kinematik und Quaternionen*, VEB Deutscher Verlag der Wissenschaften, Berlin 1960

4. R. BRACEWELL, *The Fourier Transform and its Applications*, McGraw–Hill, 2nd edition, 1986

5. TH. BÜLOW, G. SOMMER, *Algebraically Extended Representation of Multi-Dimensional Signals*, Proc. of the 10th Scandinavian Conference on Image Analysis, 559–566, 1997

6. V.M. CHERNOV, *Discrete orthogonal transforms with data representation in composition algebras*, Proc. of the 9th Scandinavian Conference on Image Analysis, 357–364, 1995

7. G.H. GRANLUND, H. KNUTSSON, *Signal Processing for Computer Vision*, Kluwer Academic Publishers, 1995

8. W.R. HAMILTON, *On quaternions, or on a new system of imaginaries in algebra*, Phil. Mag. 25, 489–495, 1844, reprinted in *The mathematical papers of Sir William Rowan Hamilton*, Vol III, *Algebra*, Cambridge University Press, London 1967

9. I.L. KANTOR, A.S. SOLODOVNIKOV, *Hypercomplex Numbers*, Springer–Verlag, New–York, 1989

10. I.R. PORTEOUS, *Clifford Algebras and the Classical Groups*, Cambridge University Press, 1995

11. G. SOMMER, E. BAYRO–CORROCHANO & TH. BÜLOW, *Geometric Algebra as a Framework for the Perception–Action Cycle*, in: Workshop on Theoretical Foundation of Computer Vision, Ed. F. Solina, Springer Verlag, Wien, 1997

# On Hestenes' Formalization for Describing Linear Image Transforms

Andrew V. Chernov and Vladimir M. Chernov

Image Processing Systems Institute of RAS,
151 Molodogvardejskaya st., IPSI RAS, 443001, Samara, Russia

e-mail: chernov@sgau.volgacom.samara.su

**Abstract.** In the framework of D.Hestenes conception of Clifford algebras a description of linear image transforms is considered. It is shown that the geometric algebra of a space with degenerate metric is an obvious and effective tool. It is stated that a lot of linear image transforms are realized with screwing (generalized rotation + shift) of this algebra.

## 1 Introduction

Clifford algebras appeared as a result of natural desire of mathematicians to include the finite-dimensional vector space into such an algebraic structure for which additional operations of vector space (inner and outer products) are expressed in terms of operations of this algebra [1], [2], [3].

The interpretation of Clifford algebras introduced by D.Hestenes (the geometric algebra, [4], [5]) was accepted immediately by some physicists due to its obviousness and an making efforts to join our interesting workshop. adequate connection with vector constructions being used in physics.

An effective application of Clifford algebras in the Computer Sciences is now based, mainly, on two ideas.

First, either informative setting of some problems of robotics, computer vision etc. has a direct physical (mechanical) nature or there exist evident physical analogues [6], [7]. For example, considering 8-dimensional geometric algebra of Euclidean space extends possibilities of four-dimensional quaternion algebra for analysis and motion simulation in robotics [8]. It is interesting to use the algebra of dual quaternions in tasks of so-called hand-eye calibration [9].

Second, the particular Clifford algebras are formally used as a convenient calculation model for data representing and increasing of effectiveness of some signal processing algorithms. For example, the data representation in the quaternion algebra or the order-2 matrices algebra allows to synthesize effective multidimensional FFT algorithms "with multioverlapping" [10], [11].

In the first case such Clifford algebra properties predominate which are connected with geometric obviousness. In the second case predominate properties connected with an existence of a Clifford algebra automorphisms group that is large enough.

Tasks of signal (image) analysis and digital processing have a clear physical character. Their solution requires a significant amount of calculations and, therefore, needs effective algorithmic support. Thus, for these tasks it is necessary to develop and use such algebraic means which have both a clear physical interpretation and calculating efficiency.

The geometric obviousness of Hestenes' conception and its successful application to physics for object local properties analysis allow to expect that its application to analysis of multidimensional signal local properties can be successful too. In authors' opinion, to develop the *local* theory of image processing in the framework of Hestenes conception, first, it is necessary to solve a quite simple task: to give an obvious description of images and their simplest (linear) transforms in terms of the geometric algebra.

Let us note that although the Hestenes formalization exists for describing some classes of 3D space linear transforms (in particular, rotations) in terms of the 8-dimensional geometric algebra, the problem considered in this work is nontrivial.

Indeed, if an image is interpreted as a subset of 3D space (a pair of arguments; brightness function), then its linear transforms are linear ones with invariant planes of arguments. In other words, coordinates of a 3D space vector associated with the image are not equivalent. This anisotropy is to be taken into account adequately under selection of the particular Clifford algebra. In authors' opinion, such an algebra is the Clifford algebra over 3D space with degenerate metric.

In the present work the Hestenes conception is developed for the case of space with degenerate metric in applying to the description of images and their linear transforms.

## 2 A review of the geometric algebra theory

Let $\mathbf{R}^3$ be a 3D real space with the basis $\{\mathbf{e}_1, \mathbf{e}_2, \mathbf{e}_3\}$. Let us consider a 8D associative $\mathbf{R}$-algebra $\Sigma$ with the basis

$$E_0 = 1, \ E_1 = \mathbf{e}_1, \ E_2 = \mathbf{e}_2, \ E_3 = \mathbf{e}_3,$$

$$E_4 = \mathbf{e}_1\mathbf{e}_2, \ E_5 = \mathbf{e}_1\mathbf{e}_3, \ E_6 = \mathbf{e}_2\mathbf{e}_3, \ E_7 = \mathbf{e}_1\mathbf{e}_2\mathbf{e}_3 \tag{1}$$

under the following rules of the basis elements multiplication:

$$\mathbf{e}_k\mathbf{e}_k = \epsilon_k \in \mathbf{R}, \ \mathbf{e}_k\mathbf{e}_j = -\mathbf{e}_j\mathbf{e}_k, \ (j \neq k; \ j, k = 1, 2, 3). \tag{2}$$

Equalities (2) induce the rules of the basis elements (1) multiplication, and further, rules of arbitrary elements multiplication:

$$A = \sum_{j=o}^{7} \alpha_j E_j \in \Sigma \ (\alpha_j \in \mathbf{R}).$$

**Definition 1.** *Let us call the algebra $\Sigma$ as a Clifford algebra of 3D space (more precisely, Ref. [2], Clifford algebra of a quadratic form*

$$Q(t) = \epsilon_1 t_1^2 + \epsilon_2 t_2^2 + \epsilon_3 t_3^2 \quad ).$$

According of the main conception of Hestenes, elements of the Clifford algebra $\Sigma$ (multivectors) can be represented in the form:

$$S = a_0 + a_1 \mathbf{e}_1 + a_2 \mathbf{e}_2 + a_3 \mathbf{e}_3$$

$$+a_{23} \mathbf{e}_2 \wedge \mathbf{e}_3 + a_{31} \mathbf{e}_3 \wedge \mathbf{e}_1 + a_{12} \mathbf{e}_1 \wedge \mathbf{e}_2 + a_{123} \mathbf{e}_1 \wedge \mathbf{e}_2 \wedge \mathbf{e}_3. \tag{3}$$

Let $A_r$ be an order-$r$ multivector (vector, bivector, trivector ); $\mathbf{a}$ be a vector. Then the relations

$$\mathbf{a} \cdot A_r = \frac{1}{2} \left( \mathbf{a} A_r + (-1)^r A_r \mathbf{a} \right), \tag{4}$$

$$\mathbf{a} \wedge A_r = \frac{1}{2} \left( \mathbf{a} A_r - (-1)^r A_r \mathbf{a} \right) \tag{5}$$

extend the concept of the usual scalar (inner) and outer products of vectors. Then the Clifford product can be written in the form called geometrical product

$$\mathbf{a} A_r = \mathbf{a} \cdot A_r + \mathbf{a} \wedge A_r,$$

and further,

$$A_s B_r = <AB>_{r+s} + <AB>_{r+s-2} + ... + <AB>_{|r-s|}, \tag{6}$$

where $<A>_r$ is an operation of taking of order-$r$ part.

**Definition 2.** *Let us call the Clifford algebra $\Sigma$ with elements representation in the form (3) and operations induced by relations (4)-(6) as a geometrical algebra of 3D space.*

## 3  Geometric algebra of images

If $\mathbf{e}_1^2 = \mathbf{e}_2^2 = \mathbf{e}_3^2 = 1$, then relations (2) define a Euclidean structure on $\mathbf{R}^3$. In this case, subalgebra $\Sigma_0^4 \subset \Sigma$ with the basis

$$\{1, \ \mathbf{e}_2 \wedge \mathbf{e}_3, \ \mathbf{e}_3 \wedge \mathbf{e}_1, \ \mathbf{e}_1 \wedge \mathbf{e}_2\} \tag{7}$$

is isomorphic to the quaternion algebra. This algebra is sufficient for describing a rigid body motion, but its dimension is not sufficient for describing arbitrary linear transforms of an affine space.

If

$$\mathbf{e}_1^2 = \mathbf{e}_2^2 = 1, \quad \mathbf{e}_3^2 = 0, \tag{8}$$

the algebra $\Sigma$ contains a subalgebra of dual quaternions. In this case, it has been used efficiently while solving applied tasks [9], [12].

In this paper one more application of Clifford algebra under the condition (8) is considered.

**Definition 3.** *Let us call the 8D Clifford algebra $\Sigma^*$ with elements representation in the form (3) under the condition (8) as Hestenes algebra of images.*

The reasons of introducing such a term will be clear below.

**Proposition 1.** *Subalgebra $\Sigma_{12}^4 \subset \Sigma^*$ with the basis*

$$\{1,\ e_1,\ e_2,\ e_1 \wedge e_2\}$$

*is isomorphic to the algebra $\mathbf{M}_2(\mathbf{R})$ of order-2 matrices.*

Proof

An arbitrary order-2 matrix $\mathbf{W}$ is presented in the form:

$$\mathbf{W} = \alpha \begin{pmatrix} 1 & 0 \\ 0 & 1 \end{pmatrix} + \beta \begin{pmatrix} 1 & 0 \\ 0 & -1 \end{pmatrix} + \gamma \begin{pmatrix} 0 & 1 \\ 1 & 0 \end{pmatrix} + \delta \begin{pmatrix} 0 & 1 \\ -1 & 0 \end{pmatrix} \tag{9}$$

with some $\alpha, \beta, \gamma, \delta \in \mathbf{R}$.

It is verified directly that the mapping:

$$\varphi(1) \longmapsto \begin{pmatrix} 1 & 0 \\ 0 & 1 \end{pmatrix}, \quad \varphi(e_1 \wedge e_2) \longmapsto \begin{pmatrix} 0 & 1 \\ -1 & 0 \end{pmatrix},$$

$$\varphi(e_1) \longmapsto \begin{pmatrix} 1 & 0 \\ 0 & -1 \end{pmatrix}, \quad \varphi(e_2) \longmapsto \begin{pmatrix} 0 & 1 \\ 1 & 0 \end{pmatrix}$$

preserves the matrix multiplication operation.

Then $\mathbf{R}$-linear extension $\Phi$ of the mapping $\varphi$ is an isomorphism of the algebra $\mathbf{M}_2(\mathbf{R})$ of order-2 matrices and $\Sigma_{12}^4$. End Proof

The algebra $\mathbf{M}_2(\mathbf{R})$ is split composition algebra [13]: there exists such a quadratic form $N(\mathbf{W})$ on the vector space of algebra (*the norm of an element* $\mathbf{W} \in \mathbf{M}_2(\mathbf{R})$ ), that

$$N(\mathbf{W}_1 \mathbf{W}_2) = N(\mathbf{W}_1)N(\mathbf{W}_2).$$

As in general case of composition algebras, the form $N(\mathbf{W})$ on $\mathbf{M}_2(\mathbf{R})$ is generated by a $\mathbf{R}$-linear anti-isomorphism $\sigma$ :

$$\sigma(\mathbf{W}_1 \mathbf{W}_2) = \sigma(\mathbf{W}_2)\sigma(\mathbf{W}_1), \quad N(\mathbf{W}) = \sigma(\mathbf{W})\mathbf{W}.$$

This anti-isomorphism for $\mathbf{M}_2(\mathbf{R})$ is the symplectic involution:

$$\sigma : \begin{pmatrix} a & b \\ c & d \end{pmatrix} \longmapsto \begin{pmatrix} d & -b \\ -c & a \end{pmatrix} \tag{10}$$

and

$$N(\mathbf{W}) = \sigma(\mathbf{W})\mathbf{W} = \mathbf{E}\det(\mathbf{W}).$$

*Proof.* **Proposition 2.** *For the matrix* **W** *represented in the form (9), the following equality is valid:*

$$N(\mathbf{W}) = (\alpha^2 - \beta^2 - \gamma^2 + \delta^2).EndProof \tag{11}$$

**Proposition 3.** *Subalgebras* $\Sigma_3^2, \Sigma_{31}^2, \Sigma_{23}^2 \subset \Sigma^*$ *with the bases*

$$\{1, \; e_3\}, \quad \{1, \; e_3 \wedge e_1\}, \quad \{1, \; e_2 \wedge e_3\},$$

*respectively, are all isomorphic to the algebra* $\Theta$ *of dual numbers [14]:*

$$\Theta = \{\omega = a + b\theta : \qquad \theta^2 = 0, \; a, \; b \; \in \mathbf{R}\}.$$

Proof

Since

$$e_3^2 = (e_3 \wedge e_1)^2 = (e_2 \wedge e_3)^2 = 0,$$

that every dual number $\omega = a + b\theta$ is canonically identified with the elements

$$w = a + be_3, \; w = a + be_3 \wedge e_1, \; w = a + be_2 \wedge e_3, \tag{12}$$

respectively. End Proof

*Proof.* **Proposition 4.** *The subalgebra* $\Sigma_{12}^2 \subset \Sigma^*$ *with the basis*

$$\{1, \; e_1 \wedge e_2\}$$

*is isomorphic to the algebra* **C** *of complex numbers.*

Proof

As in the Proposition 3 this statement follows from the relation

$$(e_1 \wedge e_2)^2 = -1.$$

The complex number $z = a + bi$ is canonically identified with the elements $Z = a + b(e_1 \wedge e_2)$. End Proof

*Proof.* **Proposition 5.** *The subalgebra* $\Sigma_0^4 \subset \Sigma^*$ *with the basis*

$$\{1, \; e_2 \wedge e_3, \; e_3 \wedge e_1, \; e_1 \wedge e_2\}$$

*is isomorphic to the algebra* **H**$^*$ *of dual quaternions [14].*

Proof

Elements of the algebra of dual quaternions can be represented in the form

$$z = (a + di) + \theta(b + ci), \quad (a, b, c, d \in \mathbf{R})$$

with multiplications rules induced by multiplications rules for the basic elements:

$$\theta^2 = 0, \quad i^2 = -1, \quad i\theta = -\theta i.$$

Since any element $Z \in \Sigma_0^4$ can be represented in the form

$$Z = a + b(\mathbf{e}_2 \wedge \mathbf{e}_3) + c(\mathbf{e}_3 \wedge \mathbf{e}_1) + d(\mathbf{e}_1 \wedge \mathbf{e}_2)$$

$$= (a + d(\mathbf{e}_1 \wedge \mathbf{e}_2)) + (\mathbf{e}_2 \wedge \mathbf{e}_3)(b + c(\mathbf{e}_1 \wedge \mathbf{e}_2)),$$

then this statement follows from relations

$$(\mathbf{e}_1 \wedge \mathbf{e}_2)^2 = -1, \quad (\mathbf{e}_2 \wedge \mathbf{e}_3)^2 = 0.$$

End Proof

Informally, in the algebra $\Sigma^*$ real, complex and dual numbers; vectors, dual quaternions and (2x2)-matrices "live together". It allows to give an effective description of image linear transforms in terms of properties of the algebra $\Sigma^*$.

Propositions 1-5 proved above allow to formulate the main theorem on structure of the algebra $\Sigma^*$ and to understand the fact why this algebra, exactly, is good means for describing linear transforms of images.

*Proof.* **Theorem 1.** *The vector space of the algebra $\Sigma^*$ is represented in the form of direct sum*

$$\Sigma^* \cong \mathbf{M}_2(\mathbf{R}) \oplus \theta \mathbf{M}_2(\mathbf{R}) \tag{13}$$

*(it is the dualisation of $\mathbf{M}_2(\mathbf{R})$ ).*
*Let $A, B \in \Sigma^*$ :*

$$A = \mathbf{A}_1 + \theta \mathbf{A}_2, \quad B = \mathbf{B}_1 + \theta \mathbf{B}_2; \quad \mathbf{A}_1, \mathbf{A}_2, \mathbf{B}_1, \mathbf{B}_2 \in \mathbf{M}_2(\mathbf{R}).$$

*Then multiplications in $\Sigma^*$ are defined by:*

$$AB = (\mathbf{A}_1 + \theta \mathbf{A}_2)(\mathbf{B}_1 + \theta \mathbf{B}_2) = \mathbf{A}_1 * \mathbf{A}_2 + \theta(\sigma(\mathbf{A}_1) * \mathbf{B}_2 + \mathbf{A}_2 * \mathbf{B}_1), \tag{14}$$

*where $\sigma$ is involution (10), $(*)$ is the operation of matrix multiplication.*

Proof

This statement follows easily from the representation of a multivector $S$ in the form;

$$S = (a_0 + a_1 \mathbf{e}_1 + a_2 \mathbf{e}_2 + a_{12} \mathbf{e}_1 \wedge \mathbf{e}_2)$$

$$+ \mathbf{e}_3(a_3 - a_{23} \mathbf{e}_2 + a_{31} \mathbf{e}_1 + a_{123} \mathbf{e}_1 \wedge \mathbf{e}_2)$$

and from Proposition 3. End Proof

Further, multivectors

$$\mathbf{Rm}(S) = a_0 + a_1 e_1 + a_2 e_2 + a_{12} e_1 \wedge e_2, \tag{15}$$

$$\mathbf{Dm}(S) = a_3 - a_{23} e_2 + a_{31} e_1 + a_{123} e_1 \wedge e_2 \tag{16}$$

are called *real* and *dual* matrix parts of $S$ and denoted as in (15),(16).

*Proof. Remark 1.* For algebras of dual numbers $\Theta$ and dual quaternions $\mathbf{H}^*$, the following expansions being analogous to (13), are valid:

$$\Theta = \mathbf{R} \oplus \theta \mathbf{R}, \qquad \mathbf{H}^* = \mathbf{C} \oplus \theta \mathbf{C},$$

as well as the multiplications rules (14), where the operation $(*)$ is replaced by real or complex multiplications and the involution $\sigma$ is replaced by involution $\sigma_R(x) = x$ for $\Theta$ or by involution $\sigma_C(x + iy) = x - iy$ for $\mathbf{H}^*$. Thus, the algebras of dual numbers, dual quaternions and $\Sigma^*$ are generated from different algebras but with the same dualization scheme.

In Ref.[9] the use of 4D dual quaternion algebra was enough, since the *main* considered transforms (rotation in a plane) were described completely in terms of complex numbers. For the task of describing linear transform of images, such main transforms are linear transforms of $(e_1 \wedge e_2)$-plane. Certainly, in this case the matrix algebra $\mathbf{M}_2(\mathbf{R})$ is the most natural structure. Thus, the proposed approach is the development of ideas of Ref.[9] for double-dimensional algebra.

## 4 Realization of image linear transforms by rotations

**Definition 4.** *Let $\mathbf{A}^3$ be a 3D affine space, $\mathbf{R}^3$ be an associated vector space, $F(x_1, x_2)$ is a real-valued function. A subset of points $\Omega \subset \mathbf{A}^3$ :*

$$\Omega = \{(x_1, x_2, x_3) : \qquad x_3 = F(x_1, x_2)\}$$

*we shall call an image.*

**Definition 5.** *A subset*

$$\Delta_\Omega = \{\mathbf{x} : \ \mathbf{x} = x_1 e_1 + x_2 e_2 + F(x_1, x_2) e_3; \quad (x_1, x_2, x_3) \in \Omega\} \subset \mathbf{R}^3$$

we shall call a *vector model* of the image $\Omega$, and the function $F(x_1, x_2)$ we shall call a *brightness*.

*Remark 2.* Any affine transform $T$ of the space $\mathbf{A}^3$ induces a transform $\mathbf{T}$ of the space $\mathbf{R}^3$ that can be presented in the form

$$\mathbf{Tx} = \mathbf{Ux} + \mathbf{b},$$

where vector $\mathbf{b}$ defines a shift of the origin, and $\mathbf{U}$ is a linear operator. For the vector model introduced in Def.4, not any transform $\mathbf{T}$ can be interpreted as an image transform: the $(e_1 \wedge e_2)$−plane is to be $\mathbf{U}$-invariant.

Let $\overset{\rightharpoonup}{\Gamma}$ be a set of such operators $\mathbf{U}$. It is easily proved that any operator $\mathbf{U} \in \Gamma$ can be presented in the form:

$$\mathbf{U} : \mathbf{x} = x_1\mathbf{e}_1 + x_2\mathbf{e}_2 + F(x_1, x_2)\mathbf{e}_3 \mapsto$$

$$\mapsto \mathbf{L}(x_1, x_2) + (px_1 + qx_2 + \mu F(x_1, x_2))\mathbf{e}_3, \tag{17}$$

where $p, q, \mu \in \mathbf{R}$; $\mathbf{L}(x_1, x_2)$ is a linear transform of the $(\mathbf{e}_1 \wedge \mathbf{e}_2)$−plane.

Let us identify the original space $\mathbf{R}^3$ with a subset of vectors of the geometric algebra $\Sigma^*$.

The operation $S \mapsto S^\sim$ of *reversion* reverses the order of vectors in any multivector in the sum (3).

The following propositions are verified immediately.

**Proposition 6.** *Let*

$$\mathbf{x} = x_1\mathbf{e}_1 + x_2\mathbf{e}_2 + F(x_1, x_2)\mathbf{e}_3, \quad \mathbf{S}(\mathbf{x}) = S^\sim \mathbf{x} S,$$

$$S = a_0 + a_1\mathbf{e}_1 + a_2\mathbf{e}_2 + a_3\mathbf{e}_3 + a_{23}\mathbf{e}_2 \wedge \mathbf{e}_3 + a_{31}\mathbf{e}_3 \wedge \mathbf{e}_1 + a_{12}\mathbf{e}_1 \wedge \mathbf{e}_2 + a_{123}\mathbf{e}_1 \wedge \mathbf{e}_2 \wedge \mathbf{e}_3.$$

*Then*

$$\mathbf{S}(\mathbf{x}) = \left[(a_0^2 + a_1^2 - a_2^2 - a_{12}^2)x_1 + 2(a_1 a_2 - a_0 a_{12})x_2\right]\mathbf{e}_1$$

$$+ \left[2(a_1 a_2 + a_0 a_{12})x_1 + (a_0^2 - a_1^2 + a_2^2 - a_{12}^2)x_2\right]\mathbf{e}_2$$

$$+ [2(a_1 a_3 - a_0 a_{31} + a_2 a_{123} + a_{23} a_{12})x_1 + 2(a_0 a_{23} + a_2 a_3 - a_1 a_{123} + a_{31} a_{12})x_2$$

$$+ (a_0^2 - a_1^2 - a_2^2 + a_{12}^2)x_3]\mathbf{e}_3$$

$$+ [2(a_0 a_2 - a_1 a_{12})x_2 + 2(a_0 a_1 + a_2 a_{12})x_1]. \tag{18}$$

¿From the relation (18) it follows, that transform of the $(\mathbf{e}_1 \wedge \mathbf{e}_2)$−plane is defined completely by real matrix part $\mathbf{Rm}(S)$ of $S$. Determinant of the matrix $\mathbf{S}_{12}$ of the $(\mathbf{e}_1 \wedge \mathbf{e}_2)$−restriction (18) is equal:

$$\det \mathbf{S}_{12} = (a_0^2 + a_{12}^2)^2 - (a_1^2 + a_2^2)^2 = N(\mathbf{Rm}(S))(a_0^2 + a_1^2 + a_2^2 + a_{12}^2).$$

Coefficients

$$(a_1 a_3 - a_0 a_{31} + a_2 a_{123} + a_{23} a_{12}),$$

$$(a_0 a_{23} + a_2 a_3 - a_1 a_{123} + a_{31} a_{12})$$

can take different values for various $\mathbf{Dm}(S)$. Thus the following statement is correct.

**Proposition 7.** *For all* $l_{11}, l_{12}, l_{22}, l_{21}, p, q \in \mathbf{R}$ *there exist such multivector* $S$ *defined by the Proposition 6, that*

$$\mathbf{S}(\mathbf{x}) = (l_{11}x_1 + l_{12}x_2)\mathbf{e}_1 + (l_{21}x_1 + l_{22}x_2)\mathbf{e}_2 + (px_1 + qx_2 + \mu x_3)\mathbf{e}_3 + \lambda(x_1, x_2), \tag{19}$$

*where* $\lambda(x_1, x_2)$ *is a linear scalar function, and*

$$\mu = \frac{l_{11}l_{22} - l_{12}l_{21}}{\|\mathbf{Rm}(S)\|^2}, \quad \|\mathbf{Rm}(S)\|^2 = a_0^2 + a_1^2 + a_2^2 + a_{12}^2. \tag{20}$$

Further, the transform (19) under conditions (20) is called *normalized linear transform* of the image.

From Propositions 6, 7 and the relation (18), the theorem on representation of transforms for vector image model follows.

**Theorem 2.** *For any normalized linear transform* $\mathbf{U}$ *of the image* $\mathbf{x}$ *there exists such a multivector* $S \in \Sigma^*$ *and linear scalar function* $\lambda(x_1, x_2)$ *that the following relation is valid:*

$$\mathbf{Ux} = \mathbf{S(x)} - \lambda(x_1, x_2). \tag{21}$$

Let us call the transform (21) as a *(generalized) rotation,* and the transform $\mathbf{Ux} + \mathbf{b}, \ \mathbf{b} \in \mathbf{R}^3$ we shall call a *screw transform.* From Theorem 2 and Remark 2 the main theorem follows.

**Theorem 3.** *Any normalized linear transform of the vector model of an image is the screw transform.*

*Remark 3.* In the case of Euclidean space the Eq. (21) with $\lambda(x_1, x_2) = 0$ describes, in particular, a physical compound rotation of a 3D object. Eight independent parameters (coordinates of the multivector $S$) are not sufficient for describing an arbitrary linear transform of the 3D space. A specificity of image transforms concerned with invariance of the Remark 3
$(\mathbf{e}_1 \wedge \mathbf{e}_2)$–plane allows to represent such a transform as the screw one.

## 5  Some important special examples

Let $S = \mathbf{Rm}(S)$, then the Eq.(18) can be rewritten in the form

$$
\begin{aligned}
\mathbf{S(x)} = &\left[(a_0^2 + a_1^2 - a_2^2 - a_{12}^2)x_1 + 2(a_1 a_2 - a_0 a_{12})x_2\right]\mathbf{e}_1 \\
&+ \left[2(a_1 a_2 + a_0 a_{12})x_1 + (a_0^2 - a_1^2 + a_2^2 - a_{12}^2)x_2\right]\mathbf{e}_2 \\
&+ (a_0^2 - a_1^2 - a_2^2 + a_{12}^2)x_3 \mathbf{e}_3 \\
&+ \left[2(a_0 a_2 - a_1 a_{12})x_2 + 2(a_0 a_1 + a_2 a_{12})x_1\right].
\end{aligned} \tag{22}
$$

**Example 1.** *(Rotation of the* $(\mathbf{e}_1 \wedge \mathbf{e}_2)$*–plane).*

Let

$$a_2 = a_1 = 0, \ a_0 = \cos\varphi, \ a_{12} = \sin\varphi,$$

then the transform

$$\mathbf{S(x)} = [x_1 \cos 2\varphi - x_2 \sin 2\varphi]\mathbf{e}_1 + [x_1 \sin 2\varphi + x_2 \cos 2\varphi,]\mathbf{e}_2 + x_3 \mathbf{e}_3$$

is the rotation of the $(\mathbf{e}_1 \wedge \mathbf{e}_2)$–plane.

**Example 2.** *(Scaling of the image).*

Let
$$a_1 a_2 - a_0 a_{12} = a_1 a_2 + a_0 a_{12} = 0,$$
then $a_0 = 0$ or $a_{12} = 0$ and $a_1 = 0$ or $a_2 = 0$.

(a). Let $a_0 = a_1 = 0$, then from Eq.(22) the relation follows:
$$\mathbf{S(x)} = x_1 k_1 \mathbf{e}_1 + x_2 k_2 \mathbf{e}_2 - x_3 k_2 \mathbf{e}_3,$$
where $(-a_2^2 - a_{12}^2) = k_1 \leq 0$, $(a_2^2 - a_{12}^2) = k_2$.

(b). Let $a_0 = a_1 = 0$, then
$$\mathbf{S(x)} = x_1 k_1 \mathbf{e}_1 + x_2 k_2 \mathbf{e}_2 - x_3 k_1 \mathbf{e}_3,$$
where $(a_1^2 - a_{12}^2) = k_1$, $(-a_1^2 - a_{12}^2) = k_2 \leq 0$.

(c). Let $a_{12} = a_2 = 0$, then
$$\mathbf{S(x)} = x_1 k_1 \mathbf{e}_1 + x_2 k_2 \mathbf{e}_2 + x_3 k_1 \mathbf{e}_3,$$
where $(a_0^2 - a_2^2) = k_1$, $(a_0^2 + a_2^2) = k_2 \geq 0$.

(d). Let $a_{12} = a_2 = 0$, then
$$\mathbf{S(x)} = x_1 k_1 \mathbf{e}_1 + x_2 k_2 \mathbf{e}_2 + x_3 k_2 \mathbf{e}_3,$$
where $(a_0^2 + a_1^2) = k_1 \geq 0$, $(a_0^2 - a_1^2) = k_2$.

**Example 3.** *(Derivation of the image).*

Let $f(x_1, x_2) = x_3$ be a differentiable function and $f(0,0) = 0$. Then the equation of the tangent plane to the surface $f(z_1, z_2) = z_3$ in the point
$$(z_1, z_2, z_3) = 0$$
looks like:
$$x_3 = \frac{\partial f(0,0)}{\partial z_1} x_1 + \frac{\partial f(0,0)}{\partial z_2} x_2.$$
Therefore from Eq.(18) under
$$a_1 = a_2 = a_{12} = a_3 = 0, \quad 2a_{31} = -\frac{\partial f(0,0)}{\partial z_1}, \quad 2a_{23} = \frac{\partial f(0,0)}{\partial z_2}$$
we obtain an equation of the tangent plane in the form:
$$\mathbf{S(x)} = x_1 \mathbf{e}_1 + x_2 \mathbf{e}_2 + \left( \frac{\partial f(0,0)}{\partial z_1} x_1 + \frac{\partial f(0,0)}{\partial z_2} x_2 \right) \mathbf{e}_3. \tag{23}$$

In the general case of arbitrary $\mathbf{z} = (z_1, z_2, f(z_1, z_2))$ the Eq. (23) is rewritten in the form:
$$\mathbf{S(x)} = S^\sim \mathbf{x} S + \mathbf{z},$$

where

$$S = \left(1 + \frac{1}{2}\frac{\partial f(z_1, z_2)}{\partial z_1}(\mathbf{e}_1 \wedge \mathbf{e}_3) + \frac{1}{2}\frac{\partial f(z_1, z_2)}{\partial z_2}(\mathbf{e}_2 \wedge \mathbf{e}_3)\right).$$

The equation of the tangent plane as a function of $(x_1, x_2)$ defines a Frechet derivative in a neighborhood of the point $\mathbf{z}$. A concept of the Frechet derivative (tangent mapping, local linearization) is a fundamental concept in the modern calculus on manifolds [15].

# 6 Conclusion

Thus, the geometric algebra of a space with degenerate metric is use as convenient means for representation of linear image transforms in the simple (screw) form.

1. As mentioned above (see Remark 1), the algebra $\Sigma^*$ considered in the paper is one more example of algebras obtained by means of dualization. The only question is: for what tasks it is expedient to consider the dualization of the quaternion algebra $\mathbf{H}^* = \mathbf{H} \oplus \theta\mathbf{H}$ ?

In our opinion, such an algebra is an adequate mathematical means, at least, in two cases.

First, quaternions characterize well rotations of the unit sphere. By means of stereographic projection, the unit sphere is identified with points of the extended complex plane $\mathbf{C} \cup \{\infty\} = \mathbf{C}^+$. Rotations of the sphere induce linear fractional transformations on $\mathbf{C}^+$. If we assume an image to be a real function of complex argument, then the algebra $\mathbf{H}^*$ is a good tool for local analysis of linear fractional transformations of images.

Second, the group of space rotations is a model of 3D projective space. Quaternions are good means for describing of such rotations. Therefore the algebra $\mathbf{H}^*$ is also a good tool for analysis of projective image transforms in the style of Ref.[16].

2. Accepting the motto "The world is locally linear", the authors do not see principal difficulties for localizing the obtained results by introducing appropriate differential structures in the space and in the Clifford algebra (see, for example, [15]). Example 3 indicates obviously of such a localization.

3. An extrapolation of the method onto the case of digital (discrete) images is significantly more difficult. The usual approach here is a conversion to continuous approximation with the following (secondary !) discretization of the obtained data. The authors consider it to be an actual task to develop the theory of Clifford algebras and the conception of geometric algebra in application to processing signals directly for discrete spaces (e.g. over rational numbers field or finite field). A classification of composition algebras over such fields is also known [17].

A development of the obtained results is an object of special research of the authors.

# 7 Acknowledgment

The grate presentation of Dr. K. Daniilidis (Ref. [9]) on ICPR '96 was turned attention by the senior author to possibility of the effective use of dual quaternions in other tasks too. The following discussions with Dr. E. Bayro-Corrochano were also very enriching.

This work was performed with partial financial support from the Russian Foundation of Fundamental Investigations (Grant 97-01-009000).

# References

1. Clifford, W.K.: Applications of Grassmann's Exstensive Algebra. Amer. J. Math. **1** (1878) 350-358
2. Artin, E.: Geometric Algebra. Interscience Publ, Inc., NY, London, 1957
3. Plymen, R., Robinson, P.: Spinors in Hilbert Space. Cambridge Univ. Press, 1994
4. Hestenes, D.: Space-Time Algebra. Gordon and Breach, NY, 1966
5. Hestenes, D., Sobczyk,G.: Clifford Algebra to Geometric Calculus. D.Redel Publ. Comp., Dordrecht, Boston, Lancaster, Tokyo, 1984
6. Hestenes, D.: New Foundation for Clasical Mechanics. D.Redel Publ. Comp., Dordrecht, Boston, Lancaster, Tokyo, 1986
7. Doran, C.J.L.: Geometric Algebra and its Application to Mathematical Physics. PhD-Thesis, University of Cambridge, 1994
8. Bayro-Corrochano, E., Sommer, G.: Object Modelling and Collision Avoidance Using Clifford Algebra. In V. Hlavac, R. Sara (eds.), Computer Analysis of Image and Pattern, Proc. CAIP '95. LNCS **970** (1995) 699-704
9. Daniilidis, K., Bayro-Corrochano, E.: The Dual Quaternion Approach to Hand-Eye Callibration. Proc. ICPR '96 **1** (1996)
10. Chernov, V.M.: Arithmetic Methods in the Theory of Discrete Orthogonal Transforms. Proceedings SPIE **2363** (1995) 134-141
11. Chernov, V.M.: Discrete Orthogonal Transforms with Data Representation in Composition Algebras. Proceedings of the 9th Scandinavian Conference on Image Analysis (1995) 357-364
12. Walker, M.W., Shao, L., Volz, R.A.: Estimating 3D-Location Using Dual Number Quaternions. CVGIP: Image Understanding **54** (1991) 358-367
13. Jacobson, N.: Structure and Representations of Jordan Algebras. Providens, R.I., 1968
14. Yaglom, I.M.: Complex Number in Geometry. Academic Press, 1968
15. Spivak, M.: Calculus on Manifolds. W.A. Benjamen Inc. 1965
16. Lasenby, J., Bayro-Corrochano, E., Lasenby, A.N., Sommer, G.: A New Methodology for Computing Invariants in Computer Vision. Proc. ICPR '96 **1** (1996)
17. Jacobson, N.: Composition Algebras and Their Automorphisms. Rend. Circ. Mat. Palermo **7** (1958) 55-80

# Fractal Basis Functions for Pattern Recognition

Władysław Skarbek[1] and Krystian Ignasiak[2]

[1] Polish-Japanese Institute of Computer Techniques
Koszykowa 86, 02-008 Warsaw, Poland, email: skarbek@ipipan.waw.pl
[2] Electronics and Information Technology Department, Warsaw Univ. of Technology

**Abstract.** Asynchronous fractal operators are designed for images in order to implement fractal applications by massive parallel asynchronous computing system. Assuming the convergence of the standard fractal operator to an element $\tilde{f}$ which is an approximation of the original image $f$, it is proved that any asynchronous deterministic realization of local fractal operators is convergent to $\tilde{f}$ and the stochastic realization converges to $\tilde{f}$ with probability one. A new class of basis functions is defined using fractal operators designed for class representatives. Applications for object recognition and image association are presented.

## 1 Introduction

Intelligent archivization and retrieval of images is one of basic goals in contemporary computer applications especially in multimedia systems [11]. One of the most efficient image compression methods is based on the notion of fractal operator [5,8]. In this paper we extend it to asynchronous, deterministic and stochastic case to point for potential parallel implementations. Moreover, we show new *intelligent* areas of fractal applications: object recognition and image association.

Let $f \in \mathcal{I}$ be an image. We assume that its domain $D = dom(f)$ is a rectangular array of $N = |D|$ pixels. Two kinds of subdomains are considered in the definition of a fractal operator $F$:

- *target domains* (shortly t-Domains) which create a partition $\Pi$ of $D$, i.e.:

$$\Pi = \{T_1, \ldots, T_a\}, \ D = \bigcup_{i=1}^{a} T_i \ ,$$

$$T_i \cap T_j = \emptyset, \ i \neq j \ ;$$

- *source domains* (shortly s-Domains) which create a cover $\Gamma$ of $D$, i.e.:

$$\Gamma = \{S_1, \ldots, S_b\}, \ D = \bigcup_{i=1}^{b} S_i \ .$$

We say that the pair $(\Pi, \Gamma)$ is a *regular pair of domain sets* if the following conditions are satisfied:

1. each t-Domain $T_i \in \Pi$ is a discrete square of size $|T_i| = 2^{t_i} \times 2^{t_i}$ pixels, where typically $t_i = 2, 3, 4$;
2. each s-Domain $S_i \in \Pi$ is a discrete square of size $|S_i| = 2^{s_i} \times 2^{s_i}$ pixels, where typically $s_i = 3, 4, 5$;
3. each s-Domain $S_i \in \Pi$ is a sum of certain t-Domains:

$$S_i = \bigcup T_{j_k} .$$

Let $U$ be a domain from $\Pi \cup \Gamma$. Then by $\chi_U$ we mean the characteristic function of $U$, i.e. for any pixel $p \in U$:

$$\chi_U(p) \doteq \begin{cases} 1 \text{ if } p \in U , \\ 0 \text{ otherwise .} \end{cases}$$

We are modelling the subimage of the image $g$ restricted to a domain $U \in \Pi \cup \Gamma$ by $g_U = g\chi_U$. Note that

$$g = \sum_{T \in \Pi} g_T \quad \text{for any } g \in \mathcal{I} \tag{1}$$

The main idea in the construction of the fractal operator $F$ for the given image $f$, is the elaboration of a *matching function* $\mu : \Pi \to \Gamma$ such that for any $T \in \Pi$, the subimage $f_{\mu(T)}$ is the most *similar* to the subimage $f_T$ and the size of $\mu(T)$ is greater than size of $T$.

In simple, but practical case the similarity is defined by *affine mapping* acting separately between domains of subimages and between ranges of subimages, i.e. between gray scale intervals.

The degree of similarity between subimages is measured by a $p$-norm $\| \cdot \|_p$, $1 \leq p \leq \infty$.

Having $\Pi$ and $\mu$, the *affine fractal operator* is defined for any $g \in \mathcal{I}$ as follows:

$$F(g) \doteq \sum_{T \in \Pi} [c_T \cdot R_T(g) + o_T \cdot \chi_T] \tag{2}$$

where

- $c_T$ is the *contrast* between subimages $g_{\mu(T)}$ and $g_T$;
- $o_T$ is the *offset* between subimages $g_{\mu(T)}$ and $g_T$;
- $R_T \doteq P_T \circ A_T$ is the *reducing mapping*;
- $A_T$ is the *averaging mapping* effectively reducing subimage $g_{\mu(T)}$ by averaging pixels from a subsquare in the bigger s-Domain $\mu(T)$ and putting the result into the corresponding pixel of the smaller t-Domain $T$. For the regular $(\Pi, \Gamma)$ the mapping $A_T = (1/4)^k A'_T$ where:
  - $T$ is of size $2^t \times 2^t$, $\mu(T)$ is of size $2^s \times 2^s$, and $k = s - t$;
  - at certain numbering of pixels in $D$ the mapping $A'_T$ can be written in matrix notation which has:
    * $N \doteq |D|$ rows and $N$ columns;
    * exactly $4^k$ ones in rows corresponding to elements from $T$;

\* only zeros in rows corresponding to elements outside of $T$;

\* only single one in columns corresponding to elements from $\mu(T)$;

\* only zeros in columns corresponding to elements outside of $\mu(T)$;

- $P_T$ is the *affine permutation mapping* on $T$, i.e. with single ones in rows and columns corresponding to elements from $T$.

## 2  Standard results on convergence

In this section a brief review of classic results is presented (cf. [5,10]).

If we take the matrix form of the mappings $R_T$ and of the characteristic functions $\chi_T$ $(T \in \Pi)$ then any affine fractal operator $F$ can be written in a vector form:

$$F g = L g + o \tag{3}$$

where $g, o$ are vectors and $L$ is $N \times N$ matrix:

$$L \doteq \sum_{T \in \Pi} c_T \cdot R_T, \quad o = \sum_{T \in \Pi} o_T \cdot \chi_T \tag{4}$$

Using matrix notation we can also represent the iterations of $F$ by matrix powers of $L$:

**Lemma 1.**

*For any natural $k > 0$:*

$$F^{\circ k}(g) = L^k g + \sum_{i=0}^{k-1} L^{i-1} o \tag{5}$$

Obviously $F$ $(F^{\circ k})$ is Lipschitzian with factor $\|F\|_p$ $(\|F^{\circ k}\|_p)$ equal to the $p$-norm of the matrix $L$ $(L^k)$. Hence we can easily prove that

**Lemma 2.**
1. *$F$ is contractive in $p$-norm if and only if $\|L\|_p < 1$;*
2. *$F$ is eventually contractive in $p$-norm, i.e. there exists $k$ such that the operator $F^{\circ k}$ is contractive if and only if there exists $k$ such that, $\|L^k\|_p < 1$.*

From properties of mapping $A_T$ it follows immediately that for the regular pair $(\Pi, \Gamma)$, in the supremum norm $\|\cdot\|_\infty$, its operator norm $\|A_T\| = 1$. Applying of the permutation $P_T$ from the left side to $A_T$ results in permutation of its rows in matrix representation. The interchange of rows does not change the supremum norm. Therefore the supremum norm of $R_T$ is equal to one too. Hence the supremum norm of $L$ can be easily derived:

**Lemma 3.**

*If the pair of domain sets $(\Pi, \Gamma)$ is regular then*

$$\|L\|_\infty = c^* \doteq \max_{T \in \Pi} |c_T| \tag{6}$$

From the lemmas 3 and 2, we get:

**Corollary 4.**

*If $c^* < 1$ then fractal operator $F$ is contractive in supremum norm.*

Frequently, the image sequence convergence is considered in other than supremum norms, i.e. in $p$-norms for finite $p$. We are going to use some classical results and the above corollary to get a convergence theorem for any $p$-norm.

From the Hölder's inequality for $N$-dimensional vectors $\boldsymbol{x}, \boldsymbol{y}$ and real exponents $p > 1, q = p/(p-1)$:

$$\sum_{i=1}^{N} |x_i y_i| \leq \|\boldsymbol{x}\|_p \|\boldsymbol{y}\|_q \tag{7}$$

we bound $p$-norm by supremum norm for any $N$-dimensional vector $\boldsymbol{x}$ and $p \geq 1$:

$$\frac{\sqrt[p]{N} \|\boldsymbol{x}\|_\infty}{N} \leq \|\boldsymbol{x}\|_p \leq \sqrt[p]{N} \|\boldsymbol{x}\|_\infty \tag{8}$$

The above inequality leads to the conclusion that the convergence of the sequence in supremum norm implies its convergence in any finite $p$-norm.

Therefore the condition $c^* < 1$ is also sufficient for the convergence of the fractal operator iterations with any $p$-norm:

**Theorem 5.**

*Let the fractal operator $F$ be given by the equation 2. If $c^* < 1$ then*

1. *there exists a unique fixed point $\tilde{f}$ of the operator $F$;*
2. *for any initial image $g_0 \in \mathcal{I}$ and for any $p$-norm $(1 \leq p \leq \infty)$:*

$$\lim_{i \to \infty} \|F^{\circ i}(g_0) - \tilde{f}\|_p = 0 .$$

## 3    Convergence in asynchronous case

In this section we consider more general definition of the fractal operator introducing local nonlinearities $\boldsymbol{a}_T$:

$$F(g) \doteq \sum_{T \in \Pi} \boldsymbol{a}_T (c_T \cdot \boldsymbol{R}_T(g) + o_T \cdot \chi_T) \tag{9}$$

The nonlinearity $\boldsymbol{a} : R^M \to R^M$ is *Lipschitzian* if there exists a real constant $c$ such that for any $\boldsymbol{x}, \boldsymbol{y} \in R^M$ the following inequality is true:

$$\|\boldsymbol{a}(\boldsymbol{x}) - \boldsymbol{a}(\boldsymbol{y})\| \leq c \cdot \|\boldsymbol{x} - \boldsymbol{y}\| \tag{10}$$

The smallest $c$ satisfying (10) is denoted by $\|\boldsymbol{a}\|$. The norm $\|\cdot\|$ considered is one of Minkovsky $l_p$ norms which are all equivalent in $R^M$. However in this paper we choose $p = \infty$, i.e. the supremum (Tchebyshev) norm.

The dynamics of the asynchronous fractal operator is defined by asynchronous nondeterministic actions of local operators changing only subimages of the image.

Suppose $g$ denotes the whole image and $g$ is composed of subimages $g_k$, $k = 1, \ldots, L$, corresponding to t-Domains. The action of $k$-th local operator can be described by the operator $F_k$ defined as follows:

$$F_k(g) \doteq \begin{bmatrix} g_1 \\ \vdots \\ g_{k-1} \\ a_k(c_k R_k g + o_k 1) \\ g_{k+1} \\ \vdots \\ g_L \end{bmatrix} \tag{11}$$

For our analysis we consider the asynchronous, deterministic mode as well, in which actions of local operators are performed in a fixed order, for instance according to their numbering. After applying all local operators, we get a global change which is modelled by a global operator $\mathcal{F}$:

$$\mathcal{F}(g) \doteq F_L(F_{L-1}(\ldots(F_1(g))\ldots)) \tag{12}$$

$\mathcal{F}$ works in the space of images and the following lemma gives a sufficient condition for the contractivity of this operator:

**Lemma 6.** *Deterministic sequential behaviour of the fractal operator has the following features:*

1. *Operator $\mathcal{F}$ defined by the equation (12) is contractive if the following condition holds:*

$$\max_{1 \leq k \leq L} (|c_k| \cdot \|a_k\|_\infty) < 1 \tag{13}$$

2. *The fixed point $\tilde{f}$, i.e. the attractor of $\mathcal{F}$ is also the fixed point of local operators $F_k$, $k = 1, \ldots, L$.*

3. *Any permutation of local operators in the definition $\mathcal{F}$ gives a global sequential deterministic operator which is also contractive and has the same attractor $\tilde{f}$.*

**Proof:** Let $g^{(0)} \doteq g$ be any image. Applying the local operator $F_i$, defined by the equation (11) to the image $g^{(i-1)}$ we obtain the image $g^{(i)}$.

Note that the result of the global operator $\mathcal{F}(g) = g^{(L)}$ and that $g^{(i)}$ can be recursively written as follows:

$$
g^{(i)} = \begin{bmatrix} a_1 \left( c_1 R_1 g^{(0)} + o_1 1 \right) \\ \vdots \\ a_i \left( c_i R_i g^{(i-1)} + o_i 1 \right) \\ g_{i+1} \\ \vdots \\ g_L \end{bmatrix} \tag{14}
$$

In order to show that $\mathcal{F}$ is Lipschitzian, let us take now two arbitrary images $g$ and $h$ and estimate the distance using the supremum norm:

$$
\|\mathcal{F}(g) - \mathcal{F}(h)\|_\infty
$$

$$
\leq \max_{1 \leq i \leq L} \left\| a_i \left( c_i R_i g^{(i-1)} + o_i 1 \right) - a_i \left( c_i R_i h^{(i-1)} + o_i 1 \right) \right\| \tag{15}
$$

$$
\leq \max_{1 \leq i \leq L} \|a_i\| \cdot |c_i| \cdot \|g^{(i-1)} - h^{(i-1)}\|
$$

We show now by induction the essential inequality:

$$
\|g^{(i-1)} - h^{(i-1)}\| \leq \|g - h\|, \ i = 1, \ldots, L \tag{16}
$$

In the inductive step we have:

$$
\|g^{(i)} - h^{(i)}\| = \|F_i(g^{(i-1)}) - F_i(h^{(i-1)})\|
$$

$$
\leq \max \left( \|g_1^{(i-1)} - h_1^{(i-1)}\|, \ldots, \|g_{i-1}^{(i-1)} - h_{i-1}^{(i-1)}\|, \right.
$$

$$
\left\| a_i \left( c_i R_i g^{(i-1)} + o_i 1 \right) - a_i \left( c_i R_i h^{(i-1)} + o_i 1 \right) \right\|,
$$

$$
\left. \|g_{i+1}^{(i-1)} - h_{i+1}^{(i-1)}\|, \ldots \right)
$$

$$
\leq \max \left( \|g - h\|, \|a_i\| \cdot |c_i| \cdot \|g_{i-1}^{(i-1)} - h_{i-1}^{(i-1)}\| \right)
$$

$$
\leq \|g - h\| \cdot \max(1, \|a_i\| \cdot |c_i|) = \|g - h\| .
$$

Using the inequality (16) in (15) we finally get the contractivity condition:

$$
\|\mathcal{F}(g) - \mathcal{F}(h)\|_\infty \leq \left( \max_{1 \leq i \leq L} \|a_i\| \cdot |c_i| \right) \|g - h\| .
$$

This concludes the proof of the point one.

It is obvious that the conclusion about the contractivity does not rely on the particular ordering of local operators in the definition of $\mathcal{F}$. Therefore the contractivity condition with the same contractivity factor holds for any permutation in the composition of local operators.

Let us observe now that each local operator $F_k$ is responsible for the change of the independent part $g_k$ of the global state $g$. Hence if $\tilde{f}$ is the fixed point of the global operator $\mathcal{F}$, the application of $F_1$ to $\tilde{f}$ must produce $\tilde{f}$. By a simple induction we can show the same relation for all local operators. This gives the proof of the point two.

The third statement of this lemma immediately follows from the second statement as each global sequential deterministic operator is the composition of the local operators.

$\square$

Let $g$ be any global state in problem space and $\epsilon$ any positive number. By the ball $B(g, \epsilon)$ we mean the set:

$$B(g, \epsilon) \doteq \{h| \, \|g - h\| \leq \epsilon\} \, .$$

Assume that in the asynchronous nondeterministic case, the probability function of the choice of the given local operator is fixed. Then the following theorem holds:

**Theorem 7.** *If the condition (13) holds then there exists a unique global state $\tilde{f}$ such that for any initial state $g$, and positive $\epsilon$, the random sequence $F_{rand(i)}(g)$ enters the ball $B(\tilde{f}, \epsilon)$ with the probability one and stays there permanently.*

**Proof:** Let us take as $\tilde{f}$ the fixed point of the global sequential deterministic operator $\mathcal{F}$.

Since the space of images, i.e. the space of global states is bounded, there exists such a $k$, that for any initial state $g$, the $k$-th iteration of the contractive operator $\mathcal{F}$ enters the ball $B(\tilde{f}, \epsilon)$.

Once the iterative sequence enters this ball it remains there. Namely by the second point of the contractivity lemma 6, $F_i(\tilde{f}) = \tilde{f}$ and if $h \in B(\tilde{f}, \epsilon)$ then we get:

$$\|F_i(h) - \tilde{f}\| = \|F_i(h) - F_i(\tilde{f})\| \leq \|F_i\| \cdot \|h - \tilde{f}\| \leq \epsilon \, .$$

It remains only to show that the probability of a random sequence of local operators which has from a certain position, local components of global operator $\mathcal{F}^{\circ k}$, is equal to one. Though technically somewhat complex it can be demonstrated that the proof of this fact can be reduced to the proof of the following proposition: *The probability of at least one success in infinite Bernoulli scheme is equal to one.* $\square$

The first advantage of the asynchronous approach which was already noticed in the program design, is RAM memory savings by about 50%. This savings follow from the fact that there is no need for the second copy of the iterated image. All local operations can be performed on the same copy of the image.

The second advantage of nondeterministic algorithm is less number of iterations which lead to the same quality image at the decoding stage. It results in time savings. The actual time reducing is image dependent, but usually the reducing is by the factor of two. For instance we have obtained the final quality of the image Lena already after five iterations while the former approach requires 10 iterations.

# 4 Fractal based recognition and association

## 4.1 Recognition by fractal distortion measure

Let $Z = \{f_1, \ldots, f_K\} \subset \mathcal{X}$ be a set of certain objects modelled in a complete metric space $\mathcal{X}$ with a metric $\rho$. Suppose that for each $i = 1, \ldots, K$ we can design contractive mapping $F_i$ such that $F_i(f_i) \approx f_i$. In our approach $F_i$ is the fractal operator.

We propose the following methodology for the recognition of $g \in X$:

$$\text{assign } g \text{ to the object } f_i \text{ if } i = \arg\min_j \rho(g, F_j(g)) \tag{17}$$

The intuition behind the above decision rule is as follows: if $g$ is a distorted version of $f_i$ then $F_i$ as nearly invariant on $f_i$ should be nearly invariant on $g$, i.e. the value of $\rho(g, F_i(g))$ should be small while for $j \neq i$ $F_j$ contracts significantly $g$ towards $f_j$ resulting in the larger value of $\rho(g, F_j(g))$.

Note that this intuition is right only for distorted versions of objects from the set $Z$. If $g$ is not such an element then this decision rule is unable to confirm it.

To deal with rejections, i.e. with negative recognition, we suggest here providing in the given application enough dense set of training objects $Z$ and looking not only for the minimum distance but also for the distribution of several least distances $\rho(g, F_j(g))$. A flat distribution of them should indicate that $g$ is not similar to any object in the set $Z$.

In order to fix the name for the above recognition methodology, considering also that the contractivity and the Banach fixed point theorem [4,3] is in its foundations, we modestly suggest to call the decision rule 17 by the *Banach decision rule*.

## 4.2 Recognition using fractal basis functions

The Banach decision rule defined above requires a competition layer in a hardware implementation what creates serious technical problems. Moreover, this rule cannot be easily extended to rejections.

In this subsection we remove above deficiencies by introducing basis functions related to fractal operators.

Namely, let again $Z = \{f_1, \ldots, f_K\} \subset \mathcal{X}$ be class representatives for objects modelled in a complete metric space $\mathcal{X}$ with a metric $\rho$. Let a contractive mapping $F_i$ be such that $F_i(f_i) \approx f_i$. Then, we can assign with each object $f_i$ a basis function $\Phi_i$ as follows:

$$\Phi_i(g) \doteq e^{-\rho(g, F_i(g))} \tag{18}$$

In case when $F_i$ is the fractal operator then the above basis functions are called *fractal basis functions*.

Similarly to radial basis functions we can design a neural network approximating characteristic functions for classes of interests.

Suppose that there is $L$ classes ($L \le K$). Then by applying recursive least square method (RLS) we find coefficients $c_{ij}$, $i = 1, \ldots, K$, $j = 1, \ldots, L$ such that the following function:

$$\hat{\psi}_j(g) \doteq \sum_{i=1}^{K} c_{ij} \Phi_i(g)$$

approximates the characteristic function $\psi_j$ of $j$-th class.

Comparing to radial basis functions ([6]) we have defined here problem dependent basis functions which fit better to local class properties than Gaussian functions. Moreover, the design process of basis functions in fractal approach appears faster than neural learning of parameters for Gaussians.

The classifier obtained using the above basis function is called the *Banach classifier*.

## 4.3 Association Methodology

By object association we mean here the associative recall performed by an associative memory of objects.

In turn, by the associative memory we mean a system which works in two phases: training and testing phase. During the training phase it acquires certain information about training objects and incorporates it into its internal state (for instance into tables of certain coefficients). During the testing phase distorted versions of training objects are presented on the input of the associative memory while original or close to original objects are sent to the output of the system.

The behaviour of the associative memory is undefined when completely different from training examples are delivered. In practice, we encounter three types of behaviour in case of *negative association*: output is wrongly associated object, output is an object which is randomly or chaotically generated, and output is a predefined *dummy* object.

Our association methodology is based on the above introduced Banach classifier.

Let $Z = \{f_1, \ldots, f_K\} \subset \mathcal{X}$ be a set of certain objects modelled in a complete metric space $\mathcal{X}$ with a metric $\rho$ which are presented to the system during the training phase. For each $i = 1, \ldots, K$ we construct a contractive operator $F_i$ such that $F_i(f_i) \approx f_i$.

If $g \in X$ is presented during the testing phase then the associative recall is made by the following rule

$$\text{output} \quad \hat{g} = \lim_{k \to \infty} F_i^{\circ k}(g) \quad \text{where} \quad i = \arg\min_j \rho(g, F_j(g)) \tag{19}$$

Hence if $g$ was classified by the Banach classifier to be assigned to $f_i$ then the result of the associative recall is obtained by the iteration of the contractive operator $F_i$ starting from $g$. Note that there is only the information about operators $F_j$ stored in the system, which is usually more compact than the data needed for original objects $f_j$.

# 5   Applications

## 5.1   Face Recognition

The face recognition problem which we consider here is an example of object recognition task, where objects are still images of human faces stored in a face database. For the solution of this problem, we apply the methodology of Banach classifier. According this general scheme, each training image is represented by its asynchronous nondeterministic fractal operator. As a distance measure the normalized mean square error NMSE was chosen.

The experiments were performed on pictures collected in the ORL face database. It includes a set of photographs taken between April 1992 and April 1994 at the Olivetti Research Laboratory in Cambridge, UK[1] There are 10 different images of each of 40 distinct subjects. For some subjects, the images were taken at different times, varying the lighting, facial expressions (open or closed eyes, smiling or not smiling) and facial details (glasses or no glasses). All the images were taken against a dark homogeneous background with the subjects in an upright, frontal position (with tolerance for some side movement). The size of each original image is 92x112 pixels, with 256 grey levels per pixel. For our experiments we have reduced each image to the resolution 11x14 by replacing each block 8x8 by its arithmetic average. This database is maintained by Ferdinando Samaria who used it in his research [9].

Primitive use of Banach classifier to this problem takes one picture ($l = 1$) per one person in the training phase. Then results are rather poor and they exhibit high dependence on the cardinality of the training set $s = |Z|$. The average results for the recognition rate are $87.8\%, 77.1\%$, and $69.1\%$ at $s = 10, 20$, and $40$ respectively. For instance when 10 pictures of 10 persons were chosen randomly for training, on average 10 pictures out of 90 tested were assigned to wrong persons.

When the number of training pictures per person is increased to $l = 5$ by random sampling, the results are significantly improved with the following

**Decision rule 1**: decide that the face image $g$ belongs to a person $p$ if the Banach classifier assigns for $g$ a face picture $f$ belonging to the person $p$.

For instance at the full database $s = 40$ on average about 11 pictures out of 200 was wrongly classified (cf. Table 1).

In the next step we consider more general situation: for the person $i$, $l_i$ face images are used for training and $t_i$ images for testing. In our experiments

---

[1] The ORL database is free of charge, see http://www.cam-orl.co.uk/facedatabase.html

| $s$ | 10 | | 20 | | 40 | |
|---|---|---|---|---|---|---|
| $l$ | R[%] | E[%] | R[%] | E[%] | R[%] | E[%] |
| 1 | 87.8 | 12.2 | 77.1 | 22.9 | 69.1 | 30.9 |
| 3 | 95.4 | 4.6 | 90.2 | 9.8 | 88.7 | 11.3 |
| 5 | 97.2 | 2.8 | 94.8 | 5.2 | 94.3 | 5.7 |

**Table 1.** Results for face recognition – decision rule 1, random choice of training and testing faces; recognition rate – column R; error rate – column E; $s$ – number of persons, $l$ – size of training set per person.

$t_i = 10 - l_i$. Denote by $n_{p,i}$ the number of faces belonging to the person $p$ which were assigned by the Banach classifier to a person $i$ when all its testing faces were given for the recognition. Now we classify persons rather than faces using the following:

**Decision rule 2**: decide that the collection of face images taken for a person $i$ belongs to a person $p$ in the database if $n_{p,i} \geq 0.5t_i$ and for all $q \neq p$ we have $n_{q,i} < n_{p,i}$, otherwise do not take any decision.

According the above decision rule, the person is recognized only if at least half of its photographs is assigned to the same person and there is no any other person in the database with the same property. Other interpretation of this rule is by voting process: $n_{p,i}$ is number of votes given by $i$ for $p$. The person with the majority of votes is the winner.

This new decision rule not only improved the recognition rate, but also by introducing the *no decision* category it reduced significantly the error rate. For instance for the whole database at random choice of $l = 5$ training faces and remaining $t = 5$ testing faces on average 0.24 persons were wrongly classified. In other words once per four trials a false substitution occurred (cf. Table 2).

| $s$ | 10 | | | 20 | | | 40 | | |
|---|---|---|---|---|---|---|---|---|---|
| $l$ | R | N | E | R | N | E | R | N | E |
| 1 | 98.0 | 2.0 | 0.0 | 77.5 | 17.5 | 5.0 | 71.3 | 21.3 | 7.4 |
| 3 | 98.0 | 2.0 | 0.0 | 96.3 | 1.2 | 2.5 | 93.1 | 6.3 | 0.6 |
| 5 | 98.0 | 0.0 | 2.0 | 97.4 | 0.2 | 2.4 | 98.1 | 1.3 | 0.6 |

**Table 2.** Results for face recognition – decision rule 2, random choice of training and testing faces; recognition rate – column R; no decision rate – column N; error rate – column E (all rates in percents); $s$ – number of persons, $l$ – size of training set per person.

The last improvement was made by replacing the random choice of training photographs by manually preselected faces. For $l = 1$ a front view was chosen, for $l = 3$ if possible left side view and right side view were added, and finally for $l = 5$ arbitrary chosen two additional views were chosen. This preselection step seems reasonable for applications in which we can afford for more than one photograph with desired poses.

For the preselected training set, we have tried experiments with both decision rules. Using the first decision rule we get 7 false substitution for $200 = 5 \cdot 40$ pictures. However, using the second decision rule results in 100% recognition already for three training faces per one person (cf. Table 3).

| | decision rule 1 | | decision rule 2 | | |
|---|---|---|---|---|---|
| $l$ | R | E | R | N | E |
| 1 | 75.3 | 24.7 | 72.5 | 17.5 | 10.0 |
| 3 | 93.2 | 6.8 | 100.0 | 0.0 | 0.0 |
| 5 | 96.5 | 3.5 | 100.0 | 0.0 | 0.0 |

**Table 3.** Results for face recognition – preselected choice of training faces; recognition rate – column R; no decision rate – column N; error rate – column E (all rates in percents); $l$ – size of training set per person.

## 5.2 Image Association

By image association we mean object association as described in the previous section where we take images as the objects.

In contrary to the face recognition experiments, in this section during the training phase, the full resolution images were used to built corresponding fractal operators. For instance 512x512 Lena image was taken. However, to accelerate the classification stage, the distance measure NMSE was calculated on 10% of randomly chosen pixels.

**Noisy Image Association** In experiments there are noisy versions of training images presented on the input of the system. The Banach classifier assigns the input to one of images which is reconstructed by the iterations of its fractal operator.

In the tables we use abbreviations for image names: L – Lena, J – Julia, B – Baboon, M – Miyake, K – Kiuchi. The symbol I denotes the column for image names which are sorted according to the distance value. We present the results only for the first three images in the ordering fixed by increasing NMSE values.

In these experiments pixel values were not normalized, i.e. they are in the interval $[0, 255]$. Therefore for instance $\sigma = 500$ means that the standard deviation of the noise equals to about double length of nominal image range.

We see from the Table 4 that the distance to Lena image is the shortest for $\sigma$ up to 500. The wrong recall is observed for enormous noise with $\sigma = 4000$. Though the classification is correct for $\sigma < 2000$ the distance differences are insignificant.

| # | $\sigma = 20$ | | $\sigma = 150$ | | $\sigma = 250$ | | $\sigma = 500$ | | $\sigma = 4000$ | |
|---|---|---|---|---|---|---|---|---|---|---|
| | I | NMSE | I | NMSE | I | NMSE | I | NMSE | I | NMSE |
| 1 | L | 28.6 | L | 96.4 | L | 99.2 | L | 99.9 | M | 99.99 |
| 2 | B | 85.9 | B | 102.5 | J | 101.4 | J | 100.5 | L | 100.00 |
| 3 | J | 94.5 | J | 102.7 | B | 101.5 | B | 100.6 | J | 100.00 |

**Table 4.** Sorted distance measures for Lena image corrupted by additive Gaussian noise with various standard deviations $\sigma$.

| # | $\sigma = 20$ | | $\sigma = 150$ | | $\sigma = 250$ | | $\sigma = 500$ | | $\sigma = 2000$ | |
|---|---|---|---|---|---|---|---|---|---|---|
| | I | NMSE | I | NMSE | I | NMSE | I | NMSE | I | NMSE |
| 1 | L | 4.6 | L | 59.2 | L | 86.0 | L | 102.7 | J | 111.0 |
| 2 | B | 79.1 | B | 95.8 | J | 104.6 | J | 108.9 | B | 111.8 |
| 3 | J | 91.2 | J | 100.0 | B | 105.3 | B | 109.6 | L | 111.9 |

**Table 5.** Sorted distance measures for Lena image corrupted by multiplicative uniform (speckle) noise for various standard deviations $\sigma$.

For the speckle noise we have more significant distance differences for the winner than in Gaussian noise case. However, the wrong classification occurs for smaller $\sigma$ (cf. Table 5). The similar very good performance has been obtained for uniform additive noise and impulsive noise (*salt and pepper*).

**Incomplete Image Association** Partial or incomplete data we get here by setting certain pixels to zero value (black color). This can be done in deterministic or in pseudo random way. In both cases the receiver of such incomplete image knows the location of missing pixels. Therefore it can approximate the missing value in the given location by taking the average of non missing values located in a vicinity. In computer experiments we have chosen as the neighbourhood, a pixel block growing from size $5 \times 5$ to the minimum size where non hidden pixel can be found. This approximation acts as a low pass filter on the input image.

The preprocessed incomplete image is next delivered to the associative memory to be completely recovered by an associative recall.

From Table 6 we see that changing up to 95% pixels to the black color in deterministic way still results in numerically significant, correct association. The same conclusion we get from experiments with randomly chosen, partial data.

## 6 Conclusions

Asynchronous fractal operators are designed for images in order to implement fractal applications by massive parallel asynchronous computing system as for instance recurrent neural networks [6,7].

Assuming the convergence of the standard fractal operator to an element $\tilde{f}$ which is an approximation of the original image $f$, it is proved that any asynchronous deterministic realization of local fractal operators is convergent to $\tilde{f}$

| # | $s = 10\%$ | | $s = 50\%$ | | $s = 90\%$ | | $s = 95\%$ | | $s = 99.7\%$ | |
|---|---|---|---|---|---|---|---|---|---|---|
| | I | NMSE | I | NMSE | I | NMSE | I | NMSE | I | NMSE |
| 1 | L | 1.4 | L | 1.5 | L | 4.1 | L | 11.9 | M | 100.2 |
| 2 | B | 77.7 | B | 76.5 | B | 75.9 | B | 78.5 | L | 102.7 |
| 3 | J | 90.5 | J | 90.0 | J | 90.1 | J | 91.9 | K | 105.3 |

**Table 6.** Sorted distance measures for incomplete Lena image with different percentage of missing pixel area $s$.

and the stochastic realization converges to $\bar{f}$ with probability one. A new class of basis functions is defined using fractal operators designed for class representatives.

Experiments confirm practical advantage of asynchronous fractal operator over synchronous one: faster decoding and significant memory savings.

Beside compression, applications of such stochastic fractal operators for face recognition and image association show their high discrimination and association power.

**Acknowledgment:**

This work was partially supported by KBN grant, no 8T11C01808.

# References

1. S. Amari, "Learning patterns and pattern sequences by self-organizing nets of threshold elements", IEEE Trans. on Computer, C-21, pp. 1197-1206, 1972.
2. S. Amari, "Neural theory of association and concept formation", Biological Cybernetics, 26, pp. 175-185, 1977.
3. S. Banach, "Sur les operations dans les ensembles abstraits et leur applications aux equations integrales", Fundamenta Mathematica, vol.3, pp. 133-181, 1922.
4. J. Dugundi and A. Granas, *Fixed point theory*, Polish Scientific Publishers, Warszawa, 1982.
5. Y. Fisher, ed., *Fractal Image Compression – Theory and Application*, Springer Verlag (1995).
6. S. Haykin, *Neural networks – A Comprehensive Foundation*, Maxwell Macmillan International (1994).
7. J. J. Hopfield, "Neural networks and physical systems with emergent collective computational abilities," *Proc. Natl. Acad. Sci. USA, Biophysics*, 79, pp. 2554-2558 (1982).
8. A. Jacquin, "Image coding based on a fractal theory of iterated contractive image transformations," *IEEE Trans. Image Processing*, 1, pp. 18-30 (1992).
9. Ferdinando Samaria and Andy Harter, "Parameterisation of a stochastic model for human face identification," Proceedings of 2nd IEEE Workshop on Applications of Computer Vision, Sarasota FL, December 1994.
10. W. Skarbek, "On convergence of affine fractal operators", *Image Processing and Communications*, 1, pp. 33-41 (1995).
11. S. W. Smoliar and H. Zhang, "Content-Based Video Indexing and Retrieval," *IEEE Multimedia*, 1, pp. 62-72, (1994).

# Trilinear Tensor: The Fundamental Construct of Multiple-view Geometry and Its Applications

Amnon Shashua

Institute of Computer Science
The Hebrew University of Jerusalem
Jerusalem, 91904, Israel
http://www.cs.huji.ac.il/~ shashua

**Abstract.** The topic of representation, recovery and manipulation of three-dimensional (3D) scenes from two-dimensional (2D) images thereof, provides a fertile ground for both intellectual theoretically inclined questions related to the algebra and geometry of the problem and to practical applications such as Visual Recognition, Animation and View Synthesis, recovery of scene structure and camera ego-motion, object detection and tracking, multi-sensor alignment, etc.

The basic materials have been known since the turn of the century, but the full scope of the problem has been under intensive study since 1992, first on the algebra of two views and then on the algebra of multiple views leading to a relatively mature understanding of what is known as "multilinear matching constraints", and the "trilinear tensor" of three or more views.

The purpose of this paper is, first and foremost, to provide a coherent framework for expressing the ideas behind the analysis of multiple views. Secondly, to integrate the various incremental results that have appeared on the subject into one coherent manuscript.

## 1   Introduction

Given that three-dimensional (3D) objects in the world are modeled by point or line sets, then their projection onto a number of distinct image planes produces point or line sets that are related by correspondences. The relationship between a 3D point/line and its 2D projections is easily described by a simple multilinear equation whose parameters consist of the camera location (viewing position).

From an algebraic standpoint, since the 3D-to-2D relationship juxtaposes the variables of object space, variables of image space and variables of viewing position, then one can isolate subsets of these variables and consider their properties:

- Matching Constraints: Given two or more views, it is possible to eliminate the object variables and obtain multilinear functions of image measurements (image variables) and (functions of) viewing variables. In other words, the existence of a correspondence set is an algebraic constraint(s) in disguise whose form becomes explicit via an elimination process and leaves us with a "shape invariant" function.

– Shape constraints: analogously, given a sufficient number of points one can eliminate the viewing variables and obtain functions of image measurements and object variables (known as indexing functions). In other words, it is possible to factor out the role of the changing viewing position and remain with a "view invariant" function.

The application aspect naturally follows the decomposition above and can be roughly divided into two classes:

– Applications for which irrelevant image variabilities are factored out: for example, Visual Recognition of a 3D object under changing viewing positions may use the Matching Constraints to create an equivalence class of the image space generated by all views of the object; or may use the Shape Constraints as an index into a shape library. In both cases, the desire is not to reconstruct the value of unknown variables (say, the shape of the object from image measurements), but rather to find a new representation that will facilitate the matching process between input image to library models. This class of applications includes also Object Tracking by means of image stabilization processing, and Image-based Rendering (a.k.a View Synthesis) of 3D objects directly from a sample of 2D images without recovering object shape.
– Reconstruction Applications: here the goal is to recover the value of unknown variables (shape or viewing positions) from the correspondence set. This line of applications is part of Photogrammetry with origins starting at the turn of the century. Both the Matching Constraints and the Shape Constraints provide simple and linear methods for achieving this goal, but non-linear iterative methods, such as the "block bundle adjustment" of Photogrammetry, are of much use as well.

One of the important ideas that has emerged in the recent years and is related to these issues is the factorization/elimination principles from which the multilinear matching constraints have arisen and consequently the discovery of the Trilinear Tensor which has emerged as the basic building block of 3D visual analysis.

The purpose of this paper is, first and foremost, to provide a coherent framework for expressing the ideas behind the analysis of multiple views. Secondly, to integrate the various incremental results that have appeared on the subject into one coherent manuscript.

We will start with the special case of Parallel Projection (Affine camera) model in order to illuminate the central ideas, and proceed from there to the general Perspective Projection (Projective camera) model and progress through the derivation of the Matching Constraints, the Trilinear Tensor and its properties, the Fundamental matrix, and applications.

## 2 N-view and n-point Geometry With an Affine Camera

The theory underlying the relationship among multiple affine views is well understood and will serve here to illuminate the goals one wishes to obtain in the general case of perspective views.

An affine view is obtained when the projecting rays emanating from a 3D object are all parallel to each other, and their intersection with an image plane forms the "image" of the object from the vantage point defined by the direction of rays. In general, we are also allowed to take pictures of pictures as well. Let the 3D world consist of $n$ points $P_1, ..., P_n$, whose homogeneous coordinates are $(X_i, Y_i, Z_i, 1)$, $i = 1, ..., n$. Consider $N$ distinct affine views $\psi_1, ..., \psi_N$. If we ignore problems of occlusion (assuming the object is transparent), then each view consists of $n$ points $p_1^j, ..., p_n^j$, $j = 1, ..., N$, with non-homogeneous coordinates $(x_i^j, y_i^j)$, $i = 1, ..., n$.

The relationship between the 3D and 2D spaces is represented by a $2 \times 4$ matrix per view:

$$p_i^j = \begin{bmatrix} a_j^\top \\ b_j^\top \end{bmatrix} P_i$$

where $a_j, b_j$ are the rows of the matrix. The goal is to recover $P_i$ (and/or the camera transformations $a_j, b_j$) from the image measurements alone (i.e., from the set of image points). If the world is undergoing only rigid transformations, then the $2 \times 3$ left principle minor of each camera transformation is a principle minor of an orthonormal matrix (rotation in space), otherwise the world may undergo affine transformations. Furthermore, one of the camera matrices may be chosen arbitrarily and, for example, set to $a = (1, 0, 0, 0)$ and $b = (0, 1, 0, 0)$. Note that even in the affine case the task is not straightforward because the camera parameters and the space coordinates (both of which are unknown) are coupled together, hence making the estimation a non-linear problem. But now consider all the measurements stacked together:

$$\begin{bmatrix} x_1^1 & x_2^1 & \cdots & x_n^1 \\ y_1^1 & y_2^1 & \cdots & y_n^1 \\ \cdot & & & \\ \cdot & & & \\ \cdot & & & \\ x_1^N & x_2^N & \cdots & x_n^N \\ y_1^N & y_2^N & \cdots & y_n^N \end{bmatrix} = \begin{bmatrix} a_1^\top \\ b_1^\top \\ \cdot \\ \cdot \\ \cdot \\ a_N^\top \\ b_N^\top \end{bmatrix} \begin{bmatrix} X_1 & \cdots & X_n \\ Y_1 & \cdots & Y_n \\ Z_1 & \cdots & Z_n \\ 1 & \cdots & 1 \end{bmatrix}.$$

Thus, we clearly see that the rank of the $2N \times n$ matrix of image measurements is at most 4 (because the two matrices on the right hand side are at most of rank 4 each). This observation was made independently by Tomasi & Kanade [28] and Ullman & Basri [30] — each focusing on a different aspect of this result. Tomasi & Kanade took this result as an algorithm for reconstruction, namely, the measurement matrix can be factored into two matrices one representing motion and the other representing shape. The factorization can be done via the well known "Singular Value Decomposition" (SVD) method of Linear Algebra and the orthogonality constraints can be employed as well in order to obtain an Euclidean reconstruction. Ullman & Basri focused on the fact that the row space of the measurement matrix is spanned by four rows, thus a view (each view is represented by two rows) can be represented as a linear combination of other views — hence the "linear combination of views" result.

To understand the importance of the rank 4 result further, consider the following. Each column of the measurement matrix is a point in a $2N$ dimensional space, we call "Joint Image Space" (JIS). Each point $P$ in the 3D world maps into a point in JIS. The rank 4 result shows that the entire 3D space lives in a 4-dimensional linear subspace of JIS. Each point in this subspace is linearly spanned by 4 points, and the coefficients of the linear combination are a function (possibly non-linear) of 3D coordinates *alone*. Therefore, the JIS represents a direct connection between 2D and 3D where the camera parameters are *eliminated* altogether. These functions are called "indexing" functions because they allow us to index into a library of 3D objects directly from the image information.

Similarly, each row of the measurement matrix is a point in a $n$ dimensional space, we call "Joint Point Space" (JPS). Each "half" view, i.e., the collection of x or y coordinates, of a set of $n$ points maps to a point in JPS. The rank 4 result shows that all the half views occupy a 4-dimensional linear subspace of JPS[1]. Each point in this subspace is linearly spanned by 4 points, and the coefficients of the linear combination are a function (possibly non-linear) of camera parameters *alone*. Therefore, the JPS represents a direct connection between 2D and camera parameters where the 3D coordinates are *eliminated* altogether. These functions are called "matching constraints" because they describe constraints (in this case linear) across image coordinates of a number of views that must hold uniformly for all points. Finally, JPS and JIS are dual spaces. Fig. 1 illustrates these concepts.

The affine camera case provides the insight of where the goals are in the general case. With perspective views (projective camera) there is an additional coupling between image coordinates and space coordinates, which implies that the subspaces of JIS and JPS are non-linear, they live in a manifold instead. In order to capture these manifolds we must think in terms of *elimination* because this is what actually has been achieved in the example above. The most important distinction between the affine and the general cases is that in the general case we focus on those coefficients that describe the manifolds (the matching constraints or the indexing functions). These coefficients form a tensor, the "trilinear tensor", and the rank 4 result holds again, but not in the JIS or JPS but in the new space of tensors. The linearity thus appears in the general case by focusing on yet another higher level space. The remainder of this paper is about that space, its definition and its relevance to the reconstruction and other image manipulation tasks.

## 3  Matching Constraints With a General Pin-hole Camera

We wish to transfer the concepts discussed above to the general pin-hole camera. In the Parallel Projection model it was easy to capture both the matching and shape constraints in a single derivation, and which applied (because they were

---

[1] Jacobs [14] elegantly shows that this subspace is decomposed into two skewed lines, i.e., a 3D model is mapped to two lines in $R^n$.

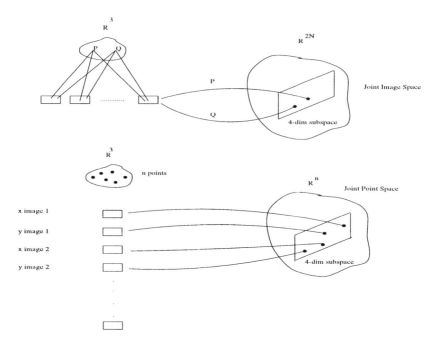

**Fig. 1.** Illustration of the "rank 4" result in the case of an affine camera. See text for explanation.

linear constraints) to generally $N$ views and $n$ points. The Pin-hole model gives rise to a slightly more complex decomposition, as described below.

A perspective view is obtained when the projecting rays emanating from a 3D object are concurrent and meet at a point known as the "center of projection". The intersection of the projecting rays with an image plane forms the "image" of the object. In general, we are also allowed to take pictures of pictures as well. Let the 3D world consist of $n$ points $P_1, ..., P_n$, whose homogeneous coordinates are $(X_i, Y_i, Z_i, 1)$, $i = 1, ..., n$. Consider $N$ distinct views $\psi_1, ..., \psi_N$. If we ignore problems of occlusion, then each view consists of $n$ points $p_1^j, ..., p_n^j$, $j = 1, ..., N$, with homogeneous coordinates $(x_i^j, y_i^j, 1)$, $i = 1, ..., n$.

The relationship between the 3D and 2D spaces is represented by a $3 \times 4$ matrix per view:

$$p_i^j \cong T_j P_i$$

where $T_j$ is the "camera matrix" and $\cong$ defines equality up to scale. In case the world undergoes rigid transformations only, then the left $3 \times 3$ principle minor of $T_j$ is orthonormal (rotation matrix), otherwise the world may undergo general projective transformations. Without loss of generality, one of the camera matrices, say $T_1$, may be chosen as $[I; 0]$ where $I$ is the $3 \times 3$ identity matrix and the fourth column is null.

Note that the major difference between the parallel projection and perspective projection models is the additional scale factor hidden in $\cong$. In case of parallel projection, the 3D-to-2D equation is *bilinear* in the unknowns (space and viewing parameters), thus a single factorization (via SVD) is sufficient for obtaining linear relations between image and viewing variables or image and space variables. However, the perspective 3D-to-2D equation is *trilinear* in the unknowns (space, viewing parameters and the scale factor), thus two steps of factorizations are needed: the first factorization will produce a bilinear structure, and the second factorization will produce a linear structure but not in the image space but in a higher level space. This will become clear in the sequel.

Consider a single point $P$ in space projected onto 4 views with camera matrices $[I; 0], T, G, H$. To simplify the indexing notation, the image points of $P$ will be denoted as $p, p', p'', p'''$ in views 1 to 4, respectively. We can eliminate the scale factors as follows. Consider $p' \cong TP$, thus

$$x' = \frac{t_1^\top P}{t_3^\top P} \tag{1}$$

$$y' = \frac{t_2^\top P}{t_3^\top P}, \tag{2}$$

where $t_i$ is the i'th row of $T$. Note that the third relation $x'/y'$ is linearly spanned by the two above, thus does not add any new information. In matrix form we obtain:

$$\begin{bmatrix} x' t_3^\top - t_1^\top \\ y' t_3^\top - t_2^\top \end{bmatrix} P = 0, \tag{3}$$

or $MP = 0$. Therefore, every view adds two rows to $M$ whose dimensions become $2N \times 4$. For $N \geq 2$ the vanishing determinant of $M$ ($|M| = 0$ because $P \neq 0$ is in the null space of $M$) provides a constraint (Matching Constraint) between the image variables and the viewing parameters — thus the space variables have been eliminated. For $N = 2$ we have exactly 1 such constraint which is bilinear in the image coordinates, for $N = 3$ we have,

$$M = \begin{bmatrix} x i_3^\top - i_1^\top \\ y i_3^\top - i_2^\top \\ x' t_3^\top - t_1^\top \\ y' t_3^\top - t_2^\top \\ x'' g_3^\top - g_1^\top \\ y'' g_3^\top - g_2^\top \end{bmatrix}, \tag{4}$$

where $i_j$ is the j'th row of $[I; 0]$ and every $4 \times 4$ minor has a vanishing determinant. Thus, there are 12 matching constraints that include all three views, which are arranged in three groups of 4: each group is obtained by fixing two rows corresponding to one of the three views. For example, the first group consists of $M_{1235}, M_{1236}, M_{1245}, M_{1246}$ where $M_{ijkl}$ is the matrix formed by taking rows $i, j, k, l$ from $M$. Each constraint has a trilinear form in image coordinates, hence they denoted as "trilinearities".

For $N = 4$ we obtain 16 constraints that include all four views (choose one row from $M$ per view) and which have a quadlinear form in image coordinates.

Clearly, the case of $N > 4$ does not add anything new (because we choose at most subsets of 4 rows at a time).

The bilinear constraint was introduced by Longuett-Higgins [15] in the context of rigid motions and later by Faugeras [6] (see, [7], with references therein) for the general projective model. The trilinearities were originally introduced by Shashua [18] and the derivation adopted here is due to Faugeras & Mourrain [8] (similar derivations also concurrently appeared in [12,29]).

We will focus on the trilinearities because (i) we will show later that the bilinear constraint arises from and is a particular case of the trilinearities, and (ii) the quadlinearities do not add any new information since they can be expressed as a linear combination of the trilinearities and the bilinear constraint [8]. The following questions are noteworthy:

1. How are the coefficients of the trilinearities (per group) arranged? We will show they are arranged as a trivalent tensor with 27 entries.
2. What are the properties of the tensor? The term "properties" contain a number of issues including (i) the geometric objects that the tensor applies onto, (ii) what do contractions (subsets of coefficients) of the tensor represent? (iii) what distinguishes this tensor from a general trivalent tensor? (iv) the connection to the bilinear constraint and other 2-view geometric constructs, and (iv) applications of the Matching Constraints and methods for 3D reconstruction from the tensor.
3. Uniqueness of the solution of the tensor from the correspondence set (image measurements) — the issue of critical configurations.
4. The relationship among tensors — factorization in Tensor space, where the rank 4 constraint we saw in the affine model resurfaces again. This issue includes the notion of "tensorial operators" and their application for rendering tasks.

The first two issues (1,2) will be addressed in the remainder of this paper — the remaining issues can be found in isolation in [21,2,22] or integrated in the full version of this manuscript [19]. We will start with a brief description of notations that will assist the reader with the technical derivations.

## 4    Primer on Tensorial Notations

We assume that the physical 3D world is represented by the 3D projective space $\mathcal{P}^3$ (object space) and its projections onto the 2D projective space $\mathcal{P}^2$ defines the image space. We use the covariant-contravariant summation convention: a point is an object whose coordinates are specified with superscripts, i.e., $p^i = (p^1, p^2, ...)$. These are called contravariant vectors. An element in the dual space (representing hyperplanes — lines in $\mathcal{P}^2$), is called a covariant vector and is represented by subscripts, i.e., $s_j = (s_1, s_2, ....)$. Indices repeated in covariant and contravariant forms are summed over, i.e., $p^i s_i = p^1 s_1 + p^2 s_2 + ... + p^n s_n$. This is known as a contraction. For example, if $p$ is a point incident to a line $s$ in $\mathcal{P}^2$, then $p^i s_i = 0$. Vectors are also called 1-valence tensors. 2-valence tensors (matrices) have two indices and the transformation they represent depends on the covariant-contravariant positioning of the indices. For example, $a_i^j$ is a mapping from points to points, and hyperplanes to hyperplanes, because $a_i^j p^i = q^j$ and $a_i^j s_j = r_i$ (in matrix form: $Ap = q$ and $A^\top s = r$); $a_{ij}$ maps points to hyperplanes; and $a^{ij}$ maps hyperplanes to points. When viewed as a matrix the row and column positions are determined accordingly: in $a_i^j$ and $a_{ji}$ the index $i$ runs over the columns and $j$ runs over the rows, thus $b_j^k a_i^j = c_i^k$ is $BA = C$ in matrix form. An outer-product of two 1-valence tensors (vectors), $a_i b^j$, is a 2-valence tensor $c_i^j$ whose $i, j$ entries are $a_i b^j$ — note that in matrix form $C = ba^\top$. An n-valence tensor described as an outer-product of $n$ vectors is a rank-1 tensor. The definition of the rank of a tensor is an obvious extension of the definition of the rank of a matrix: A rank-1 n-valence tensor is described as the outer product of n vectors; the rank of an n-valence tensor is the *smallest* number of rank-1 n-valence tensors with sum equal to the tensor. For example, a rank-1 trivalent tensor is $a_i b_j c_k$ where $a_i, b_j$ and $c_k$ are three vectors. The rank of a trivalent tensor $\alpha_{ijk}$ is the smallest $r$ such that,

$$\alpha_{ijk} = \sum_{s=1}^{r} a_{is} b_{js} c_{ks} \tag{5}$$

## 5  The Trilinear Tensor

Consider a single point $P$ in space projected onto 3 views with camera matrices $[I; 0], T, G$ with image points $p, p', p''$ respectively. Note that $P = (x, y, 1, \lambda)$ for some scalar $\lambda$. Consider $T = [A; v']$ where $A$ is the $3 \times 3$ principle minor of $T$ and $v'$ is the fourth column of $T$. Consider $p' \cong TP$ and eliminate the scale factor as was done previously:

$$x' = \frac{t_1^\top P}{t_3^\top P} = \frac{a_1^\top p + \lambda v_1'}{a_3^\top p + \lambda v_3'} \tag{6}$$

$$y' = \frac{t_2^\top P}{t_3^\top P} = \frac{a_2^\top p + \lambda v_2'}{a_3^\top p + \lambda v_3'}, \tag{7}$$

where $a_i$ is the i'th row of $A$. These two equations can be written more compactly as follows:

$$\lambda s'^\top v' + s'^\top Ap = 0 \tag{8}$$

$$\lambda s''^\top v' + s''^\top Ap = 0 \tag{9}$$

where $s' = (-1, 0, x)$ and $s'' = (0, -1, y)$. Yet in a more compact form consider $s', s''$ as row vectors of the matrix

$$s_j^\mu = \begin{bmatrix} -1 & 0 & x' \\ 0 & -1 & y' \end{bmatrix}$$

where $j = 1, 2, 3$ and $\mu = 1, 2$. Therefore, the compact form we obtain is described below:

$$\lambda s_j^\mu v'^j + p^i s_j^\mu a_i^j = 0, \tag{10}$$

where $\mu$ is a free index (i.e., we obtain one equation per range of $\mu$).

Similarly, let $G = [B; \boldsymbol{v}'']$ for the third view $p'' \cong GP$ and let $r_k^\rho$ be the matrix,

$$r_k^\rho = \begin{bmatrix} -1 & 0 & x'' \\ 0 & -1 & y'' \end{bmatrix}$$

And likewise,

$$\lambda r_k^\rho v''^k + p^i r_k^\rho b_i^k = 0, \tag{11}$$

where $\rho = 1, 2$ is a free index. We can eliminate $\lambda$ from equations 10 and 11 and obtain a new equation:

$$(s_j^\mu v'^j)(p^i r_k^\rho b_i^k) - (r_k^\rho v''^k)(p^i s_j^\mu a_i^j) = 0,$$

and after grouping the common terms:

$$p^i s_j^\mu r_k^\rho (v'^j b_i^k - v''^k a_i^j) = 0,$$

and the term in parenthesis is a trivalent tensor we call the *trilinear tensor*:

$$\boxed{T_i^{jk} = v'^j b_i^k - v''^k a_i^j. \qquad i, j, k = 1, 2, 3} \tag{12}$$

And the tensorial equations (the 4 trilinearities) are:

$$\boxed{p^i s_j^\mu r_k^\rho T_i^{jk} = 0}, \tag{13}$$

Hence, we have four trilinear equations (note that $\mu, \rho = 1, 2$). In more explicit form, these trilinearities look like:

$$x'' T_i^{13} p^i - x'' x' T_i^{33} p^i + x' T_i^{31} p^i - T_i^{11} p^i = 0,$$
$$y'' T_i^{13} p^i - y'' x' T_i^{33} p^i + x' T_i^{32} p^i - T_i^{12} p^i = 0,$$
$$x'' T_i^{23} p^i - x'' y' T_i^{33} p^i + y' T_i^{31} p^i - T_i^{21} p^i = 0,$$
$$y'' T_i^{23} p^i - y'' y' T_i^{33} p^i + y' T_i^{32} p^i - T_i^{22} p^i = 0.$$

Since every corresponding triplet $p, p', p''$ contributes four linearly independent equations, then seven corresponding points across the three views uniquely determine (up to scale) the tensor $T_i^{jk}$. Equation 12 was first introduced in [18] and the tensorial derivation leading to Equation 13 was first introduced in [20].

The trilinear tensor has been well known in disguise in the context of Euclidean line correspondences and was not identified at the time as a tensor but as a collection of three matrices (a particular contraction of the tensor as we shall see later) [25,26,32]. The link between the two and the generalization to projective space was identified later by Hartley [10,11].

Before we delve further on the properties of the trilinear tensor, we can readily identify the first application — called *image transfer* in Photogrammetric circles or a.k.a *image reprojection*. Image transfer is the task of predicting the location of $p''$ from the corresponding pair $p, p'$ given a small number of basis matching triplets $p_i, p'_i, p''_i$. This task can be readily achieved using the geometry of two views, simply by intersecting epipolar lines (we will later discuss these concepts) — as long as the three camera centers are not on a line, however. With the trilinearities one can achieve a general result:

**Proposition 1.** *A triplet of points* $p, p', p''$ *is in correspondence iff the four trilinear constraints are satisfied.*

The implication is simple: take 7 triplets $p_i, p'_i, p''_i$, $i = 1, ..., 7$ and recover linearly the coefficients of the tensor (for each $i$ we have 4 linearly independent equations for the tensor). For any new pair $p, p'$ extract the coordinates of $p''$ directly from the trilinearities. This will always work without singularities. In practice, due to errors in image measurements and outliers, one uses more advanced techniques for recovering the tensor (cf. [3,16,5]) and exploits further algebraic constraints among its coefficients [9].

# 6    Properties of the Tensor

The first striking property of the tensor is that it is an object that operates on both point and line correspondences. This becomes readily apparent from Equation 13 that simply tells us that the tensor operates on a point $p$, on a line passing through $p'$, and on a line passing through $p''$. To see why this is so, consider $s_j^\mu p'^j = 0$ which means that $s_j^1$ and $s_j^2$ are two lines coincident with $p'$ (lines and points in projective plane are duals of one another, thus their scalar product vanishes when they are coincident). Since any line $s_j$ passing through $p'$ can be described as a linear combination of the lines $s_j^1$ and $s_j^2$, and any linear combination of two trilinearities is also a trilinearity (i.e. vanishes on $p, p', p''$), and since the same argument holds for $r_k^1$ and $r_k^2$, we have that:

$$p^i s_j r_k \mathcal{T}_i^{jk} = 0 \qquad (14)$$

where $s_j$ is *some* line through $p'$ and $r_k$ is *some* line through $p''$. In other words,

**Proposition 2.** *A trilinearity represents a correspondence set of a point in the reference image and two lines (not necessarily corresponding) passing through the matching points in the remaining two images, i.e., is a point-line-line configuration. Analogously, in 3D space the configuration consists of a line-plane-plane, where the line is the optical ray of the reference image and the planes are defined by the optical centers and the image lines mentioned above.*

Figure 2 provides a pictorial description of the geometry represented by a trilinearity. The lines in the four trilinearities in Equation 13 are simply the horizontal and vertical scan lines of the image planes — we will call this representation of the trilinearities the *canonical* representation because with it each trilinearity is represented by the minimal number of non-vanishing coefficients (12 instead of

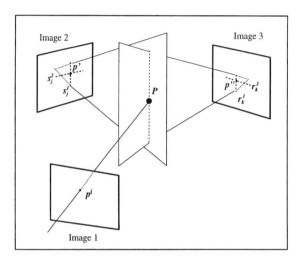

**Fig. 2.** Each of the four trilinear equations describes a matching between a point $p$ in the first view, some line $s_j^\mu$ passing through the matching point $p'$ in the second view and some line $r_k^\rho$ passing through the matching point $p''$ in the third view. In space, this constraint is a meeting between a ray and two planes (Figure adopted from [2]).

27). The line-plane-plane description was first introduced by Faugeras & Mourrain [8] using Grassmann-Cayley algebra, and the analogous point-line-line description was introduced earlier in the context of Euclidean motion by Spetsakis & Aloimonos [26]. The derivation above, however, is the most straightforward one because it simply comes for free by observing how Equation 13 is organized. Finally, by similar observation one can see that a triplet of matching lines provides 2 trilinearities[2], thus 13 triplets of matching lines are sufficient for solving (linearly) for the tensor.

Before we continue further consider the applications of the point-line-line property. The first application is straightforward, whereas the second requires some elaboration. Consider a polyhedral object, like a rooftop of a house. Say the corner of the roof is visible in the first image, but is occluded in the remaining images (second and third). Thus the image measurements consist of a point-line-line arrangement, and is sufficient for providing a constraint for camera motion (the tensor). In other words, using the remarkable property of the tensor operating on both points and lines one can enrich the available feature space significantly (see for example, [3] for an application of this nature).

The second application, naturally related, is the issue of estimating Structure and Motion directly from image spatio-temporal derivatives, rather than through explicit correspondence set (points or lines). For example, the trilinear constraint (Equation 14) can be replaced by a "model-based brightness constraint" by

---

[2] two contractions with covariant vectors leaves us with a covariant vector, thus three matching lines provide two linear equations for the tensor elements.

having the lines $s_j$ and $r_k$ become:

$$s_j = \begin{pmatrix} -I_x \\ -I_y \\ I'_t + xI_x + yI_y \end{pmatrix} \qquad r_k = \begin{pmatrix} -I_x \\ -I_y \\ I''_t + xI_x + yI_y \end{pmatrix} \qquad (15)$$

where $I_x, I_y$ are the spatial derivatives at location $(x, y)$ and $I'_t, I''_t$ are the temporal derivatives (the image difference) between the reference image and image two and three, respectively. Hence, every pixel with a non-vanishing gradient contributes one linear constraint for camera motion. Stein & Shashua [27] provide the details and an elaborate experimental setup and also show that there are a few subtleties (and open problems) that make a successful implementation of "direct estimation" quite challenging.

## 6.1 Tensor Contractions

We have discussed so far the manner by which the tensor operates on geometrical entities of points and lines and the applications arising from it. We now turn our attention to similar properties of *subsets* of the tensor arising from contractions of the tensor to bivalent tensors (matrices). There are two kinds of contractions, the first yielding the well known three matrices of line geometry, and the second provides something new in the form of homography matrices. We will start with the latter contraction.

Consider the matrix arising from the contraction,

$$\delta_k T_i^{jk} \qquad (16)$$

which is a $3 \times 3$ matrix, we denote by $E$, obtained by the linear combination $E = \delta_1 T_i^{j1} + \delta_2 T_i^{j2} + \delta_3 T_i^{j3}$ (which is what is meant by a contraction), and $\delta_k$ is an *arbitrary* covariant vector. Clearly, if $\delta_k = r_k$ then $E$ maps $p$ onto $p'$ because $p^i r_k T_i^{jk} \cong p'^j$ (or $Ep \cong p'$). The question of interest, therefore, is whether $E$ has any *general* meaning? The answer is affirmative with the details described below.

Recall that the projection of the space point $P$ onto the second image satisfies $p' \cong TP$ where $T = [A; v']$. Let the three camera centers be denoted by $C, C', C''$ of the first, second and third views respectively, i.e., $TC' = 0$. The tensor operates on the ray $\overline{CP}$ and two planes one for each image. For the second image, choose the plane $CC'P$, known as the epipolar plane, which is the plane passing through the two camera centers and the space point $P$. This plane intersects the second image at a point, known as the epipole, which is exactly $v'$. Clearly, the line $s_j$ is simply the epipolar line $p' \times v'$ defined by the vector product of $p'$ and $v'$ (see Figure 3). In the third image, since $\delta_k$ is arbitrary, we have a plane that does not contain $p''$. Let the plane defined by the point $C''$ and the line $\delta_k$ in the third image plane be denoted by $\pi$. Since $\pi$ does not necessarily contain $p''$, then the intersection of $\pi$ with the epipolar plane $CC'P$ is some point $\tilde{P}$ on the ray $\overline{CP}$. Clearly, the projection of $\tilde{P}$ onto the second

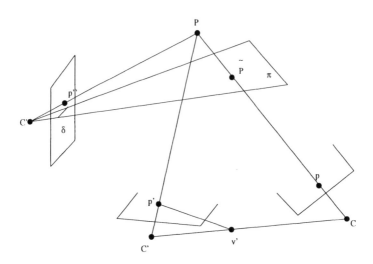

**Fig. 3.** The contraction $\delta_k \mathcal{T}_i^{jk}$ is a homography matrix due to the plane $\pi$ determined by the center $C''$ and the line $\delta_k$. See text for details.

image is a point $\tilde{p}'$ on the epipolar line (the epipolar line is the projection of the ray $\overline{CP}$ onto the second image). Hence,

$$p^i \delta_k \mathcal{T}_i^{jk} \cong \tilde{p}'^j \tag{17}$$

or $Ep \cong \tilde{p}'$. In other words, the matrix $E$ is a 2D projective transformation (a *homography*) from the first to the second image planes via the plane $\pi$, i.e., the concatenation of the mappings (i) from first image onto $\pi$, and (ii) the mapping from $\pi$ onto the second image. Stated formally,

**Proposition 3.** *The contraction $\delta_k \mathcal{T}_i^{jk}$ for some arbitrary $\delta_k$ is a homography matrix from image one onto image two determined by the plane containing the third camera center $C''$ and the line $\delta_k$ in the third image plane. Generally, the rank of $E$ is 3. Likewise, the contraction $\delta_j \mathcal{T}_i^{jk}$ is a homography matrix between the reference image and the third image.*

Clearly, one can generate up to three distinct homography matrices because $\delta_k$ is spanned by three covariant vectors. Let the *standard* contractions be identified by selecting $\delta_k$ be $(1, 0, 0)$ or $(0, 1, 0)$ or $(0, 0, 1)$, thus the three homography matrices are $\mathcal{T}_i^{j1}, \mathcal{T}_i^{j2}$ and $\mathcal{T}_i^{j3}$, and we denote them by $E_1, E_2, E_3$ respectively. Thus, $E_1, E_2$ are associated with the planes passing through the horizontal and vertical scan-lines around the origin $(0, 0)$ of the third image (and of course containing the center $C''$), and $E_3$ is associated with the plane parallel to the third image plane. The matrices $E_1, E_2, E_3$ were first introduced by Shashua & Werman [24] where further details can be found therein.

The applications of the standard homography contractions include 3D reconstruction of structure and camera motion. The camera motion is simply

$T = [E; \boldsymbol{v'}]$ where $E$ is one of the standard contractions or a linear combination of them (we will discuss the recovery of $\boldsymbol{v'}$ later). Similarly, any two homography matrices, say $E_1, E_2$ define a projective invariant $\kappa$ defined by,

$$p' \cong E_1 p + \kappa E_2 p.$$

More details on 3D reconstruction from homography matrices can be found in [17,23]. We will encounter further applications of the standard homography contractions later in the paper.

Finally, the contractions $\mathcal{T}_1^{jk}, \mathcal{T}_2^{jk}$ and $\mathcal{T}_3^{jk}$ are the three matrices used by [25,32] to study the structure from motion problem from line correspondences (see [11], for more details).

## 6.2 The Bilinear Constraint

We wish to reduce the discussion back to the context of two views. We saw in Section 3 that the bilinear and trilinear constraints all arise from the same principle of vanishing determinants of $4 \times 4$ minors of $M$. The question of interest is what form do the coefficients of the bilinear constraint take, and how is that related to the trilinear tensor?

Starting from Equation 12, repeated for convenience below,

$$\mathcal{T}_i^{jk} = v'^j b_i^k - v''^k a_i^j$$

we will consider the case of two views as a *degenerate* instance of $\mathcal{T}_i^{jk}$ in the following way. Instead of three distinct images, we have two distinct images and the third image is a replica of the second image. Thus, the two camera matrices are $[A; \boldsymbol{v'}]$ and again $[A; \boldsymbol{v'}]$. Substituting $A$ instead of $B$ and $\boldsymbol{v'}$ instead of $\boldsymbol{v''}$ in Equation 12, we obtain a new trivalent tensor of the form:

$$\mathcal{F}_i^{jk} = v'^j a_i^k - v'^k a_i^j. \tag{18}$$

The tensor $\mathcal{F}_i^{jk}$ follows the same contraction properties as $\mathcal{T}_i^{jk}$. For example, the point-line-line property is the same with the exception that the two lines are in the same image:

$$p^i s_j' s_k'' \mathcal{F}_i^{jk} = 0,$$

where $s_j'$ and $s_k''$ are any two lines (say the horizontal and vertical scan lines) that intersect at $p'$. The standard contractions apply here as well: $\delta_k \mathcal{F}_i^{jk}$ is a homography matrix from image one onto image two due to the plane $\pi$ that passes through the camera center $C'$ and the line $\delta_k$ in the second image — but now, since $\pi$ contains $C'$, it is a rank 2 homography matrix instead of rank 3.

In closer inspection one can note that 9 of the 27 elements of $\mathcal{F}_i^{jk}$ vanish and the remaining 18 are divided into two sets which differ only in sign, i.e., 9 of those elements can be arranged in a matrix $F$ and the other 9 in $-F$, where $F$ satisfies $p'^\top F p = 0$ and $F = [\boldsymbol{v'}]_x A$ where $[\boldsymbol{v'}]_x$ is the skew-symmetric matrix of vector products ($[\boldsymbol{v'}]_x u = \boldsymbol{v'} \times u$); and the contraction $\delta_k \mathcal{F}_i^{jk}$ is the matrix $[\delta]_x F$.

The matrix $F$ is known as the "Fundamental" matrix [15,6], and the constraint $p'^\top F p = 0$ is the (and only) bilinear constraint. Further details on $\mathcal{F}_i^{jk}$ and its properties can be found in [1].

Finally, given the three standard homography matrices, $E_1, E_2, E_3$, one can readily obtain $F$ from the following constraint:

$$E_j^\top F + F^\top E_j = 0$$

which yields 18 linear equations of rank 8 for $F$. Similarly, cross products between columns of two homographies provide epipolar lines which can be used to recover the epipole $v'$ — or simply recover $F$ and then $F^\top v' = 0$ will provide a solution for $v'$.

# 7 Discussion

We have presented the foundations for a coherent and integrated treatment of Multiple View Geometry whose main analysis vehicle is the "trilinear tensor" which captures in a very simple and straightforward manner the basic structures associated with this problem of research.

We have left several issues out of the scope of this paper, and some issues are still an open problem. The issues we left out include (i) uniqueness of solution — the issue of critical configurations [22], (ii) properties of the tensor manifold — relation among tensors across many views, tensorial operators and applications for rendering [21,2], and (iii) Shape Constraints which are the dual of the multilinear matching constraints [31,4,13]. Issues that are still open include the tensorial structure behind the quadlinear matching constraints, the tensor governing the shape constraints and its properties.

# References

1. S. Avidan and A. Shashua. Unifying two-view and three-view geometry. Technical report, Hebrew University of Jerusalem, November 1996.
2. S. Avidan and A. Shashua. View synthesis in tensor space. In *Proceedings of the IEEE Conference on Computer Vision and Pattern Recognition*, Puerto Rico, June 1997.
3. P. Beardsley, P. Torr, and A. Zisserman. 3D model acquisition from extended image sequences. In *Proceedings of the European Conference on Computer Vision*, April 1996.
4. S. Carlsson. Duality of reconstruction and positioning from projective views. In *Proceedings of the workshop on Scene Representations*, Cambridge, MA., June 1995.
5. R. Deriche, Z. Zhang, Q.T. Luong, and O.D. Faugeras. Robust recovery of the epipolar geometry for an uncalibrated stereo rig. In *Proceedings of the European Conference on Computer Vision*, pages 567–576, Stockholm, Sweden, May 1994. Springer-Verlag, LNCS 800.

6. O.D. Faugeras. What can be seen in three dimensions with an uncalibrated stereo rig? In *Proceedings of the European Conference on Computer Vision*, pages 563–578, Santa Margherita Ligure, Italy, June 1992.

7. O.D. Faugeras. Stratification of three-dimensional vision: projective, affine and metric representations. *Journal of the Optical Society of America*, 12(3):465–484, 1995.

8. O.D. Faugeras and B. Mourrain. On the geometry and algebra of the point and line correspondences between N images. In *Proceedings of the International Conference on Computer Vision*, Cambridge, MA, June 1995.

9. O.D. Faugeras and T. Papadopoulo. A nonlinear method for estimating the projective geometry of three views. Submitted, June 1997.

10. R. Hartley. Lines and points in three views — a unified approach. In *Proceedings of the ARPA Image Understanding Workshop*, Monterey, CA, November 1994.

11. R. Hartley. A linear method for reconstruction from lines and points. In *Proceedings of the International Conference on Computer Vision*, pages 882–887, Cambridge, MA, June 1995.

12. A. Heyden. Reconstruction from image sequences by means of relative depths. In *Proceedings of the International Conference on Computer Vision*, pages 1058–1063, Cambridge, MA, June 1995.

13. M. Irani and P. Anandan. Parallax geometry of pairs of points for 3D scene analysis. In *Proceedings of the European Conference on Computer Vision*, LNCS 1064, pages 17–30, Cambridge, UK, April 1996. Springer-Verlag.

14. D.W. Jacobs. Matching 3D models to 2D images. *International Journal of Computer Vision*, 21(1/2):123–153, January 1997.

15. H.C. Longuet-Higgins. A computer algorithm for reconstructing a scene from two projections. *Nature*, 293:133–135, 1981.

16. Torr P.H.S., Zisserman A., and Murray D. Motion clustering using the trilinear constraint over three views. In *Workshop on Geometrical Modeling and Invariants for Computer Vision*. Xidian University Press., 1995.

17. A. Shashua. Projective structure from uncalibrated images: structure from motion and recognition. *IEEE Transactions on Pattern Analysis and Machine Intelligence*, 16(8):778–790, 1994.

18. A. Shashua. Algebraic functions for recognition. *IEEE Transactions on Pattern Analysis and Machine Intelligence*, 17(8):779–789, 1995.

19. A. Shashua. Trilinear tensor: The fundamental construct of multiple-view geometry and its applications. Submitted for journal publication, June 1997.

20. A. Shashua and P. Anandan. The generalized trilinear constraints and the uncertainty tensor. In *Proceedings of the ARPA Image Understanding Workshop*, Palm Springs, CA, February 1996.

21. A. Shashua and S. Avidan. The rank4 constraint in multiple view geometry. In *Proceedings of the European Conference on Computer Vision*, Cambridge, UK, April 1996.

22. A. Shashua and S.J. Maybank. Degenerate $n$ point configurations of three views: Do critical surfaces exist? Technical Report TR 96-19, Hebrew University of Jerusalem, November 1996.

23. A. Shashua and N. Navab. Relative affine structure: Canonical model for 3D from 2D geometry and applications. *IEEE Transactions on Pattern Analysis and Machine Intelligence*, 18(9):873–883, 1996.

24. A. Shashua and M. Werman. Trilinearity of three perspective views and its associated tensor. In *Proceedings of the International Conference on Computer Vision*, June 1995.

25. M.E. Spetsakis and J. Aloimonos. Structure from motion using line correspondences. *International Journal of Computer Vision*, 4(3):171–183, 1990.

26. M.E. Spetsakis and J. Aloimonos. A unified theory of structure from motion. In *Proceedings of the ARPA Image Understanding Workshop*, 1990.

27. G. Stein and A. Shashua. Model based brightness constraints: On direct estimation of structure and motion. In *Proceedings of the IEEE Conference on Computer Vision and Pattern Recognition*, Puerto Rico, June 1997.

28. C. Tomasi and T. Kanade. Shape and motion from image streams – a factorization method. *International Journal of Computer Vision*, 9(2):137–154, 1992.

29. B. Triggs. Matching constraints and the joint image. In *Proceedings of the International Conference on Computer Vision*, pages 338–343, Cambridge, MA, June 1995.

30. S. Ullman and R. Basri. Recognition by linear combination of models. *IEEE Transactions on Pattern Analysis and Machine Intelligence*, PAMI-13:992—1006, 1991. Also in M.I.T AI Memo 1052, 1989.

31. D. Weinshall, M. Werman, and A. Shashua. Duality of multi-point and multi-frame geometry: Fundamental shape matrices and tensors. In *Proceedings of the European Conference on Computer Vision*, LNCS 1065, pages 217–227, Cambridge, UK, April 1996. Springer-Verlag.

32. J. Weng, T.S. Huang, and N. Ahuja. Motion and structure from line correspondences: Closed form solution, uniqueness and optimization. *IEEE Transactions on Pattern Analysis and Machine Intelligence*, 14(3), 1992.

# Algebraic and Geometric Tools to Compute Projective and Permutation Invariants

Gabriella CSURKA[1] and Olivier FAUGERAS[2]

[1] INRIA Rhône-Alpes, 655 Av. de l'Europe, 38330 Montbonnot Saint Martin, France
[2] INRIA Sophia-Antipolis, 2004 Route des Lucioles, 06902 Sophia Antipolis, France

**Abstract.** This paper studies the computation of projective invariants in pairs of images from uncalibrated cameras, and presents a detailed study of the projective and permutation invariants for configurations of points and/or lines. We give two basic computational approaches, one algebraic and one geometric, and also the relations between the invariants computed by different approaches. In each case, we show how to compute invariants in projective space assuming that the points and lines have already been reconstructed in an arbitrary projective basis, and also, how to compute them directly from image coordinates in a pair of views using only point and line correspondences and the fundamental matrix. Finally, we develop combinations of those projective invariants which are insensitive to permutations of the geometric primitives of each of the basic configurations.

## Introduction

Various visual or visually-guided robotics tasks may be carried out using only a projective representation which show the importance of projective informations at different steps in the perception-action cycle. We can mention here the obstacle detection and avoidance [13], goal position prediction for visual servoing [7] or 3D object tracking [12]. More recently it has been shown, both theoretically and experimentally, that under certain conditions an image sequence taken with an uncalibrated camera can provide 3-D Euclidean structure as well. The latter paradigm consists in recovering projective structure first and then upgrading it into Euclidean structure [14,4]. Additionally, we believe that computing structure without explicit camera calibration is more robust than using calibration because we need not make any (possibly incorrect) assumptions about the Euclidean geometry (remembering that calibration is itself often erroneous).

All these show the importance of the projective geometry as well in computer vision than in robotics, and the various applications show that projective informations can be useful at different steps in the perception-action cycle. Still the study of every geometry is based on the study of properties which are invariant under the corresponding group of transformations, the projective geometry is characterized by the projective invariants.

This paper is dedicated to the study of the various configurations of points and/or lines in 3D space; it gives algebraic and geometric methods to compute projective invariants in the space and/or directly from image measurements. In the 3D case, we will suppose that we have an arbitrary three-dimensional projective representation of the object obtained by explicit projective reconstruction.

In the image case, we will suppose that the only information we have is the image measurements and the epipolar geometry of the views and we will compute three-dimensional invariants without any explicit reconstruction.

First we show that arbitrary configurations of points and/or lines can be decomposed into minimal sub-configurations with invariants, and these invariants characterize the the original configuration. This means that it is sufficient to study only these minimal configurations (six points, four points and a line, three points and two lines, and four lines). For each configuration we show how to compute invariants in 3D projective space and in the images, using both algebraic and geometric approaches.

As these invariants generally depend on the order of the points and lines, we will also look for features which are both projective and permutation invariants.

## 1   Projective Invariants

**Definition 1.** *Suppose that* $\mathbf{p}$ *is a vector of parameters characterizing a geometric configuration and* $\mathcal{T}$ *is a group of linear transformations, such that* $\rho \mathbf{y} = \mathbf{T} \mathbf{x}$, $\mathbf{T} \in \mathcal{T}$, *where* $\mathbf{x}$ *are* $\mathbf{y}$ *are vectors of homogeneous coordinates. A function* $I(\mathbf{p})$ *is invariant under the action of the group* $\mathcal{T}$ *if* $I(\mathbf{T}(\mathbf{p})) = I(\mathbf{p})$, $\forall \mathbf{T} \in \mathcal{T}$, *where* $I(\mathbf{T}(\mathbf{p}))$ *is the value of* $I$ *after the transformation* $\mathbf{T}$.

If $I_1(\mathbf{p}), \ldots I_n(\mathbf{p})$ are $n$ invariants, any $f(\mathbf{p}) = f(I_1(\mathbf{p}), \ldots I_n(\mathbf{p}))$ is also invariant. So if we have several invariants, it is possible that not all of them are functionally independent. The maximum number of independent invariants for a configuration is given by the following proposition [5,6]:

**Proposition 2.** *If* $\mathcal{S}$ *is the space parameterizing a given geometric configuration (for example six points, four lines) and* $\mathcal{T}$ *is a group of linear transformations, the number of functionally independent invariants of a configuration* $\mathbf{p}$ *of* $\mathcal{S}$ *under the transformations of* $\mathcal{T}$ *is:*

$$N_{inv} = dim(\mathcal{S}) - (dim(\mathcal{T}) - min_{\mathbf{p} \in \mathcal{S}}(dim(\mathcal{T}_{\mathbf{p}}))), \qquad (1)$$

*where* $\mathcal{T}_{\mathbf{p}}$ *is the isotropy sub-group of the configuration* $\mathbf{p}$ *defined as* $\mathcal{T}_{\mathbf{p}} = \{\mathbf{T} \in \mathcal{T} \mid \mathbf{T}(\mathbf{p}) = \rho \mathbf{p}\}$.

Generally, $min_{\mathbf{p} \in \mathcal{S}}(dim(\mathcal{T}_{\mathbf{p}})) = 0$. However, certain types of configurations have non trivial isotropy sub-groups. For example, for Euclidean transformations, the distance between two 3D points is an invariant and $dim(\mathcal{S}) - dim(\mathcal{T}) = 3 + 3 - 6 = 0$. Consequently, $min_{\mathbf{p} \in \mathcal{S}}(dim(\mathcal{T}_{\mathbf{p}})) \neq 0$. Indeed, the sub-group of rotations about the axis defined by two points, leaves both of the points fixed.

The most commonly studied transformation groups in computer vision are the Euclidean, affine and projective transformations. Since we want to work with weakly calibrated cameras, we will use projective transformations. As projective transformations of 3D space have $16 - 1 = 15$ parameters, points 3 and lines 4, using proposition **2**, we can easily see that we need at least six points, four points and a line, two points and three lines, three points and two lines or four lines to produce some invariants. We will say that these configurations are *minimal*.

Taking a non-minimal configuration of points and lines, we can decompose it into several minimal configurations. It is easy to show that the invariants of the sub configurations characterize the invariants of the original configuration. This means that we only need to be able to compute invariants for minimal configurations. For example, consider a configuration of seven points denoted $\mathbf{M}_{i,i=1..7}$. From proposition **2** there are $3 \times 7 - 15 = 6$ independent invariants. To obtain a set of six independent invariants characterizing the configuration it is sufficient to compute the three independent invariants $\lambda_{i,i=1..3}$ of the configuration $\mathbf{M}_{i,i=1..6}$ and the three independent invariants $\mu_{i,i=1..3}$ of the configuration $\mathbf{M}_{i,i=2..7}$.

We only discuss configurations of points and lines, not planes. Invariants of configurations of planes and/or lines can be computed in the same way as those of points and/or lines, by working in the dual space [11,2]. For the same reasons, we do not need to consider in detail configurations of two points and three lines. Indeed, these configurations define six planes which correspond to configurations of six points in the dual space [2]. The other four minimal configurations we will study in detail are: six points, four points and a line , three points and two lines and four lines.

## 1.1 Projective Invariants Using Algebraic Approach

Consider eight points in space represented by their homogeneous coordinates $\mathbf{M}_{i,i=1..8} = (x_i, y_i, z_i, t_i)^{\mathsf{T}}$, and compute the following ratio of determinants:

$$I = \frac{[\mathbf{M}_1\,\mathbf{M}_2\,\mathbf{M}_3\,\mathbf{M}_4][\mathbf{M}_5\,\mathbf{M}_6\,\mathbf{M}_7\,\mathbf{M}_8]}{[\mathbf{M}_{\dot{1}}\,\mathbf{M}_{\dot{2}}\,\mathbf{M}_{\dot{3}}\,\mathbf{M}_{\dot{4}}][\mathbf{M}_{\dot{5}}\,\mathbf{M}_{\dot{6}}\,\mathbf{M}_{\dot{7}}\,\mathbf{M}_{\dot{8}}]}, \tag{2}$$

where $\dot{k}$ denotes the value $\sigma(k)$ for an arbitrary permutation $\sigma$ of $\{1, 2, \ldots, 8\}$.

The invariant $I$ can be computed also from a pair of images using only image measurements and the fundamental matrix using the Grassmann-Cayley, also called the double algebra as below [1,3]:

$$I = \frac{(\beta_{12-34}^{\mathsf{T}}\mathbf{F}\alpha_{12-34})(\beta_{56-78}^{\mathsf{T}}\mathbf{F}\alpha_{56-78})(\overline{\beta_{\dot{1}}^{\mathsf{T}}\mathbf{F}\alpha_{\dot{2}}})(\overline{\beta_{\dot{3}}^{\mathsf{T}}\mathbf{F}\alpha_{\dot{4}}})(\overline{\beta_{\dot{5}}^{\mathsf{T}}\mathbf{F}\alpha_{\dot{6}}})(\overline{\beta_{\dot{7}}^{\mathsf{T}}\mathbf{F}\alpha_{\dot{8}}})}{(\beta_{\dot{1}\dot{2}-\dot{3}\dot{4}}^{\mathsf{T}}\mathbf{F}\alpha_{\dot{1}\dot{2}-\dot{3}\dot{4}})(\beta_{\dot{5}\dot{6}-\dot{7}\dot{8}}^{\mathsf{T}}\mathbf{F}\alpha_{\dot{5}\dot{6}-\dot{7}\dot{8}})(\overline{\beta_1^{\mathsf{T}}\mathbf{F}\alpha_2})(\overline{\beta_3^{\mathsf{T}}\mathbf{F}\alpha_4})(\overline{\beta_5^{\mathsf{T}}\mathbf{F}\alpha_6})(\overline{\beta_7^{\mathsf{T}}\mathbf{F}\alpha_8})} \tag{3}$$

where, $\alpha_{i_1i_2-i_3i_4} = \alpha_{i_1}\alpha_{i_2} \wedge \alpha_{i_3}\alpha_{i_4}$, $\beta_{i_1i_2-i_3i_4} = \beta_{i_1}\beta_{i_2} \wedge \beta_{i_3}\beta_{i_4}$ and $(\overline{\beta_i^{\mathsf{T}}\mathbf{F}\alpha_j})$ stands for the expression $sign(\beta_j^{\mathsf{T}}\mathbf{F}\alpha_i)\sqrt{-(\beta_j^{\mathsf{T}}\mathbf{F}\alpha_i)(\beta_i^{\mathsf{T}}\mathbf{F}\alpha_j)}$.

Note that if we change the bases in the images such that $\alpha' = \mathbf{H}_1\alpha$ and $\beta' = \mathbf{H}_2\beta$ are the image coordinates in the new bases, the corresponding fundamental matrix is $\mathbf{F}' = \mathbf{H}_2^{-\mathsf{T}}\mathbf{F}\mathbf{H}_1^{-1}$, so the quantities $\beta'^{\mathsf{T}}_j\mathbf{F}'\alpha'_i = \beta_j^{\mathsf{T}}\mathbf{H}_2\mathbf{H}_2^{-\mathsf{T}}\mathbf{F}\mathbf{H}_1^{-1}\mathbf{H}_1\alpha_i = \beta_j^{\mathsf{T}}\mathbf{F}\alpha_i$ are independent of the bases chosen.

## 1.2 Six Point Configurations

Now consider a configuration of six points $\mathbf{A}_{i,i=1..6}$ in 3D projective space. From proposition **2**, there are 3 independent invariants for this configuration. Using

(2) and (3) we can deduce the following three invariants:

$$I_1 = \frac{[\mathbf{A_1\,A_3\,A_2\,A_5}][\mathbf{A_3\,A_4\,A_2\,A_6}]}{[\mathbf{A_1\,A_3\,A_2\,A_6}][\mathbf{A_3\,A_4\,A_2\,A_5}]} = \frac{(\beta_{13-25}^{\mathsf{T}}\mathbf{F}\alpha_{13-25})(\beta_{34-26}^{\mathsf{T}}\mathbf{F}\alpha_{34-26})}{(\beta_{13-26}^{\mathsf{T}}\mathbf{F}\alpha_{13-26})(\beta_{34-25}^{\mathsf{T}}\mathbf{F}\alpha_{34-25})}$$

$$I_2 = \frac{[\mathbf{A_1\,A_2\,A_3\,A_5}][\mathbf{A_1\,A_4\,A_3\,A_6}]}{[\mathbf{A_1\,A_2\,A_3\,A_6}][\mathbf{A_1\,A_4\,A_3\,A_5}]} = \frac{(\beta_{12-35}^{\mathsf{T}}\mathbf{F}\alpha_{12-35})(\beta_{14-36}^{\mathsf{T}}\mathbf{F}\alpha_{14-36})}{(\beta_{12-36}^{\mathsf{T}}\mathbf{F}\alpha_{12-36})(\beta_{14-35}^{\mathsf{T}}\mathbf{F}\alpha_{14-35})}$$

$$I_3 = \frac{[\mathbf{A_2\,A_3\,A_1\,A_5}][\mathbf{A_2,A_4,A_1,A_6}]}{[\mathbf{A_2\,A_3\,A_1\,A_6}][\mathbf{A_2,A_4,A_1,A_5}]} = \frac{(\beta_{23-15}^{\mathsf{T}}\mathbf{F}\alpha_{23-15})(\beta_{24-16}^{\mathsf{T}}\mathbf{F}\alpha_{24-16})}{(\beta_{23-16}^{\mathsf{T}}\mathbf{F}\alpha_{23-16})(\beta_{24-15}^{\mathsf{T}}\mathbf{F}\alpha_{24-15})}$$

To show that they are independent, change coordinates so that $\mathbf{A}_{i,i=1..5}$ become a standard basis. Denote the homogeneous coordinates of $\mathbf{A}_6$ in this basis by $(p, q, r, s)^{\mathsf{T}}$. Computing $I_{j,j=1..3}$ in this basis we obtain $I_1 = \frac{p}{s}$, $I_2 = \frac{q}{s}$ and $I_3 = \frac{r}{s}$. These invariants are clearly independent.

Alternatively, one can also take a geometric approach to compute six point invariants. The basic idea is to construct cross ratios using the geometry of the configuration. In this case we will give two different methods that compute independent projective invariants.

The first method constructs a pencil of planes using the six points. Taking two of the points, for example $\mathbf{A}_1$ and $\mathbf{A}_2$, the four planes defined by $\mathbf{A}_1, \mathbf{A}_2$ and $\mathbf{A}_{k,k=3..6}$ belong to the pencil of planes through the line $\mathbf{A}_1\mathbf{A}_2$. Their cross ratio is an invariant of the configuration. Taking other pairs of points for the axis of the pencil gives further cross ratios. The relation between these and $I_{j,j=1..3}$ is:

$$\begin{aligned}
\{\mathbf{A}_2\mathbf{A}_3\mathbf{A}_1, \mathbf{A}_2\mathbf{A}_3\mathbf{A}_4; \mathbf{A}_2\mathbf{A}_3\mathbf{A}_5, \mathbf{A}_2\mathbf{A}_3\mathbf{A}_6\} &= I_1 \\
\{\mathbf{A}_1\mathbf{A}_3\mathbf{A}_2, \mathbf{A}_1\mathbf{A}_3\mathbf{A}_4; \mathbf{A}_1\mathbf{A}_3\mathbf{A}_5, \mathbf{A}_1\mathbf{A}_3\mathbf{A}_6\} &= I_2 \quad\quad (4)\\
\{\mathbf{A}_1\mathbf{A}_2\mathbf{A}_3, \mathbf{A}_1\mathbf{A}_2\mathbf{A}_4; \mathbf{A}_1\mathbf{A}_2\mathbf{A}_5, \mathbf{A}_1\mathbf{A}_2\mathbf{A}_6\} &= I_3
\end{aligned}$$

The second method [5] consists in constructing six coplanar points from the six general ones and computing invariants in the projective plane. For example, if take the plane $\mathbf{A}_1\mathbf{A}_2\mathbf{A}_3$ and cut it by the three lines $\mathbf{A}_4\mathbf{A}_5$, $\mathbf{A}_4\mathbf{A}_6$ and $\mathbf{A}_5\mathbf{A}_6$ obtaining the intersections $\mathbf{M}_1$, $\mathbf{M}_2$ and $\mathbf{M}_3$ coplanar with $\mathbf{A}_{i,i=1.3}$. Five coplanar points, for example $\mathbf{A}_{i,i=1..3}, \mathbf{M}_{i,i=1,2}$, give two cross ratios $\lambda_1, \lambda_2$. Any other set of five, for example $\mathbf{A}_{i,i=1..3}, \mathbf{M}_{i=1,3}$, gives two further cross ratios $\lambda_3, \lambda_4$ but only three of the four cross ratios are independent. Indeed $(\lambda_4 - \lambda_2)(\lambda_3 - 1) - (\lambda_3 - \lambda_1)(\lambda_4 - 1) = 0$. The relation between $\lambda_{i,i=1..3}$ and $I_{j,j=1..3}$ are $\lambda_1 = \frac{I_1}{I_3}$, $\lambda_2 = \frac{I_2}{I_3}$ and $\lambda_3 = \frac{I_1-1}{I_3-1}$.

So we have several methods of computing geometric invariants in 3D space. To do this, we need an arbitrary projective reconstruction of the points. However, the invariants can be also be computed directly from the images by using the fact that the cross ratio is preserved under perspective projections.

First consider the case of the pencil of planes. We know that the cross ratio of four planes of a pencil is equal to the cross ratio of their four points of intersections with an arbitrary transversal line. So if we are able to find the image of the intersection point of a line and a plane we can also compute the required cross ratio in the image. The coplanar point method uses the same principle, computing the intersection of a line and a plane from image measurements.

We want to compute the images of the intersection of a line $\mathbf{A}_3\mathbf{A}_4$ and a plane $\mathbf{A}_1\mathbf{A}_2\mathbf{A}_5$ given only the projections of the five points in two images $\mathbf{a}_{i,i=1..5}$ and

$\mathbf{a}'_{i,i=1..5}$ and the fundamental matrix between the two images. Take a point $\mathbf{p}$ in the first image and a point $\mathbf{p}'$ in the second one. These points are the projections we are looking for if and only if $\mathbf{p}' = \mathbf{Fp} \times (\mathbf{a}'_3 \times \mathbf{a}'_4)$ and $p$ verify (details in [2]):

$$
\begin{cases}
\mathbf{p} \cdot (\mathbf{a}_3 \times \mathbf{a}_4) = 0 \\
\left((\mathbf{a}'_2 \times \mathbf{a}'_5) \times (\mathbf{a}'_1 \times (\mathbf{Fp} \times (\mathbf{a}'_3 \times \mathbf{a}'_4)))\right)^{\mathsf{T}} \mathbf{F}\left((\mathbf{a}_2 \times \mathbf{a}_5) \times (\mathbf{a}_1 \times \mathbf{p})\right) = 0
\end{cases}
$$

The first equation is linear. The second one is quadratic, but it is shown in [2] that it can be decomposed into two linear components, one of which is irrelevant (zero only when $\mathbf{p}$ belongs the epipolar line of $\mathbf{a}'_1$). So finally we obtain two linear equations which give the solution for $\mathbf{p}$ and $\mathbf{p}' = \mathbf{Fp} \times (\mathbf{a}'_3 \times \mathbf{a}'_4)$.

Another way to compute the intersection of a plane and a line from image measurements is to use the homography induced between the images by the plane. Let us denote the homography induced by the 3D plane $\mathbf{\Pi}$ by $\mathbf{H}$. The image $\mathbf{n}'$ of the intersection $\mathbf{N}$ of a line $\mathbf{L}$ and the plane $\mathbf{\Pi}$ is then $\mathbf{n}' = \mathbf{l}' \times \mathbf{H}^{-\mathsf{T}}\mathbf{l}$.

To compute the homography $\mathbf{H}$ of the plane $\mathbf{A}_1\mathbf{A}_2\mathbf{A}_5$, we use the fact that $\rho_i\mathbf{a}'_i = \mathbf{H}_5\mathbf{a}_i$ for $i \in \{1,2,5\}$ and $\rho_e\mathbf{e}_{21} = \mathbf{H}_5\mathbf{e}_{12}$. We denote $\mathbf{a}_i = (x_i, y_i, t_i)^{\mathsf{T}}$, $\mathbf{a}'_i = (x'_i, y'_i, t'_i)^{\mathsf{T}}$ for $i \in \{1,2,5\}$, $\mathbf{e}_{12} = (x_e, y_e, t_e)^{\mathsf{T}}$, $\mathbf{e}_{21} = (x'_e, y'_e, t'_e)^{\mathsf{T}}$ and the coefficients of the matrix $\mathbf{H}$ by $h_{ij}$. Then for each $j \in \{1,2,5,e\}$, we have:

$$
\begin{pmatrix}
x_j t'_j & y_j t'_j & t_j t'_j & 0 & 0 & 0 & x_j x'_j & y_j x'_j & t_j x'_j \\
0 & 0 & 0 & x_j t'_j & y_j t'_j & t_j t'_j & x_j y'_j & y_j y'_j & t_j y'_j
\end{pmatrix}
\begin{pmatrix} h_{11} \\ \vdots \\ h_{33} \end{pmatrix} = 0
\tag{5}
$$

As $\mathbf{H}$ is defined only up to a scale factor, it has only eight independent degrees of freedom and can be computed from the eight equations of (5).

## 1.3 Configurations of One Line and Four Points

Denoting the four points by $\mathbf{A}_{i,i=1..4}$ and the line by $\mathbf{L}$ we obtain (cf. (2) and (3)) the following invariant:

$$
I = \frac{[\mathbf{L}\,\mathbf{A}_1\,\mathbf{A}_3][\mathbf{L}\,\mathbf{A}_2\,\mathbf{A}_4]}{[\mathbf{L}\,\mathbf{A}_1\,\mathbf{A}_4][\mathbf{L}\,\mathbf{A}_2\,\mathbf{A}_3]} = \frac{(\beta_{0-13}^{\mathsf{T}}\mathbf{F}\alpha_{0-13})(\beta_{0-24}^{\mathsf{T}}\mathbf{F}\alpha_{0-24})(\beta_1^{\mathsf{T}}\mathbf{F}\alpha_4)(\beta_2^{\mathsf{T}}\mathbf{F}\alpha_3)}{(\beta_{0-14}^{\mathsf{T}}\mathbf{F}\alpha_{0-14})(\beta_{0-23}^{\mathsf{T}}\mathbf{F}\alpha_{0-23})(\beta_1^{\mathsf{T}}\mathbf{F}\alpha_3)(\beta_2^{\mathsf{T}}\mathbf{F}\alpha_4)}
\tag{6}
$$

where $[\mathbf{L}\,\mathbf{A}_i\,\mathbf{A}_j] = [\mathbf{P}\,\mathbf{Q}\,\mathbf{A}_i\,\mathbf{A}_j]$ for $\mathbf{P}$ and $\mathbf{Q}$ arbitrary two distinct points on the line $\mathbf{L}$, $\alpha_i, \beta_i, \mathbf{l}, \mathbf{l}'$ are the projections of $\mathbf{A}_{i,i=1..4}$ and $\mathbf{L}$ in the two images, $\alpha_{0-ij} = \mathbf{l} \times (\alpha_j \times \alpha_i)$ and $\beta_{0-ij} = \mathbf{l}' \times (\beta_j \times \beta_i)$.

Using the geometric approach, the four planes $\mathbf{L}\mathbf{A}_{i,i=1..4}$ belong to a pencil so they define an invariant cross ratio $\lambda$. Another approach, given by Gros in [5] is to consider the four planes defined by the four possible triplets of points and cut the line $\mathbf{L}$ with them. This gives another cross ratio $\lambda'$ for the configuration. Of course, we only have one independent invariant, so there are relations between $I$, $\lambda$ and $\lambda'$. Indeed, we have $I = \lambda$ and $\frac{1}{\lambda} + \frac{1}{\lambda'} = 1$.

The method of computing $\lambda$ and $\lambda'$ directly from the images is basically the same as for the configuration of six points .

### 1.4 Configurations of Three Points and Two Lines

The following two ratios are independent invariants for configurations of three points $\mathbf{A}_{i,i=1..3}$ and two lines $\mathbf{L}_{k,k=1,2}$ in 3D:

$$I_1 = \frac{[\mathbf{L}_1\,\mathbf{A}_1\,\mathbf{A}_2][\mathbf{L}_2\,\mathbf{A}_1\,\mathbf{A}_3]}{[\mathbf{L}_1\,\mathbf{A}_1\,\mathbf{A}_3][\mathbf{L}_2\,\mathbf{A}_1\,\mathbf{A}_2]} = \frac{(\beta_{1-12}^{\mathsf{T}}\mathbf{F}\alpha_{1-12})(\beta_{2-13}^{\mathsf{T}}\mathbf{F}\alpha_{2-13})}{(\beta_{1-13}^{\mathsf{T}}\mathbf{F}\alpha_{1-13})(\beta_{2-12}^{\mathsf{T}}\mathbf{F}\alpha_{2-12})}$$

$$I_2 = \frac{[\mathbf{L}_1\,\mathbf{A}_1\,\mathbf{A}_2][\mathbf{L}_2\,\mathbf{A}_2\,\mathbf{A}_3]}{[\mathbf{L}_1\,\mathbf{A}_2\,\mathbf{A}_3][\mathbf{L}_2\,\mathbf{A}_1\,\mathbf{A}_2]} = \frac{(\beta_{1-12}^{\mathsf{T}}\mathbf{F}\alpha_{1-12})(\beta_{2-23}^{\mathsf{T}}\mathbf{F}\alpha_{2-23})}{(\beta_{1-23}^{\mathsf{T}}\mathbf{F}\alpha_{1-23})(\beta_{2-12}^{\mathsf{T}}\mathbf{F}\alpha_{2-12})}$$

These invariants can also be obtained as follows. Cut the two lines with the plane $\mathbf{A}_1\mathbf{A}_2\mathbf{A}_3$ defined by the three points to give $\mathbf{R}_1$ and $\mathbf{R}_2$. The five coplanar points define a pair of invariants, for example $\lambda_1 = \{\mathbf{A}_1\mathbf{A}_2, \mathbf{A}_1\mathbf{A}_3; \mathbf{A}_1\mathbf{R}_1, \mathbf{A}_1\mathbf{R}_2\} = I_1$ and $\lambda_2 = \{\mathbf{A}_2\mathbf{A}_1, \mathbf{A}_2\mathbf{A}_3; \mathbf{A}_2\mathbf{R}_1, \mathbf{A}_2\mathbf{R}_2\} = I_2$.

Another way to compute a pair of independent invariants for this configuration is to consider the three planes $\mathbf{L}_1\mathbf{A}_{i,i=1..3}$ and the plane $\mathbf{A}_1\mathbf{A}_2\mathbf{A}_3$ and cut them by $\mathbf{L}_2$ This gives a cross ratio $\lambda_1' = \frac{(\lambda_2-1)}{\lambda_2}$. Changing the role of $\mathbf{L}_1$ and $\mathbf{L}_2$ gives another cross ratio $\lambda_2' = \frac{(\lambda_2-1)}{\lambda_1}$. The cross ratios $\lambda_{i,i=1,2}$ and $\lambda_{i,i=1,2}'$ can be computed directly in the images in the same way as the cross ratios of the configuration of six points (finding images of intersections of lines and planes).

### 1.5 Configurations of Four Lines

Consider four lines $\mathbf{L}_{i,i=1..4}$ in 3D projective space. This configuration has 16 parameters, so naively we might expect to have $16 - 15 = 1$ independent invariant. However, the configuration has a 1D isotropy subgroup, so there are actually two independent invariants [6,5,2].

The existence of two independent cross ratios can also be shown geometrically. Assume first that the lines are in general position, in the sense that they are skew and none of them can be expressed as a linear combination of three others. Consider the first three lines $\mathbf{L}_{i,i=1..3}$. As no two of them are coplanar, there is a one parameter family $\mathcal{K}$ of lines meeting all three of them. $\mathcal{K}$ sweeps out a quadric surface in space, ruled by the members of $\mathcal{K}$ and also by a complementary family of generators $\mathcal{L}$, to which $\mathbf{L}_1$, $\mathbf{L}_2$ and $\mathbf{L}_3$ belong [10,6]. Members of $\mathcal{K}$ are mutually skew, and similarly for $\mathcal{L}$, but each member of $\mathcal{K}$ intersect each member of $\mathcal{L}$ exactly once. Another property of generators is that all members of each family can be expressed as a linear combination of any three of them. By our independence assumptions, the fourth line $\mathbf{L}_4$ does not belong to either family (if $\mathbf{L}_4$ belonged to $\mathcal{L}$ it would be a linear combination of the $\mathbf{L}_{i,i=1..3}$ and if $\mathbf{L}_4$ was in $\mathcal{K}$ it would cut each of the lines $\mathbf{L}_{i,i=1..3}$). Hence, $\mathbf{L}_4$ cuts the surface in two real or imaginary points, $\mathbf{A}_4$ and $\mathbf{B}_4$ (these may be identical if $\mathbf{L}_4$ is tangent to the surface). For each point of the surface there is a unique line of each family passing through it. Denote the lines of $\mathcal{K}$ passing through $\mathbf{A}_4$ and $\mathbf{B}_4$ by $\mathbf{T}_1$ and $\mathbf{T}_2$ respectively. Let these lines cut the $\mathbf{L}_{i,i=1..3}$ in $\mathbf{A}_{i,i=1..3}$ and $\mathbf{B}_{i,i=1..3}$ respectively. In this way, we obtain two cross ratios:

$$\lambda_1 = \{\mathbf{A}_1, \mathbf{A}_2; \mathbf{A}_3, \mathbf{A}_4\} \qquad \text{and} \qquad \lambda_2 = \{\mathbf{B}_1, \mathbf{B}_2; \mathbf{B}_3, \mathbf{B}_4\} \qquad (7)$$

Before continuing, consider the various degenerate cases.

– Provided that the first three lines are mutually skew, they still define the two families $\mathcal{K}$ and $\mathcal{L}$. The fourth line can then be degenerate in the following ways:
  - If $\mathbf{L}_4$ intersects one ($\mathbf{L}_i$) or two ($\mathbf{L}_i, \mathbf{L}_j$) of the three lines, we have the same solution as in the general case with $\mathbf{A}_4 = \mathbf{A}_i$ or $\mathbf{A}_4 = \mathbf{A}_i$ and $\mathbf{B}_4 = \mathbf{B}_j$ respectively.
  - If $\mathbf{L}_4 \in \mathcal{K}$ intersects all three lines $\mathbf{L}_{i,i=1..3}$, the two transversals $\mathbf{T}_1$ and $\mathbf{T}_2$ are equal to $\mathbf{L}_4$ and the cross ratios are no longer defined.
  - If $\mathbf{L}_4 \in \mathcal{L}$ (linear combination of the lines $\mathbf{L}_{i,i=1..3}$) every line belonging to $\mathcal{K}$ cuts the four lines $\mathbf{L}_{i,i=1..4}$ and there is only one characteristic cross ratio.
– Given two pairs of coplanar lines, say $\mathbf{L}_1, \mathbf{L}_2$ and $\mathbf{L}_3, \mathbf{L}_4$, there are two transversals intersecting all four lines. One is the line between the intersection of $\mathbf{L}_1$ and $\mathbf{L}_2$ and the intersection of $\mathbf{L}_3$ and $\mathbf{L}_4$. In this case the cross ratio $\lambda_1$ equals 1. The other is the line of intersection of the planes defined by $\mathbf{L}_1\mathbf{L}_2$ and $\mathbf{L}_3\mathbf{L}_4$ which cut $\mathbf{L}_{i,i=1..4}$ defining a second cross ratio [5].
– If three of the lines are coplanar, say $\mathbf{L}_{i,i=1..3}$, the plane $\mathbf{\Pi}$ containing them intersects $\mathbf{L}_4$ in a point $\mathbf{A}_4$. If the lines $\mathbf{L}_{i,i=1..3}$ belong to a pencil with center $\mathbf{B} \neq \mathbf{A}_4$, the lines $\mathbf{L}_{i,i=1..3}$ and $\mathbf{B}\mathbf{A}_4$ define one cross ratio. Otherwise, we have no invariants.
– Finally, if all of the lines lie in the same plane $\mathbf{\Pi}$, the possible cases are that they belong to a pencil and define one cross ratio or they do not belong to a pencil and there are no invariants.

As in the preceding cases, we can also compute the invariants using the algebraic approach:

$$\begin{aligned} I_1 &= \frac{[\mathbf{L}_1\,\mathbf{L}_2][\mathbf{L}_3\,\mathbf{L}_4]}{[\mathbf{L}_1\,\mathbf{L}_4][\mathbf{L}_2\,\mathbf{L}_3]} = \frac{(\beta_{1-2}^\mathsf{T}\mathbf{F}\alpha_{1-2})\,(\beta_{3-4}^\mathsf{T}\mathbf{F}\alpha_{3-4})}{(\beta_{1-4}^\mathsf{T}\mathbf{F}\alpha_{1-4})\,(\beta_{2-3}^\mathsf{T}\mathbf{F}\alpha_{2-3})} \\ I_2 &= \frac{[\mathbf{L}_1\,\mathbf{L}_3][\mathbf{L}_2\,\mathbf{L}_4]}{[\mathbf{L}_1\,\mathbf{L}_4][\mathbf{L}_2\,\mathbf{L}_3]} = \frac{(\beta_{1-3}^\mathsf{T}\mathbf{F}\alpha_{1-3})\,(\beta_{2-4}^\mathsf{T}\mathbf{F}\alpha_{2-4})}{(\beta_{1-4}^\mathsf{T}\mathbf{F}\alpha_{1-4})\,(\beta_{2-3}^\mathsf{T}\mathbf{F}\alpha_{2-3})} \end{aligned} \tag{8}$$

where $\mathbf{l}_i$ and $\mathbf{l}_i'$ are the projections of the $\mathbf{L}_{i,i=1..4}$ in the images, $\beta_{i-j} = \mathbf{l}_j' \times \mathbf{l}_i'$ and $\alpha_{i-j} = \mathbf{l}_j \times \mathbf{l}_i$. The relation between $I_{i,i=1,2}$ and $\lambda_{j,j=1,2}$ is $I_1 = 1 - \lambda_1 - \lambda_2 + \lambda_1\lambda_2$ and $I_2 = \lambda_1\lambda_2$ and conversely, given $I_1$ and $I_2$, $\lambda_1$ and $\lambda_2$ are the solutions of the equation $\lambda^2 - (1 - I_1 + I_2)\lambda + I_2 = 0$. As $I_1$ and $I_2$ are real, $\lambda_1$ and $\lambda_2$ are either real or a complex conjugate pair.

**Computation of $\lambda_1$ and $\lambda_2$ in 3D:** First, we suppose that we know a projective representation of the lines in some projective basis. They can be represented by their Plücker coordinates $\mathbf{L}_i = (l_1^{(i)}, l_2^{(i)}, l_3^{(i)}, l_4^{(i)}, l_5^{(i)}, l_6^{(i)})_{i=1..4}^\mathsf{T}$ [10].

We look for lines $\mathbf{T}$ that intersect the $\mathbf{L}_{i,i=1..4}$. If $\mathbf{T} = (p_1, p_2, p_3, p_4, p_5, p_6)^\mathsf{T}$, it represents a transversal line if and only if:

$$\begin{cases} l_4^{(i)}p_1 + l_5^{(i)}p_2 + l_6^{(i)}p_3 + l_1^{(i)}p_4 + l_2^{(i)}p_5 + l_3^{(i)}p_6 = 0, & \forall i = 1..4 \\ p_1p_4 + p_2p_5 + p_3p_6 = 0. \end{cases}$$

This is a homogeneous system in the six unknowns $p_{i,i=1..6}$, containing four linear equations and one quadratic one. It always has two solutions: $\mathbf{T}_k =$

$(p_1^k, p_2^k, p_3^k, p_4^k, p_5^k, p_6^k)_{k=1,2}^\top$ defined up to a scale factor, which may be real or complex conjugates. These lines cut the $\mathbf{L}_{i,i=1..4}$ in $\mathbf{A}_{i,i=1..3}$ and $\mathbf{B}_{i,i=1..3}$ which gives $\lambda_1$ and $\lambda_2$ ((7)).

**Computation of $\lambda_1$ and $\lambda_2$ from a Pair of Views:** Define $\mathbf{F}$ to be the fundamental matrix and $\mathbf{l}_{i,i=1..4} = (u_i, v_i, t_i)$ and $\mathbf{l}'_{i,i=1..4} = (u'_i, v'_i, t'_i)$ to be the projections of the lines $\mathbf{L}_{i,i=1..4}$ into the images. Consider a line $\mathbf{l}$ in the first image, with intersections $\mathbf{x}_i = \mathbf{l}_i \times \mathbf{l}$ with the lines $\mathbf{l}_{i,i=1..4}$. If the intersections $\mathbf{x}'_i = \mathbf{F}\mathbf{x}_i \times \mathbf{l}'_i$ of the epipolar lines with the $\mathbf{l}'_{i,i=1..4}$ in the second image all lie on some line $\mathbf{l}'$, then $\mathbf{l}$ and $\mathbf{l}'$ are the images of a transversal line $\mathbf{L}$. The condition of this can be written:

$$\begin{cases} [\mathbf{x}'_1\ \mathbf{x}'_2\ \mathbf{x}'_3] = [\mathbf{F}(\mathbf{l}_1 \times \mathbf{l}) \times \mathbf{l}'_1 \quad \mathbf{F}(\mathbf{l}_2 \times \mathbf{l}) \times \mathbf{l}'_2 \quad \mathbf{F}(\mathbf{l}_3 \times \mathbf{l}) \times \mathbf{l}'_3] = 0 \\ [\mathbf{x}'_1\ \mathbf{x}'_2\ \mathbf{x}'_4] = [\mathbf{F}(\mathbf{l}_1 \times \mathbf{l}) \times \mathbf{l}'_1 \quad \mathbf{F}(\mathbf{l}_2 \times \mathbf{l}) \times \mathbf{l}'_2 \quad \mathbf{F}(\mathbf{l}_4 \times \mathbf{l}) \times \mathbf{l}'_4] = 0 \end{cases}$$

This system of equations is difficult to deal with directly. To simplify the computation we make a change of projective basis, so that the new basis satisfies $(\mathbf{e}'_i)^T \mathbf{F} \mathbf{e}_i = 0$, for $i = 1..4$ and $\mathbf{F}\mathbf{e}_3 = \mathbf{F}^T \mathbf{e}'_3 = 0$. The fundamental matrix then becomes $\mathbf{F} = \begin{pmatrix} 0 & 1 & 0 \\ -1 & 0 & 0 \\ 0 & 0 & 0 \end{pmatrix}$ and the two pencils of epipolar lines take the form $\mathbf{m}_i = \mathbf{m}'_i = (1, \theta_i, 0)^\top$ [2].

Using this parameterization, and after simplifications [2] we obtain:

$$\lambda^k = \frac{((l_{13})^1 + \theta_1^k (l_{13})^2)\,((l_{24})^1 + \theta_2^k (l_{24})^2)}{((l_{23})^1 + \theta_2^k (l_{23})^2)\,((l_{14})^1 + \theta_1^k (l_{14})^2)}$$

where $(\theta_1^1, \theta_2^1)$ and $(\theta_1^2, \theta_2^2)$, reals or complex conjugates, are the significant[1] solutions of the system:

$$\begin{cases} ((l_{23})^1 + \theta_2 (l_{23})^2)((l'_{13})^1 + \theta_1 (l'_{13})^2) - ((l_{13})^1 + \theta_1 (l_{13})^2)((l'_{23})^1 + \theta_2 (l'_{23})^2) = 0 \\ ((l_{24})^1 + \theta_2 (l_{24})^2)((l'_{14})^1 + \theta_1 (l'_{14})^2) - ((l_{14})^1 + \theta_1 (l_{14})^2)((l'_{24})^1 + \theta_2 (l'_{24})^2) = 0 \end{cases}$$

with[2] $\mathbf{l}_{ij} = (v_i - v_j,\ u_j - u_i,\ u_i v_j - u_j v_i)^\top$ and $\mathbf{l}'_{ij} = (v'_i - v'_j,\ u'_j - u'_i,\ u'_i v'_j - u'_j v'_i)^\top$.

## 2 Projective and Permutation Invariants

The invariants of the previous sections depend on the ordering of the underlying point and line primitives. However, if we want to compare two configurations, we often do not know the correct relative ordering of the primitives. If we work with order dependent invariants, we must compare all the possible permutations of the elements. For example, if $I_{k,k=1..3}$ are the three invariants of a set of points $\mathbf{A}_{i,i=1..6}$ and we would like to know if another set $\mathbf{B}_{i,i=1..6}$ is projectively equivalent, we must compute the triplet of invariants for each of the 6!=720 possible permutations of the $\mathbf{B}_i$. This is a very costly operation, which can be avoided if we look for permutation-invariant invariants.[3]

---

[1] The other two solutions are $(1, 0, 0)^\top$ and $(0, 1, 0)^\top$.

[2] With our parameterization we can consider without lost of generality that $t_i = t'_i = 1$.

[3] We generalize here the work of Meer et al. [8] and Morin [9] on the invariants of five coplanar points to 3D configurations of points and lines.

**Definition 3.** *A function $I$ is a projective and permutation invariant for a con-figuration formed by elements $x_1, \ldots, x_k$ if and only if we have for all projective transformations $\mathbf{T}$ and for all permutations $\tau \in \mathcal{S}_k$*

$$I(x_1, \ldots, x_k) = I(x_{\tau(1)}, \ldots, x_{\tau(k)}) = I(\mathbf{T}(x_{\tau(1)}), \ldots, \mathbf{T}(x_{\tau(k)})). \qquad (9)$$

The projective invariants we computed were all expressed as cross ratios, which are not permutation invariants. As the group of permutations $\mathcal{S}_4$ has 24 elements, there are potentially 24 possible cross ratios of four collinear points. However, not all of these are distinct, because the permutations $\tau_1 = (1 \leftrightarrow 2)$ and $\tau_3 = (3 \leftrightarrow 4)$ have the same effect on the cross ratio. We know that all permutations can be written as products of transpositions of adjoining elements. For $\mathcal{S}_4$ there are three such transpositions $\tau_1 = (1 \leftrightarrow 2)$, $\tau_2 = (2 \leftrightarrow 3)$ and $\tau_3 = (3 \leftrightarrow 4)$. The effect of $\tau_1$ and $\tau_3$ is $F_1(\lambda) = \frac{1}{\lambda}$ and that of $\tau_2$ is $F_2(\lambda) = 1 - \lambda$. If we compute all the possible combinations of these functions, we obtain the following six values: $\lambda_1 = \lambda$, $\lambda_2 = \frac{1}{\lambda}$, $\lambda_3 = 1 - \lambda$, $\lambda_4 = \frac{1}{1-\lambda}$, $\lambda_5 = \frac{\lambda}{\lambda-1}$ and $\lambda_6 = \frac{\lambda-1}{\lambda}$.

To obtain permutation invariant cross ratios of four elements, it is sufficient to take an arbitrary symmetric function of the $\lambda_{i,i=1..6}$. The simplest two symmetric functions $J_0(\lambda) = \sum_{i=1}^{6} \lambda_i = 3$ and $J_0'(\lambda) = \prod_{i=1}^{6} \lambda_i = 1$ are not interesting because they give constant values. Taking the second order basic symmetric functions $J_1(\lambda) = \sum_{i=1}^{6} \lambda_i^2$ and $J_2(\lambda) = \sum_{i,j \in \{1..6\}, i<j} \lambda_i \lambda_j$ we obtain unbounded functions. Further invariants can be generated by taking combinations of the former ones $(J_0, J_0', J_1$ and $J_2)$, for example $J(\lambda) = \frac{J_1 J_0}{J_1 - J_2}$ proposed in [8] bounded between 2 and 2.8 or $G(\lambda) = -\frac{J_1}{J_2}$ bounded between 2 and 14. These functions are characterized by the fact that they have the same value for each of the six arguments $\lambda_{i,i=1..6}$.

Let us see what happens in the case of six points in projective space. We saw that the invariants of the points $\mathbf{A}_{i,i=1..6}$ given by (4) correspond to cross ratios of pencils of planes. Define: $I_{\tau(1)\tau(2)\tau(3)\tau(4)\tau(5)\tau(6)} = \{\mathbf{A}_{\tau(1)}\mathbf{A}_{\tau(2)}\mathbf{A}_{\tau(3)}, \mathbf{A}_{\tau(1)}\mathbf{A}_{\tau(2)}\mathbf{A}_{\tau(4)}; \mathbf{A}_{\tau(1)}\mathbf{A}_{\tau(2)}\mathbf{A}_{\tau(5)}, \mathbf{A}_{\tau(1)}\mathbf{A}_{\tau(2)}\mathbf{A}_{\tau(6)}\}$. It is easy to see that $I_1 = I_{231456}$, $I_2 = I_{132456}$ and $I_3 = I_{123456}$.

We are interested in the effect of permutations of points on the invariant $\lambda = I_{123456}$. The first remark is that if we interchange the first two elements the value is the same: $I_{123456} = I_{213456}$, which is obvious from a geometric viewpoint as $\mathbf{A}_1 \mathbf{A}_2 \mathbf{A}_k$ and $\mathbf{A}_2 \mathbf{A}_1 \mathbf{A}_k$ represent the same plane.

If we fix the first two points and permute the last four, we permute the four planes giving the cross ratio. So if we apply one of the above symmetric functions, for example $J$, the results will be invariant. Using the following proposition proved in [2] we can find all the possible values for $J(I_{\tau(1)\tau(2)\tau(3)\tau(4)\tau(5)\tau(6)})$.

**Proposition 4.** *An arbitrary permutation $\tau$ of $\mathcal{S}_6$ can be written as a product of four permutations $\tau = \pi \pi_{1k} \pi_{2n} \pi_{1m}$, where $\pi_{ij} = (i \leftrightarrow j)$, $k \in \{1, 2\}$, $1 \leq m < n \leq 6$ and $\pi$ is a permutation of the last four elements.*

Denote the value of $I_{\tau(1)\tau(2)\tau(3)\tau(4)\tau(5)\tau(6)}$ by $\tau(I_{123456})$. We have then $\tau(I_{123456}) = \pi(\pi_{1k}(\pi_{2n}(\pi_{1m}(I_{123456}))))$ and applying (for example) $J$, we obtain:

$$J(\tau(I_{123456})) = J(\pi(\pi_{1k}(\pi_{2n}(\pi_{1m}(I_{123456}))))) = J(\pi_{1k}(\pi_{2n}(\pi_{1m}(I_{123456})))),$$

because $\pi$ changes only the order of the four planes giving the cross ratio. But $\pi_{1k}$ has no effect on the value of the cross ratio: $\pi_{11}$ is the identity and $\pi_{12}$ interchanges the first two elements but leaves the planes invariant, so $J(\tau(I_{123456})) = J((\pi_{2n}(\pi_{1m}(I_{123456}))))$.

Consequently, we obtain only $C_6^2 = 15$ different possible values instead of $6! = 720$. These will be denoted by $J_{mn} = J((\pi_{2n}(\pi_{1m}(I_{123456}))))$. It is easy to see that $J_{mn} = J(I_{mni_1i_2i_3i_4})$ where $1 \leq m < n \leq 6$ and $\{i_1, i_2, i_3, i_4\} = \{1, 2, 3, 4, 5, 6\} \setminus \{m, n\}$. The geometric meaning of this is that we fix the pair of points giving the axis of the pencil of planes.

As a function of $I_1, I_2$ and $I_3$, the 15 values are:

$$J(I_1)\ J(\tfrac{I_2}{I_3})\ J(\tfrac{I_2-1}{I_3-1})\ J(\tfrac{I_3(I_2-1)}{I_2(I_3-1)}) \qquad J(\tfrac{I_3-I_1}{I_3-I_2})$$
$$J(I_2)\ J(\tfrac{I_1}{I_3})\ J(\tfrac{I_1-1}{I_3-1})\ J(\tfrac{I_3(I_1-1)}{I_1(I_3-1)}) \qquad J(\tfrac{I_2(I_3-I_1)}{I_1(I_3-I_2)}) \qquad (10)$$
$$J(I_3)\ J(\tfrac{I_1}{I_2})\ J(\tfrac{I_1-1}{I_2-1})\ J(\tfrac{I_2(I_1-1)}{I_1(I_2-1)})\ J(\tfrac{(I_2-1)(I_3-I_1)}{(I_1-1)(I_3-I_2)})$$

Note that the values of $I_1, I_2, I_3$ depend on the order of the points, but any other ordering gives the same values in different ordering. For this reason, we sort the 15 values after computing them to give a vector that characterizes the configuration independently of the order of underlying points.

Now, consider the configuration of a line and four unordered points. This case is a very simple case because we have a cross ratio, so it is enough to take the value of the function $J(I)$ where $I$ is the invariant given by (6). Similarly to compute permutation invariants of four lines in space, we apply $J$ to the pair of cross ratios $\lambda_1$ and $\lambda_2$ given in Sect. 1.5. It turn out that the order of the two sets of intersection points of the four lines with the two transversals change in the same way under permutations of the lines. When $\lambda_1$ and $\lambda_2$ are complex conjugates, $J(\lambda_1)$ and $J(\lambda_2)$ are also complex conjugates, but if we want to work with real invariants it is sufficient to take $J_1 = J(\lambda_1) + J(\lambda_2)$ and $J_2 = J(\lambda_1)J(\lambda_2)$.

The fourth basic configuration is the set of three points and two lines. We saw that we can compute invariants for this configuration using the invariants of five coplanar points. In this case it is sufficient to apply the results of [8], which show that if $I_1$ and $I_2$ are a pair of invariants for five coplanar points we can take the sorted list with elements: $J(I_1)$, $J(I_2)$, $J(\tfrac{I_1}{I_2})$, $J(\tfrac{I_2-1}{I_1-1})$ and $J(\tfrac{I_1(I_2-1)}{I_2(I_1-1)})$, to obtain a permutation invariant. But this means that we make mixed permutations of points and lines, which is unnecessary because when we want to compare two such configurations we have no trouble distinguishing lines from points. When we have five points in the plane $\mathbf{A}_1\mathbf{A}_2\mathbf{A}_3$ we will require that the center points of the considered pencils of lines be chosen among $\mathbf{A}_{i,i=1..3}$ and not among the intersection points $\mathbf{R}_{j,j=1,2}$ (Sect. 1.4). In this way we have the cross ratios of the form $\lambda_1 = \{\mathbf{A}_1\mathbf{A}_2, \mathbf{A}_1\mathbf{A}_3; \mathbf{A}_1\mathbf{R}_1, \mathbf{A}_1\mathbf{R}_2\} = I_1$. If we interchange $\mathbf{A}_1$ and $\mathbf{A}_2$, we obtain $I_2$. Interchanging $\mathbf{A}_2$ and $\mathbf{A}_3$ gives $\tfrac{1}{I_1}$ and interchanging $\mathbf{A}_1$ and $\mathbf{A}_3$ we obtain $\tfrac{I_1}{I_2}$. The permutation of the two lines (i.e. $\mathbf{R}_1$ and $\mathbf{R}_2$) gives $\tfrac{1}{I_1}$. Hence, applying $J$ we find that the sorted list of elements $J(I_1)$, $J(I_2)$ and $J(\tfrac{I_1}{I_2})$ is a projective and permutation invariant for this configuration.

## 2.1 Some Ideas Concerning More Complex Configurations

Consider a configuration of $N > 6$ points[4] From proposition **2**, this configuration has $3N - 15 = 3(N-5)$ independent invariants. Denote the $N$ points by $\mathbf{A}_{i,i=1..N}$ and the $3(N-5)$ invariants of this configuration by $\lambda_i^k{}_{,i=1..3,k=1..N-5}$, where $\lambda_1^k, \lambda_2^k, \lambda_3^k$ are the invariants of the sub-configurations $\mathbf{A}_1, \ldots \mathbf{A}_5, \mathbf{A}_{k+5}$.

As the invariants $\lambda_i^k$ depend on the order of the points, we try to generalize the above approach to the $N$ point case. First note that the invariants are cross ratios of four planes of a pencil $I_{ijklmn} = \{\mathbf{A}_i\mathbf{A}_j\mathbf{A}_k, \mathbf{A}_i\mathbf{A}_j\mathbf{A}_l; \mathbf{A}_i\mathbf{A}_j\mathbf{A}_m, \mathbf{A}_i\mathbf{A}_j\mathbf{A}_n\}$, where $i,j,k,l,m,n \in \{1, \ldots, N\}$. Consequently, for a given set $\{i,j,k,l,m,n\} \subset \{1, 2, \ldots, N\}$, we have $C_6^2 = 15$ different values $J(I_{ijklmn})$. But, there are $C_N^6$ subsets of six points, so we have $C_N^6 \cdot C_6^2$ different values which can be computed from the $3(N-5)$ independent invariants $\lambda_i^k$.

We show how to obtain these values in the case of 7 points, but the approach is the same for $N > 7$. Denote the $3(7-5) = 6$ independent invariants of this configuration by $\lambda_1^1 = I_{231456}$, $\lambda_2^1 = I_{132456}$, $\lambda_3^1 = I_{123456}$, $\lambda_1^2 = I_{231457}$, $\lambda_2^2 = I_{132457}$ and $\lambda_3^2 = I_{123457}$.

The $C_7^6 \cdot C_6^2 = 105$ values can be obtained by calculating for each subset of six points $\{i,j,k,l,m,n\} \subset \mathcal{M} = \{1,2,3,4,5,6,7\}$ such that $i < j < k < l < m < n$ the values $I_1 = I_{jkilmn}$, $I_2 = I_{ikjlmn}$ and $I_3 = I_{ijklmn}$ in function of $\lambda_i^j$ :

| subsets | $\mathcal{M} \setminus \{7\}$ | $\mathcal{M} \setminus \{6\}$ | $\mathcal{M} \setminus \{5\}$ | $\mathcal{M} \setminus \{4\}$ | $\mathcal{M} \setminus \{3\}$ | $\mathcal{M} \setminus \{2\}$ | $\mathcal{M} \setminus \{1\}$ |
|---|---|---|---|---|---|---|---|
| $I_1$ | $\lambda_1^1$ | $\lambda_1^2$ | $\dfrac{\lambda_1^2}{\lambda_1^1}$ | $\dfrac{1-\lambda_1^2}{1-\lambda_1^1}$ | $\dfrac{\lambda_3^1(\lambda_1^2-\lambda_3^2)}{\lambda_3^2(\lambda_1^1-\lambda_3^1)}$ | $\dfrac{\lambda_2^1(\lambda_1^2-\lambda_2^2)}{\lambda_2^2(\lambda_1^1-\lambda_2^1)}$ | $\dfrac{\lambda_1^1(\lambda_2^2-\lambda_1^2)}{\lambda_1^2(\lambda_2^1-\lambda_1^1)}$ |
| $I_2$ | $\lambda_2^1$ | $\lambda_2^2$ | $\dfrac{\lambda_2^2}{\lambda_2^1}$ | $\dfrac{1-\lambda_2^2}{1-\lambda_2^1}$ | $\dfrac{\lambda_3^1(\lambda_2^2-\lambda_3^2)}{\lambda_3^2(\lambda_2^1-\lambda_3^1)}$ | $\dfrac{\lambda_2^1(\lambda_3^2-\lambda_2^2)}{\lambda_2^2(\lambda_3^1-\lambda_2^1)}$ | $\dfrac{\lambda_1^1(\lambda_3^2-\lambda_1^2)}{\lambda_1^2(\lambda_3^1-\lambda_1^1)}$ |
| $I_3$ | $\lambda_3^1$ | $\lambda_3^2$ | $\dfrac{\lambda_3^2}{\lambda_3^1}$ | $\dfrac{1-\lambda_3^2}{1-\lambda_3^1}$ | $\dfrac{\lambda_3^1(1-\lambda_3^2)}{\lambda_3^2(1-\lambda_3^1)}$ | $\dfrac{\lambda_2^1(1-\lambda_2^2)}{\lambda_2^2(1-\lambda_2^1)}$ | $\dfrac{\lambda_1^1(1-\lambda_1^2)}{\lambda_1^2(1-\lambda_1^1)}$ |

and then apply (10). Finally, we sort the resulting list of 105 elements.

## Conclusion

We have presented a detailed study of the projective and permutation invariants of configurations of points, lines and planes. These invariants can be used as a basis for indexing, describing, modeling and recognizing polyhedral objects from perspective images.

The invariants of complex configurations can be computed from those of minimal configurations into which they can be decomposed. So it was sufficient to treat only the case of minimal configurations. Also, in projective space there is a duality between points and planes that preserves cross ratios, so configurations of planes or planes and lines can be reduced to point and point-line ones.

For each configuration we gave several methods to compute the invariants. There are basically two approaches - algebraic and geometric - and in each case we showed the relations between the resulting invariants. We analyzed the

---

[4] We will consider only the case of points. The other cases can be handled similarly, but the approach is more complex and will be part of later work.

computation of these invariants in 3D space assuming that the points and lines had already been reconstructed in an arbitrary projective basis, and we also gave methods to compute them directly from correspondences in a pair of images. In the second case the only information needed is the matched projections of the points and lines and the fundamental matrix.

Finally, for each basic configuration we also gave permutation and projective invariants, and suggested ways to treat permutation invariance for more complicated configurations.

**Acknowledgment:** We would like to thank Bill Triggs for his carefully reading of the draft of this manuscript.

# References

1. S. Carlsson. Multiple image invariants using the double algebra. In J. Mundy and A. Zissermann, editors, *Proceeding of the* DARPA–ESPRIT *workshop on Applications of Invariants in Computer Vision, Azores, Portugal*, pages 335–350, October 1993.
2. G. Csurka. *Modelisation projective des objets tridimensionnels en vision par ordinateur*. Thèse de doctorat, Université de Nice – Sophia Antipolis, April 1996.
3. G. Csurka and O. Faugeras. Computing three-dimensional projective invariants from a pair of images using the Grassmann-Cayley algebra. *Image and Vision Computing*, 1997. to appear.
4. F. Devernay and O. Faugeras. From projective to euclidean reconstruction. In *Proceedings of the Conference on Computer Vision and Pattern Recognition, San Francisco, California, USA*, June 1996. to appear.
5. P. Gros. 3D projective invariants from two images. In *Proceeding of the* DARPA–ESPRIT *workshop on Applications of Invariants in Computer Vision, Azores, Portugal*, pages 65–85, October 1993.
6. R. Hartley. Invariants of lines in space. In *Proceedings of* DARPA *Image Understanding Workshop*, pages 737–744, 1993.
7. R. Horaud, F. Dornaika, and B. Espiau. Visually guided object grasping. IEEE *Transactions on Robotics and Automation*, 1997. submitted.
8. P. Meer, S. Ramakrishna, and R. Lenz. Correspondence of coplanar features through $p^2$-invariant representations. In *Proceedings of the 12th International Conference on Pattern Recognition, Jerusalem, Israel*, pages A–196–202, 1994.
9. L. Morin. *Quelques contributions des invariants projectifs à la vision par ordinateur*. Thèse de doctorat, Institut National Polytechnique de Grenoble, January 1993.
10. J.G. Semple and G.T. Kneebone. *Algebraic Projective Geometry*. Oxford Science Publication, 1952.
11. C. Tisseron. *Géométries affine, projective et euclidienne*. HERMANN,Paris, 1988.
12. U. Uenohara and T. Kanade. Geometric invariants for verification in 3-d object tracking. In *Proceedings of the* IEEE/RSJ *International Conference on Intelligent Robots and Systems, Osaka, Japan*, volume II, pages 785–790, November 1996.
13. C. Zeller and O. Faugeras. Applications of non-metric vision to some visual guided tasks. In *Proceedings of the 12th International Conference on Pattern Recognition, Jerusalem, Israel*, 1994.
14. A. Zisserman, P.A. Beardsley, and I.D. Reid. Metric calibration of a stereo rig. In *Workshop on Representation of Visual Scenes, Cambridge, Massachusetts, USA*, pages 93–100, June 1995.

# A Unified Language for Computer Vision and Robotics

Eduardo Bayro-Corrochano[*], Joan Lasenby[‡]

[*] Computer Science Institute, Christian Albrechts University, Kiel, Germany.
email: edb@informatik.uni-kiel.de
[‡]Department of Engineering, Trumpington Street, Cambridge CB2 1PZ.
email: jl@eng.cam.ac.uk

**Abstract.** Geometric algebra is an universal mathematical language which provides very comprehensive techniques for analyzing the complex geometric situations occurring in artificial Perception Action Cycle systems. In the geometric algebra framework such a system is both easier to analyze and to control in real time computations. This paper describes the application of rotors and motors for tasks involving the algebra of the 3D kinematics. Using purely geometric derivations and the constraints for point and line correspondences in n-views projective invariants are computed and the projective depth is discussed in terms of the generalized cross-ratio.

**Categories**: Clifford algebra; geometric algebra; robotics; hand-eye calibration; computer vision; projective invariants; projective depth.

## 1 Introduction

Biological and artificial intelligent systems show particular behaviour according to their situation and environment. They exist in a manifold structure where the time is distinguished as an axis rotated orthogonally from any spatial axis. Survival depends on the system's interrelated perceptive and active capabilities. This observable dependency can be delimited in a cycle of success. Within a Perception Action Cycle (PAC) a system interacts with its environment via visual and non-visual sensors for learning or accomplishing its task. Since each component of PAC systems requires different mathematical techniques, the construction of artificial PAC systems demands the fusion of the fields of signal theory, computer vision, robotics and neural computing in a framework with a powerful representation capability and a strong geometric basis. Currently, different mathematical systems are routinely employed for each part of the cycle. Each of these systems is limited in its applicability to one part of the cycle, making communication between different processes very difficult. Clearly, our ability to control the PAC would be considerably enhanced if a single mathematical language were employed throughout. In this paper the authors propose to use *geometric algebra* to analyze and construct algorithms for each phase of the PAC [29]. Geometric algebra is a coordinate-free approach to geometry based on the algebras of Grassmann [13] and Clifford [9]. Some preliminary applications of geometric algebra in the field of computer vision, robotics, neural computing and low level signal processing have already been given [1, 21, 3, 19, 2, 4]. For a more complete introduction see [14] and for other brief summaries see [16, 1, 3].

The next section will give the basics for the modelling of a work space for the projective space in terms of geometric algebra. The geometric algebra for the 3D kinematics is explained in section three. As two typical cases of robotics the motion of a multi-link and the hand-eye calibration are presented in Section four. The discussion of interesting issues of computer vision like the computation of projective invariants using n-views is presented in Section 7 and the recovery of projective depth is analyzed in Section 8. Finally, in the conclusion section the relevance of the geometric algebra framework for artificial PAC systems is discussed.

## 2   The 4D geometric algebra for the projective space

In a geometric algebra $\mathcal{G}_{p,q,r}$ we identify $p$ and $q$ as the dimensions of the maximal subspaces with positive and negative signatures, respectively (the signature of a vector $a$ is positive, negative or zero according as $a^2 > 0, < 0, = 0$). It is important for real applications to regard the signature of the modeled space to facilitate the computations. In the case of $\mathcal{G}_{3,0,0}$ we are adopting the standard Euclidean signature for the ordinary space, $E^3$, this forces to adopt the same signature for the 4-dimensional space $\mathcal{G}_{1,3,0}$ which we associate with the projective space $P^3$. This is spanned with the following basis

$$\underbrace{1}_{scalar}, \quad \underbrace{\gamma_k}_{4\ vectors}, \quad \underbrace{\gamma_2\gamma_3, \gamma_3\gamma_1, \gamma_1\gamma_2, \gamma_4\gamma_1, \gamma_4\gamma_2, \gamma_4\gamma_3,}_{6\ bivectors} \quad \underbrace{i\gamma_k}_{4\ pseudovectors}, \quad \underbrace{i}_{pseudoscalar} \quad (1)$$

where $\gamma_4^2 = +1$, $\gamma_k^2 = -1$ for k=1,2,3. The pseudoscalar is $i = \gamma_1\gamma_2\gamma_3\gamma_4$ with

$$i^2 = (\gamma_1\gamma_2\gamma_3\gamma_4)(\gamma_1\gamma_2\gamma_3\gamma_4) = -(\gamma_3\gamma_4)(\gamma_3\gamma_4) = -1. \quad (2)$$

The fourth basis vector $\gamma_4$ can be seen also as selected direction or *projective split* [3] in 4D. The basis element $\gamma_4$ helps to associate multivectors of the 4D space with multivectors of the 3D space. The role and use of the projective split for a variety of problems involving the algebra of incidence can be found in [3].

## 3   The 4D geometric algebra for 3D kinematics

One alternative to model the work space for the robotic field could simply the geometric algebra $\mathcal{G}_{3,0,0}$ of the 3D space. Since general displacements are non-linear transformations it would be more beneficial if we compute in a higher dimensional space. That is why the authors chose the special 4D algebra of the motors $\mathcal{G}_{3,0,1}^+$ as an efficient framework for 3D kinematics.

### 3.1   Motors

Clifford introduced the biquaternions with the name motors which is the abbreviation of "moment and vector"[10] for the algebra of 3D kinematics. Motors are dual numbers with the necessary condition of $i^2 = 0$. They can be found in a special even geometric algebra $\mathcal{G}_{3,0,1}^+$ which here will be called the algebra of the motors. For detailed discussion of the role of dual, double and complex numbers in geometric algebra see [2]. Actually the algebra of the motors is a subalgebra of $\mathcal{G}_{3,0,1}$ for the 4D space with a similar basis presented in (1) with the difference that $\gamma_k^2 = 1$ for k=1,2,3 and $\gamma_4^2 = 0$ and that the pseudoscalar $i = \gamma_1\gamma_2\gamma_3\gamma_4$ squares to zero.

The important role that the motors play as a linear transformation is that they can absorb the translation component of a rigid motion. Let us explain this in more detail. Since in $R^3$ the simple translation is a nonlinear transformation the general displacement will be too. Unfortunately the displacement in $R^3$ can not be represented as a 3x3 matrix transformation. The way how we can go around is embedding $R^3$ in the $R^4$ space. In this 4D space the general displacements will take place in the hyperplane $X_4 = 1$. For example the motion of any point $\mathbf{p}$ can be now expressed compactly using a 4x4 homogeneous transformation matrix, i.e.

$$\begin{bmatrix} \mathbf{p}' \\ 1 \end{bmatrix} = \begin{bmatrix} \mathbf{R} & \mathbf{t} \\ 0^T & 1 \end{bmatrix} \begin{bmatrix} \mathbf{p} \\ 1 \end{bmatrix}. \tag{3}$$

Recall that a general displacement can be also expressed in terms of dual matrix, i.e.

$$\begin{bmatrix} R & \mathbf{t} \\ 0^T & 1 \end{bmatrix} \equiv R + i[\mathbf{t}]_x R \tag{4}$$

where $R$ is a rotation matrix and $[\mathbf{t}]_x$ the tensorial notation of the antisymmetric matrix representing the translation. We will show below that this transformation is equivalent to a motor, see equation (9) . Note that the homogeneous coordinates are similar to the ones used in the geometric algebra of the projective space $\mathcal{G}_{1,3,0}$, however if we want to compute using motors, which requires that the pseudoscalar squares to zero, we are compelled to switch to the motor algebra or $\mathcal{G}_{3,0,1}^+$.

The algebra of the motors has the basis

$$\underbrace{1}_{scalar} \quad , \quad \underbrace{\gamma_2\gamma_3, \gamma_3\gamma_1, \gamma_1\gamma_2, \gamma_4\gamma_1, \gamma_4\gamma_2, \gamma_4\gamma_3}_{6 \;\; bivectors} \quad , \quad \underbrace{i}_{pseudoscalar} \tag{5}$$

where $i^2 = 0$. A motor $\boldsymbol{M}_g$ in general is

$$\boldsymbol{M}_g = \boldsymbol{m}\boldsymbol{M} \tag{6}$$

where $\boldsymbol{m} = a + ib$ is a dual number for dilation and $\boldsymbol{M}$ is a unit motor. From now on all equations referring motors use the unit motor $\boldsymbol{M}$. A basic geometric interpretation of a unit motor $M$ can be given using the sum of two non-coplanar lines expressed in terms of dual bivector basis, i.e.

$$
\begin{aligned}
M &= X_1 X_2 + X_3 X_4 = (X_1 \cdot X_2 + X_1 \wedge X_2) + (X_3 \cdot X_4 + X_3 \wedge X_4) \\
&= (a_0 + a_1 \gamma_2 \gamma_3 + a_2 \gamma_3 \gamma_1 + a_3 \gamma_1 \gamma_2) + i(b_0 + b_1 \gamma_2 \gamma_3 + b_2 \gamma_3 \gamma_1 + b_3 \gamma_3 \gamma_2) \\
&= R + iR',
\end{aligned}
\tag{7}
$$

This tells that a motor can be seen also as a dual rotor or dual quaternion. Let us now analyze the motor equation (7). If the lines are non-coplanar the motor represents a general displacement or rigid motion and it is exact equivalent to a screw [10], else being coplanar they build a new line which can be seen as a *degenerated* motor. Thus, it is also convenient if the translation is expressed as a sort of a rotor which might be called translator

$$
T = e^{\frac{1}{2}ti} = 1 + i\frac{t}{2},
\tag{8}
$$

where $t = t_1 \gamma_2 \gamma_3 + t_2 \gamma_3 \gamma_1 + t_3 \gamma_1 \gamma_2$. The motor in terms of a translator reads

$$
M = R + iR' = R + i\frac{t}{2}R = (1 + i\frac{t}{2})R = TR.
\tag{9}
$$

The translator can be seen simply as the representation of a rotation plane displaced from the reference origin by $t$ and with the same orientation of the vector $t$. The vector $t$ can be also expressed in terms of the rotors using

$$
R'\tilde{R} = (\frac{t}{2}R)\tilde{R}
\tag{10}
$$

therefore

$$
t = 2R'\tilde{R}
\tag{11}
$$

where the multiplication is a geometric product.

The norm of a motor $M$ is defined as follows

$$
|M| = M\tilde{M} = TR\tilde{R}\tilde{T} = (1 + i\frac{t}{2})R\tilde{R}(1 - i\frac{t}{2}) = 1 + i\frac{t}{2} - i\frac{t}{2} = 1, \tag{12}
$$

where $\tilde{M}$ is the conjugate motor and 1 is the identity. The combination of two rigid motions can be expressed using two motors. The resultant motor describes the overall displacement, namely

$$
M_c = M_a M_b = (R_a + iR'_a)(R_b + iR'_b) = R_c + iR'_c.
\tag{13}
$$

Note that pure rotations combine multiplicatively and the dual parts, containing the translation components, combine additively.

## 3.2 Representing points, lines and planes in 4D

The special algebra of motors $\mathcal{G}_{3,0,1}^{+}$ has a bivector basis which in 4D span the line space. Thus let us start with the line using this bivector basis. Assume two points $\boldsymbol{X}_1 = (X_{11}, X_{12}, X_{13}, 1)$ and $\boldsymbol{X}_2 = (X_{21}, X_{22}, X_{23}, 1)$ lying on the hyperplane $X_4 = 1$ and belonging to this line. The line can be defined simply as an outer product of these points, i.e.

$$
\begin{aligned}
l_d = \boldsymbol{X}_1 \wedge \boldsymbol{X}_2 =\ & (X_{12}X_{23} - X_{13}X_{22})\gamma_2\gamma_3 + (X_{13}X_{21} - X_{11}X_{23})\gamma_3\gamma_1 + .. + \\
& (X_{11}X_{22} - X_{12}X_{21})\gamma_1\gamma_2 + (X_{21} - X_{11})\gamma_4\gamma_1 + .. + \\
& (X_{22} - X_{12})\gamma_4\gamma_2 + (X_{23} - X_{13})\gamma_4\gamma_3.
\end{aligned}
\tag{14}
$$

Since this equation consists only of bivectors, it can be expressed straightforward in terms of the bivector basis, namely

$$
\begin{aligned}
l_d &= (L^{23}\gamma_2\gamma_3 + L^{31}\gamma_3\gamma_1 + L^{12}\gamma_1\gamma_2) + (L^{41}\gamma_4\gamma_1 + L^{42}\gamma_4\gamma_2 + L^{43}\gamma_4\gamma_3) \\
&= (L^{23}\gamma_2\gamma_3 + L^{31}\gamma_3\gamma_1 + L^{12}\gamma_1\gamma_2) + i(L^{41}\gamma_2\gamma_3 + L^{42}\gamma_3\gamma_1 + L^{43}\gamma_1\gamma_2).
\end{aligned}
\tag{15}
$$

Note that this is equivalent to the line expression using Plücker coordinates. The real part can be seen as the line direction denoted as a vector $\boldsymbol{n}$ and the dual part as the moment which is nothing else as the cross product between $\boldsymbol{n}$ and any vector $\boldsymbol{q}$ touching the line, i.e.

$$
l_d = \boldsymbol{n} + i(\boldsymbol{n} \times \boldsymbol{q}) = \boldsymbol{n} + i\boldsymbol{m},
\tag{16}
$$

where $\boldsymbol{n} \times \boldsymbol{q} = -i\boldsymbol{n} \wedge \boldsymbol{q}$. This line representation using dual numbers is easier to understand and to manipulate algebraically and it is fully equivalent to the one in terms of Plücker coordinates.

For the case of the point representation, embedding a 3D point expressed as a vector $\boldsymbol{x}$ on the hyperplane $X_4 = 1$, the point $\boldsymbol{q}$ equation in $\mathcal{G}_{3,0,1}^{+}$ reads

$$
q = 1 + x_1\gamma_4\gamma_1 + x_2\gamma_4\gamma_2 + x_3\gamma_4\gamma_3 = 1 + i(x_1\gamma_2\gamma_3 + x_2\gamma_3\gamma_1 + x_3\gamma_1\gamma_2) = 1 + i\boldsymbol{x}.
\tag{17}
$$

Now, resorting to the duality principle we use the dual of the scalar i.e the pseudoscalar times $d$ and the dual of the bivector basis to write straightforwardly the plane equation, i.e

$$
\phi = n_1\gamma_2\gamma_3 + n_2\gamma_3\gamma_1 + n_3\gamma_1\gamma_2 + id = \boldsymbol{n} + id
\tag{18}
$$

where $\boldsymbol{n}$ stands for the normal of the plane and $d$ for the distance of the plane to the origin.

## 4 Modelling the 3D Motion of Points, Lines and Planes

In this section we will present the modelling of the 3D motion of basic geometric entities using rotors and motors. We will see below in the case of the hand-eye calibration that is preferable to use motors for computing the rotation and translation of an unknown rigid motion simultaneously. Because using the rotor approach we compute the translation decoupled of the rotation increasing therefore the inaccuracy. Let us now consider the modelling of the motion of points, lines and planes in both $R^3$ and $R^4$.

In $\mathcal{G}_{3,0,0}$ a line can be described in terms of any couple of points lying on the line, i.e. $\boldsymbol{x} = \theta \boldsymbol{p}_1 + \boldsymbol{p}_2$. The motion equation of the line is then the same as for the point equation(20). In the algebra of the motors $\mathcal{G}_{3,0,1}^+$ we expressed the line as equation (16) and proceed as follows

$$l_a = \boldsymbol{n}_a + i\boldsymbol{m}_a = \boldsymbol{M}l_b\tilde{\boldsymbol{M}} = \boldsymbol{R}\boldsymbol{n}_b\tilde{\boldsymbol{R}} + i(\boldsymbol{R}\boldsymbol{n}_b\tilde{\boldsymbol{R}}' + \boldsymbol{R}'\boldsymbol{n}_b\tilde{\boldsymbol{R}} + \boldsymbol{R}\boldsymbol{m}_b\tilde{\boldsymbol{R}}) \qquad (19)$$

The 3D motion of a point $\boldsymbol{x}$ in $\mathcal{G}_{3,0,0}$ has the equation

$$\boldsymbol{x}' = \boldsymbol{R}\boldsymbol{x}\tilde{\boldsymbol{R}} + \boldsymbol{t}. \qquad (20)$$

In case of the algebra of the motors $\mathcal{G}_{3,0,1}^+$ we use the point representation of equation (17)

$$\boldsymbol{M}(1 + i\boldsymbol{x})\overline{\tilde{\boldsymbol{M}}} = \boldsymbol{T}\boldsymbol{R}(1 + i\boldsymbol{x})\tilde{\boldsymbol{R}}\boldsymbol{T} = 1 + i(\boldsymbol{R}\boldsymbol{x}\tilde{\boldsymbol{R}} + \boldsymbol{t}). \qquad (21)$$

The expression $\overline{\tilde{\boldsymbol{M}}} = \tilde{\boldsymbol{R}}\boldsymbol{T}$ was found independently by the authors and it is similar to the one presented by W. Blaschke [5]. Yet in general all our motor equations explain directly that motor expressions consist of the successive action of rotors and translators. Now, when dealing with the motion of planes as it is shown below again we apply from the right $\overline{\tilde{\boldsymbol{M}}}$. This is because the plane is the dual of the point. Since the algebra of the motors has a bivector basis which span the line space and if we use this basis for representing points (geometrically of a lower dimension) and for planes (one dimension higher) we require $\overline{\tilde{\boldsymbol{M}}}$ instead of simply $\tilde{\boldsymbol{M}}$ as a sort of compensation for this asymmetry.

For the plane in $\mathcal{G}_{3,0,0}$ we use a multivector representation of the formula of Hesse, i.e. $\boldsymbol{H} = d + \boldsymbol{n}$. Note that this multivector consists of a scalar and a vector. Any point lying on this plane fulfills $\boldsymbol{x}\cdot\boldsymbol{n} - d = 0$. Using this we can now write the motion of the plane

$$\boldsymbol{H}' = (\boldsymbol{R}\boldsymbol{x}\tilde{\boldsymbol{R}} + \boldsymbol{t})\cdot(\boldsymbol{R}\boldsymbol{n}\tilde{\boldsymbol{R}}) + (\boldsymbol{R}\boldsymbol{n}\tilde{\boldsymbol{R}}). \qquad (22)$$

Since $(\boldsymbol{R}\boldsymbol{x}\tilde{\boldsymbol{R}})\cdot(\boldsymbol{R}\boldsymbol{n}\tilde{\boldsymbol{R}}) = \boldsymbol{x}\cdot\boldsymbol{n}$, this becomes $\boldsymbol{H}' = \boldsymbol{x}\cdot\boldsymbol{n} + \boldsymbol{R}\boldsymbol{n}\tilde{\boldsymbol{R}} + \boldsymbol{t}\cdot(\boldsymbol{R}\boldsymbol{n}\tilde{\boldsymbol{R}})$ which can be finally written as

$$\boldsymbol{H}' = \boldsymbol{R}\boldsymbol{H}\tilde{\boldsymbol{R}} + (\boldsymbol{R}\boldsymbol{H}\tilde{\boldsymbol{R}})\cdot\boldsymbol{t}. \qquad (23)$$

The motion of a plane in $\mathcal{G}_{3,0,1}^+$ can be seen as the motion of the dual of the point, thus using the expression of equation (18) the motion equation of the plane is

$$\boldsymbol{M}(\boldsymbol{n} + id)\overline{\tilde{\boldsymbol{M}}} = \boldsymbol{T}\boldsymbol{R}(\boldsymbol{n} + id)\tilde{\boldsymbol{R}}\boldsymbol{T} = \boldsymbol{R}\boldsymbol{n}\tilde{\boldsymbol{R}} + i(d + (\boldsymbol{R}\boldsymbol{n}\tilde{\boldsymbol{R}})\cdot\boldsymbol{t}). \qquad (24)$$

## 4.1 Application 1: movement of a robot arm

Let us consider a complicated robot mechanism in terms of a system of linked $n$-bars. This calls for an optimization approach to find out its configuration during a smooth movement. Let us analyze the problem first in 3D space and then in the 4D space.

In 3D using the geometric algebra $\mathcal{G}_{3,0,0}$ we define reference-frames attached to each turnable join. Any connected two bars, the $j-th$ bar and the $j+1$-bar, are referred simple by the relative position of the $j-th$ join with the $j-1$-th join and the next bar moved by its own join. This can be simply expressed by

$$x_{j+1} = R_{j-1}x_{j-1}\tilde{R}_{j-1}+R_jx'_j\tilde{R}_j = R_{j-1}x_{j-1}\tilde{R}_{j-1}+R_jR_{j-1}x_j\tilde{R}_{j-1}\tilde{R}_j \quad (25)$$

where $R_j$ is the rotor applied to the reference frame attached to the $j$-join and $x'$ is the translation from reference frame $j-1$ to reference frame $j$ which corresponds simply to the length of the $j$-bar. The equation for the position of end effector considering the whole mechanism reads

$$x = R_1x_1\tilde{R}_1 + R_2x'_2\tilde{R}_2 + R_3x'_3\tilde{R}_3 + ... + R_nx'_n\tilde{R}_n. \quad (26)$$

Now for the 4D space we will use the algebra of the motors $\mathcal{G}^+_{3,0,1}$. In the 4D the equation (25) reads

$$x'_{j+1} = R_jT_jR_{j-1}x'_{j-1}\tilde{R}_{j-1}T_j\tilde{R}_j, \quad (27)$$

where $x'_{j-1} = (1+ix_{j-1})$ and $x'_{j+1} = (1+ix_{j+1})$ referred to their own coordinate systems. Assuming that all the robot bars are of the same length $x$, then $T_1 = (1 + ix) = T_2 = T_3 = ... = T_n$. Equation (26) for the whole mechanism in 4D is now

$$x'_n = R_{n-1}...T_3R_2T_2R_1x'_1\tilde{R}_1T_2\tilde{R}_2T_3...\tilde{R}_{n-1}. \quad (28)$$

Comparing the 4D and 3D expressions, the formers are linear and more simple. This can be exploited when computing any robot arm motion, i.e. we can for a particular motion simplify the equation (28) by canceling some redundant or conjugate translators and rotors. This happens for instance when some degree of freedom of the robot arm joins remains for this particular motion invariant. Recall that the rotors and translators are bivectors and you can commute and associate them without acting the sign.

## 4.2 Application 2: Motors for hand-eye calibration as a case of motion of lines

The well known hand-eye equation firstly formulated by Shiu and Ahmad [27] and Tsai and Lenz [28] reads

$$\mathbf{AX} = \mathbf{XB} \quad (29)$$

where $\mathbf{A} = \mathbf{A}_1 \mathbf{A}_2^{-1}$ and $\mathbf{B} = \mathbf{B}_1 \mathbf{B}_2^{-1}$ express the elimination of the transformation hand-base to world. Here matrices are represented in bold. The geometry of the hand-eye system is depicted in Figure 1. From the expression (29) the following matrix equation and a vector equation can be derived $\mathbf{R_A R_X} = \mathbf{R_X R_B}$ and $(\mathbf{R}_A - \mathbf{I})\mathbf{t}_X = \mathbf{R}_X \mathbf{t}_B - \mathbf{t}_A$. Most of the approaches estimate first the rotation matrix decoupled from the translation. The problem requires at least two motions with rotations having not parallel axes [28]. Horaud and Dornaika [18] showed the instability of the computation of the $\mathbf{A_i}$ matrices given the projective matrices $\mathbf{M}_i = \mathbf{CA}_i = (\mathbf{CR}_{A_i} \mathbf{Ct}_{A_i})$. Let us assume that the matrix of the intrinsic parameters $\mathbf{C}$ remains constant during motions and that one extrinsic calibration $\mathbf{A_2}$ is known. Introducing $\mathbf{N}_i = \mathbf{CR}_{A_i}$ and $\mathbf{n}_i = \mathbf{Ct}_{A_i}$ and replacing $\mathbf{X} = \mathbf{A_2 Y}$, we get now as the hand-eye unknown $\mathbf{Y}$. Thus the equation (29) can be reformulated as $\mathbf{A}_2^{-1} \mathbf{A}_1 \mathbf{Y} = \mathbf{YB}$. Now if $\mathbf{A}_2^{-1} \mathbf{A}_1$ is written as a function of the projection parameters it is possible to get an expression fully independent of the intrinsic parameters $\mathbf{C}$, i.e.

$$\mathbf{A}_2^{-1} \mathbf{A}_1 = \begin{bmatrix} \mathbf{N}_2^{-1} \mathbf{N}_1 & \mathbf{N}_2^{-1}(\mathbf{n}_1 - \mathbf{n}_2) \\ 0^T & 1 \end{bmatrix} = \begin{bmatrix} \mathbf{R} & \mathbf{t} \\ 0^T & 1 \end{bmatrix}. \tag{30}$$

Taking into consideration the selected matrices and relations, this result allows

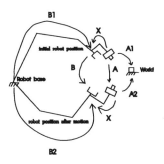

**Fig. 1.** Abstraction of the hand-eye system.

anyway to consider the formulation of the hand-eye problem again with the standard equation (29) which can be solved in terms of motors as

$$M_A M_X = M_X M_B \tag{31}$$

where $M_A = A + iA'$, $M_B = B + iB'$ and $M_X = R + iR'$. According to the congruence theorem of Chen [7] in this kind of problem the rotation angle and pitch of $M_A$ and $M_B$ remain invariant through out all the hand movements. Thus the consideration of this information can be neglected. It suffices to regard the rotation axis of the involved motors, i.e. the previous equation is reduced as the motion of the line axis of the hand towards the line axis of the camera. For that we can use the equation (19) for the computation of the real and dual components of $l_A$, i.e.

$$l_A = a + ia' = Rb\tilde{R} + i(Rb\tilde{R}' + Rb'\tilde{R} + R'b\tilde{R}). \qquad (32)$$

After some simple manipulations according the relation $\tilde{R}R' + \tilde{R}'R = 0$ we get the following matrix

$$\begin{bmatrix} a - b & [a+b]_\times & 0_{3\times 1} & 0_{3\times 3} \\ a' - b' & [a'+b']_\times & a - b & [a+b]_\times \end{bmatrix} \begin{bmatrix} R \\ R' \end{bmatrix} = 0 \qquad (33)$$

where the matrix - we will call $S$ - is a $6 \times 8$ matrix and the vector of unknowns $(R^T, R'^T)$ is 8-dimensional. More technical details about the foundations and implementation of this approach can be found in [12, 2].

## 5   Projective Invariants

In the last two decades invariant theory captured also the attention of the computer vision community. This interest in invariants results from their usefulness in tasks like reconstruction, object recognition and hand-eye calibration. These are some examples of a much wider spectrum of invariants arising in a PAC system. In this section we will show the power of geometric algebra by computing a well known invariant which results when we consider six 3D points $P_i$, $i = 1, .., 6$ in general position, represented by vectors $\{x_i, X_i\}$ in $E^3$ and $R^4$ respectively.

### 5.1   Projective invariants using 2 uncalibrated cameras

3D projective invariants can be formed from these points, and an example of such an invariant is

$$Inv = \frac{[X_1 X_2 X_3 X_4][X_4 X_5 X_2 X_6]}{[X_1 X_2 X_4 X_5][X_3 X_4 X_2 X_6]}. \qquad (34)$$

It will be highly desirable to compute the brackets $[X_i X_j X_k X_l]$ simply in terms of **image coordinates** of points $P_i$, $P_j$, $P_k$, $P_l$, in order to compute this invariant straightforwardly. Carlsson [6] discussed the computation of such invariants from a pair of images in terms of the image coordinates and the fundamental matrix, $F$, using the dual algebra. Subsequent work by Csurka and Faugeras [11] discussed corrections to Carlsson's expressions by including a series of scale factors. In contrast using geometric algebra we benefit of the duality principle and the projective split which allows us to simplify enormously the algebraic manipulation of the equations. Consider the scalar $S_{1234}$ formed from the bracket of 4 points

$$S_{1234} = [X_1 X_2 X_3 X_4] = (X_1 \wedge X_2 \wedge X_3 \wedge X_4)I_4^{-1} = (X_1 \wedge X_2) \wedge (X_3 \wedge X_4)I_4^{-1}. \qquad (35)$$

The quantities $(\mathbf{X}_1 \wedge \mathbf{X}_2)$ and $(\mathbf{X}_3 \wedge \mathbf{X}_4)$ represent the line joining points $P_1$ and $P_2$, and $P_3$ and $P_4$. Let us represent the optical centres of both cameras by $\boldsymbol{a}_0$ and $\boldsymbol{b}_0$ and their image planes by $\{\boldsymbol{a}_1, \boldsymbol{a}_2, \boldsymbol{a}_3\}$ and $\{\boldsymbol{b}_1, \boldsymbol{b}_2, \boldsymbol{b}_3\}$. Let the projection of points $\{P_i\}$ through the centres of projection onto the image planes be given by the vectors $\{\boldsymbol{a}_i'\}$ and $\{\boldsymbol{b}_i'\}$ which are ordinary vectors in $E^3$. The representations of these vectors in $R^4$ will be $\mathbf{A}_i, \mathbf{B}_i, \mathbf{A}_i', \mathbf{B}_i'...$, etc.

In [3] it is shown that the bracket of these 4 points (in $R^4$) can be equated as

$$S_{1234} = [\mathbf{X}_1\mathbf{X}_2\mathbf{X}_3\mathbf{X}_4] \equiv [\mathbf{A}_0\mathbf{B}_0\mathbf{A}_{1234}'\mathbf{B}_{1234}']. \tag{36}$$

This is achieved by the process of splitting up the bracket into two parts, $\mathbf{X}_1 \wedge \mathbf{X}_2$ and $\mathbf{X}_3 \wedge \mathbf{X}_4$ and then expressing each of these lines (bivectors) as the meet of two planes (trivectors). During this algebraic computation, since we are working in $R^4$, we are effectively ignoring any scale factors due to the arbitrary choices of the $\gamma_4$ components. Thus, when we take ratios of brackets to form our invariants we must ensure that, if we want to express the brackets in the form of equation (36), the same decomposition of $\mathbf{X}_i \wedge \mathbf{X}_j$ must occur in the numerator and denominator so that these arbitrary factors cancel. In the case of $Inv$, we have

$$Inv = \frac{\{(\mathbf{X}_1\wedge\mathbf{X}_2)\wedge(\mathbf{X}_3\wedge\mathbf{X}_4)\}I_4{}^{-1}\{(\mathbf{X}_4\wedge\mathbf{X}_5)\wedge(\mathbf{X}_2\wedge\mathbf{X}_6)\}I_4{}^{-1}}{\{(\mathbf{X}_1\wedge\mathbf{X}_2)\wedge(\mathbf{X}_4\wedge\mathbf{X}_5)\}I_4{}^{-1}\{(\mathbf{X}_3\wedge\mathbf{X}_4)\wedge(\mathbf{X}_2\wedge\mathbf{X}_6)\}I_4{}^{-1}}. \tag{37}$$

Expanding the bracket in equation (36) by expressing the intersection points in terms of the $\mathbf{A}$'s and $\mathbf{B}$'s ($\mathbf{A}_i' = \alpha_{ij}\mathbf{A}_j$ and $\mathbf{B}_i' = \beta_{ij}\mathbf{B}_j$) and defining a matrix $\tilde{\boldsymbol{F}}$ such that

$$\tilde{F}_{ij} = [\mathbf{A}_0\mathbf{B}_0\mathbf{A}_i\mathbf{B}_j] \tag{38}$$

and vectors $\boldsymbol{\alpha}_{1234} = (\alpha_{1234,1}, \alpha_{1234,2}, \alpha_{1234,3})$ and $\boldsymbol{\beta}_{1234} = (\beta_{1234,1}, \beta_{1234,2}, \beta_{1234,3})$ it is easy to see that we can write $S_{1234} = \boldsymbol{\alpha}^T{}_{1234}\tilde{\boldsymbol{F}}\boldsymbol{\beta}_{1234}$ [6]. The ratio

$$Inv = \frac{(\boldsymbol{\alpha}^T{}_{1234}\tilde{\boldsymbol{F}}\boldsymbol{\beta}_{1234})(\boldsymbol{\alpha}^T{}_{4526}\tilde{\boldsymbol{F}}\boldsymbol{\beta}_{4526})}{(\boldsymbol{\alpha}^T{}_{1245}\tilde{\boldsymbol{F}}\boldsymbol{\beta}_{1245})(\boldsymbol{\alpha}^T{}_{3426}\tilde{\boldsymbol{F}}\boldsymbol{\beta}_{3426})} \tag{39}$$

is therefore seen to be an invariant. Note that equation (39) is invariant whatever values of the $\gamma_4$ components of the vectors $\mathbf{A}_i, \mathbf{B}_i, \mathbf{X}_i$ etc. are chosen. A confusion arises if we attempt to express the $Inv$ of Eq. (39) in terms of what we actually observe, i.e. the 3D image coordinates and the fundamental matrix calculated from these image coordinates. To avoid that it is necessary to transfer the computations of Eq. (39) carried out in $R^4$ to 3D. Let us explain now this procedure.

A point $P_i$ will be projected onto a point in image plane 1, say $\boldsymbol{a}_i'$, which can be written as

$$\boldsymbol{a}_i' = \boldsymbol{a}_1 + \lambda_i(\boldsymbol{a}_2 - \boldsymbol{a}_1) + \mu_i(\boldsymbol{a}_3 - \boldsymbol{a}_1) = \delta_{i1}\boldsymbol{a}_1 + \delta_{i2}\boldsymbol{a}_2 + \delta_{i3}\boldsymbol{a}_3 \tag{40}$$

so that $\sum_{j=1}^{3} \delta_{ij} = 1$. Similarly, we have $\boldsymbol{b}'_i = \epsilon_{i1}\boldsymbol{b}_1 + \epsilon_{i2}\boldsymbol{b}_2 + \epsilon_{i3}\boldsymbol{b}_3$ (so that $\sum_{j=1}^{3} \epsilon_{ij} = 1$). Using the projective split we can now write the $\alpha_{ij}$'s and $\beta_{ij}$'s in terms of the $\delta_{ij}$'s and $\epsilon_{ij}$'s:

$$\alpha_{ij} = \frac{\mathbf{A}'_i \cdot \gamma_4}{\mathbf{A}_j \cdot \gamma_4} \delta_{ij} \qquad \beta_{ij} = \frac{\mathbf{B}'_i \cdot \gamma_4}{\mathbf{B}_j \cdot \gamma_4} \epsilon_{ij} \tag{41}$$

The 'fundamental' matrix $\tilde{F}$ is such that $\boldsymbol{\alpha}^T_i \tilde{F} \boldsymbol{\beta}_i = 0$, if $\boldsymbol{\alpha}_i$ and $\boldsymbol{\beta}_i$ are the vectors of coefficients of the points in planes 1 and 2 produced by the same world point $P_i$. Now, given more than eight pairs of corresponding observed points in the two planes, $(\boldsymbol{\delta}_i, \epsilon_i)$, $i = 1, .., 8$, we can form an 'observed' fundamental matrix $F$ such that

$$\boldsymbol{\delta}^T_i F \epsilon_i = 0. \tag{42}$$

This $F$ can be found by some method such as the Longuet-Higgins 8-point algorithm [23] or, more correctly, by some method which gives an $F$ which has the true structure [24]. Therefore, if we define $\tilde{F}$ by

$$\tilde{F}_{kl} = (\mathbf{A}_k \cdot \gamma_4)(\mathbf{B}_l \cdot \gamma_4) F_{kl} \tag{43}$$

then it follows from equations (41) that

$$\alpha_{ik} \tilde{F}_{kl} \beta_{il} = (\mathbf{A}'_i \cdot \gamma_4)(\mathbf{B}'_i \cdot \gamma_4) \delta_{ik} F_{kl} \epsilon_{il}. \tag{44}$$

If $F$ is estimated then an $\tilde{F}$ defined as in equation (43) will also act as a fundamental matrix in $R^4$.

Now let us look again at the invariant $Inv$. According to the above, we can write the invariant as

$$Inv = \frac{(\boldsymbol{\delta}^T_{1234} F \epsilon_{1234})(\boldsymbol{\delta}^T_{4526} F \epsilon_{4526})\phi_{1234}\phi_{4526}}{(\boldsymbol{\delta}^T_{1245} F \epsilon_{1245})(\boldsymbol{\delta}^T_{3426} F \epsilon_{3426})\phi_{1245}\phi_{3426}} \tag{45}$$

where $\phi_{pqrs} = (\mathbf{A}'_{pqrs} \cdot \gamma_4)(\mathbf{B}'_{pqrs} \cdot \gamma_4)$. We see therefore that the ratio of the $\boldsymbol{\delta}^T F \epsilon$ terms which resembles the expression for the invariant in $R^4$, but uses only the observed coordinates and the estimated fundamental matrix, will not be an invariant. Instead, we need to include the factors $\phi_{1234}$ etc., which do not cancel. It is relatively easy to show [20] that these factors can be formed as follows. Since $\boldsymbol{a}'_3$, $\boldsymbol{a}'_4$ and $\boldsymbol{a}'_{1234}$ are collinear we can write $\boldsymbol{a}'_{1234} = \mu_{1234}\boldsymbol{a}'_4 + (1 - \mu_{1234})\boldsymbol{a}'_3$. Then, by expressing $\mathbf{A}'_{1234}$ as the intersection of the line joining $\mathbf{A}'_1$ and $\mathbf{A}'_2$ with the plane through $\mathbf{A}_0, \mathbf{A}'_3, \mathbf{A}'_4$ we can projective split and equate terms to give

$$\frac{(\mathbf{A}'_{1234} \cdot \gamma_4)(\mathbf{A}'_{4526} \cdot \gamma_4)}{(\mathbf{A}'_{3426} \cdot \gamma_4)(\mathbf{A}'_{1245} \cdot \gamma_4)} = \frac{\mu_{1245}(\mu_{3426} - 1)}{\mu_{4526}(\mu_{1234} - 1)}. \tag{46}$$

The values of $\mu$ are readily obtainable from the images. The factors $\mathbf{B}'_{pqrs} \cdot \gamma_4$ are found in a similar way so that if $\boldsymbol{b}'_{1234} = \lambda_{1234}\boldsymbol{b}'_4 + (1 - \lambda_{1234})\boldsymbol{b}'_3$ etc., the overall expression for the invariant becomes

$$Inv = \frac{(\delta^T_{1234}\boldsymbol{F}\epsilon_{1234})(\delta^T_{4526}\boldsymbol{F}\epsilon_{4526})}{(\delta^T_{1245}\boldsymbol{F}\epsilon_{1245})(\delta^T_{3426}\boldsymbol{F}\epsilon_{3426})}\frac{\mu_{1245}(\mu_{3426}-1)}{\mu_{4526}(\mu_{1234}-1)}\cdot\frac{\lambda_{1245}(\lambda_{3426}-1)}{\lambda_{4526}(\lambda_{1234}-1)}. \quad (47)$$

Thus, to summarize, given the coordinates of a set of 6 corresponding points in the two image planes (where these 6 points are projections from arbitrary world points but with the assumption that they are not coplanar) we can form 3D projective invariants provided we have some estimate of $\boldsymbol{F}$. A more detailed discussion on this issue you can find in [22].

## 6 Projective Structure Using n Uncalibrated Cameras

In this section we present the application of cross-ratio [20] for computing the *projective depth* discovered by Sashua [26]. This can be easily calculated using the cross-ratio of projected points lying on an epipolar line of any of the n cameras. This relation remains constant also for the ratio of the segments of an optical ray delimited by a tetrahedron or reference frame as is depicted in the Figure 2.

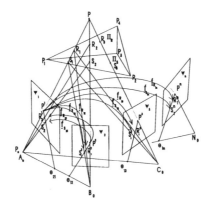

**Fig. 2.** Invariant projective depth using n uncalibrated cameras.

### 6.1 Homomorphic transformations

Consider a world point $P$ and 4 other distinct points $P_i, i = 1, 2, 3, 4$ defining a tetrahedron. Let $\Pi_R = P_1 \wedge P_3 \wedge P_4$ and $\Pi_S = P_1 \wedge P_2 \wedge P_3$ and assume $P$ does not lie on either of these two planes – see figure 2. Let $\mathbf{R}_i$ and $\mathbf{S}_i$ be the intersections of the line joining the optical centre of the $i$th camera with point $P$ with the planes $\Pi_R$ and $\Pi_S$, e.g. $\mathbf{R}_1 = \Pi_R \vee (\mathbf{A}_0 \wedge P)$. Let $\mathbf{R}_i^n$ and $\mathbf{S}_i^n$ be the projections of the points $\mathbf{R}_i$ and $\mathbf{S}_i$ onto the $n$th image planes – e.g. $\mathbf{R}_1^2 = (\mathbf{B}_0 \wedge \mathbf{R}_1) \vee (\mathbf{B}_1 \wedge \mathbf{B}_2 \wedge \mathbf{B}_3)$ etc. Note that $\mathbf{R}_i^i$ and $\mathbf{S}_i^i$ are simply the projections of the world point $P$ onto the $i$th image plane, e.g. $\mathbf{R}_1^1 = \mathbf{S}_1^1 = (\mathbf{A}_0 \wedge P) \vee (\mathbf{A}_1 \wedge \mathbf{A}_2 \wedge \mathbf{A}_3)$. Let us call the $i$th image plane $\psi_i$.

In order to compute a cross-ratio which will be defined later, we must be able to calculate the image coordinates of $\mathbf{R}_i^n, \mathbf{S}_i^n$. We can do this by finding the *homomorphic* transformations or homographies relating projected points in one image plane to the projected points in another. Consider the homography between image planes $\psi_i$ and $\psi_j$ due to the plane $\Pi_S$. If the projections of $\mathbf{P}_1, \mathbf{P}_2, \mathbf{P}_3$ onto $\psi_i$ and $\psi_j$ are $\{\mathbf{P}_k^i\}$ and $\{\mathbf{P}_k^j\}$, for $k = 1, 2, 3$, then the linear function $\underline{f}_{ij}^S$ representing this transformation must satisfy

$$\underline{f}_{ij}^S(\mathbf{P}_k^i) = \mathbf{P}_k^j \quad \text{for } k = 1, 2, 3. \tag{48}$$

Here we are working in $R^3$ so that the non-linear projective transformations in $\mathcal{E}^2$ (plane to plane) become linear – the above linear-function representation is outlined in [21]. Similarly, the corresponding homography due to the plane $\Pi_R$ is represented by the linear function $\underline{f}_{ij}^R$ given by

$$\underline{f}_{ij}^R(\mathbf{P}_k^i) = \mathbf{P}_k^j \quad \text{for } k = 1, 3, 4. \tag{49}$$

If four point correspondences from each plane are known then these linear functions can be recovered up to a scale factor by simple linear techniques. Since the homographies must map the epipole in one image plane onto the epipole in the other, we can choose the epipoles as the fourth point if these are known; $\underline{f}_{ij}^R(\mathbf{E}_{ji}) = \mathbf{E}_{ij}$ etc.

## 6.2 Computing the projective depth

The fundamental projective invariant in 1D is the cross-ratio. We can form a cross-ratio from the collinear points $\mathbf{P}, \mathbf{R}_1, \mathbf{S}_1, \mathbf{A}_0$, namely

$$\rho = \frac{(\mathbf{R}_1 \wedge \mathbf{A}_0)I_2^{-1}}{(\mathbf{P} \wedge \mathbf{A}_0)I_2^{-1}} \frac{(\mathbf{P} \wedge \mathbf{S}_1)I_2^{-1}}{(\mathbf{R}_1 \wedge \mathbf{S}_1)I_2^{-1}}. \tag{50}$$

The cross-ratio $\rho$ will be invariant when projected onto any other image plane. Consider this cross-ratio in the image plane of the second camera;

$$\rho = \frac{(\mathbf{R}_1^2 \wedge \mathbf{E}_{12})I_2^{-1}}{(\mathbf{P}^2 \wedge \mathbf{E}_{12})I_2^{-1}} \frac{(\mathbf{P}^2 \wedge \mathbf{S}_1^2)I_2^{-1}}{(\mathbf{R}_1^2 \wedge \mathbf{S}_1^2)I_2^{-1}}. \tag{51}$$

If we know the linear functions $\underline{f}_{12}^S, \underline{f}_{12}^R$, then we can write this ratio as

$$\rho = \frac{(\underline{f}_{12}^R(\mathbf{P}^1) \wedge \mathbf{E}_{12})I_2^{-1}}{(\mathbf{P}^2 \wedge \mathbf{E}_{12})I_2^{-1}} \frac{(\mathbf{P}^2 \wedge \underline{f}_{12}^S(\mathbf{P}^1))I_2^{-1}}{(\underline{f}_{12}^R(\mathbf{P}^1) \wedge \underline{f}_{12}^S(\mathbf{P}^1))I_2^{-1}}, \tag{52}$$

and the general form for the $i$-camera and $j$-camera

$$\rho = \left( \frac{(\underline{f}_{ij}^R(\mathbf{P}^i) \wedge \mathbf{E}_{ij})I_2^{-1}}{(\mathbf{P}^j \wedge \mathbf{E}_{ij})I_2^{-1}} \right) \left( \frac{(\mathbf{P}^j \wedge \underline{f}_{ij}^S(\mathbf{P}^i))I_2^{-1}}{(\underline{f}_{ij}^R(\mathbf{P}^i) \wedge \underline{f}_{ij}^S(\mathbf{P}^i))I_2^{-1}} \right). \tag{53}$$

The term in the right bracket is termed *projective depth* in [26]. If we have a number of views available then, in this framework, a more robust estimate of $k$ would be given by

$$k = \frac{1}{n} \sum_{(i \neq j)} \frac{(\mathbf{P}^j \wedge \underline{f}^S_{ij}(\mathbf{P}^i)) I_2^{-1}}{(\underline{f}^R_{ij}(\mathbf{P}^i) \wedge \underline{f}^R_{ij}(\mathbf{P}^i)) I_2^{-1}}, \tag{54}$$

where $n$ is the number of estimates used.

Finally according with $k$ the reconstruction of 3D coordinates of points is straightforward [26]. Considering the relations $\mathbf{P} \cong \mathbf{R}_1 + k\mathbf{S}_1$ and that of the fifth point (mapped onto the focal center $\mathbf{A}_0$) $\mathbf{P}_0 \cong \mathbf{R}_1 + k'\mathbf{S}_1$ (scaled so that k'=1), it can be seen that the depth $k$ is actually an invariant up to uniform scale. Using this $k$ we can reconstruct for each five points in general position (three lying on $\Pi_1$ and three lying on $\Pi_2$) the 3D coordinates of a point $\mathbf{P}_i$.

## 7 Conclusions

This paper presented the Clifford algebra in its geometric interpretation as a common language for the treatment of problems of robotics and computer vision. The authors focused in the geometry of 3D and 4D spaces which are necessary for the representation and manipulation of basic geometric entities required in those areas. In the first field we analyze the 3D and 4D modelling of motion in complicated mechanisms and in the hand-eye calibration problem. A 3D motion or general displacement is a nonlinear transformation, but linear if it is represented in 4D. This motivates us to use the 4D geometric algebra to solve in a linear manner problems involving 3D motions. An very interesting example is the hand-eye calibration. This requires a nonlinear solution strategy if it is treated in 3D. However when the 3D representation is extended to the 4D space using the motor algebra the problem of computing of the unknown motion becomes linear. In the second part of the paper it is shown how geometric algebra can be used for the algebra of incidence useful in the projective space. For intersections of planes, lines etc. and for the discussion of projective transformations we find that we do not need to invoke the standard concepts or machinery of classical projective geometry, all that is needed is the idea of the *projective split* relating the quantities in $R^4$ to quantities in our 3D world and the algebra of incidence. The case of computing invariants using n uncalibrated cameras is analyzed. Here the duality principle and projective split help to reduce the complexity of the computation. Finally using the cross-ratio of points lying in the epipolar lines of n cameras the projective reconstruction is addressed. The authors believe that the geometric algebra approach opens a new way to deal with problems in computer vision and differs with the standard approaches substantially due to its powerful algebraic and geometric properties.

Since PAC systems require different mathematical techniques for visual and motor signal processing, the construction of PAC systems demands of a framework where should be possible the fusion of the fields of signal theory, computer vision, robotics and neural computing. It have been seen by the problems treated in this paper that geometric algebra indeed has a powerful representation capability and a strong geometric basis. That is why the authors believe that it is a competitive language to provide a unified approach for the design and implementation of real time PAC systems. At last it is easy to identify that the disparate mathematical techniques used nowadays in PAC are simply special cases of the wider class of mathematics provided by geometric algebra.

**Acknowledgment**

The authors are thankful to Gerald Sommer for his valuable suggestions and for the enriching disscussions. Eduardo Bayro-Corrochano is supported by the Deutsche Forschungsgemeinschaft project SO 320-2-1 "Geometrische Algebra ein Repräsentationsrahmen für den Wahrnehmungs-Handlungs-Zyklus" and Joan Lasenby by a Royal Society University Research Fellowship.

# References

1. Bayro-Corrochano, E. and Lasenby, J. 1995. Object modelling and motion analysis using Clifford algebra. In Proceedings of Europe-China Workshop on *Geometric Modeling and Invariants for Computer Vision*, Ed. Roger Mohr and Wu Chengke, Xi'an, China, April, pp. 143-149.

2. Bayro-Corrochano, E., Daniilidis, K. and Sommer, G. 1997. Hand-Eye calibration in terms of motion of lines using Geometric Algebra. In *Proc. of the 10th Scandinavian Conference on Image Analysis SCIA '97*, Lappeenranta, Finland, June 9-11, Vol.I , pp. 397-404.

3. Bayro-Corrochano E., Lasenby J., Sommer G. Geometric Algebra: A framework for computing point and line correspondences and projective structure using n uncalibrated cameras *IEEE Proceedings of ICPR'96 Viena, Austria*, Vol. I, pages 334-338, August, 1996.

4. Bayro-Corrochano E., Buchholz S., Sommer G. 1996. Selforganizing Clifford neural network *IEEE ICNN'96 Washington, DC*, June, pp. 120-125.

5. Blaschke, W. 1960. Kinematik und Quaternionen. VEB Deutscher Verlag der Wissenschaften, Berlin 1960.

6. Carlsson, S. 1994. The Double Algebra: and effective tool for computing invariants in computer vision. *Applications of Invariance in Computer Vision*, Lecture Notes in Computer Science 825; Proceedings of 2nd-joint Europe-US Workshop, Azores, October 1993. Eds. Mundy, Zisserman and Forsyth. Springer-Verlag.

7. Chen H. A screw motion approach to uniqueness analysis of head-eye geometry. In *IEEE Conf. Computer Vision and Pattern Recognition*, pages 145–151, Maui, Hawaii, June 3-6, 1991.

8. Chou J.C.K. and Kamel M. Finding the position and orientation of a sensor on a robot manipulator using quaternions. *Intern. Journal of Robotics Research*, 10(3):240-254, 1991.

9. Clifford, W.K. 1878. Applications of Grassmann's extensive algebra. *Am. J. Math.* 1: 350–358.

10. Clifford., W.K. 1873. Preliminary sketch of bi-quaternions. *Proc. London Math. Soc.*, 4:381–395.

11. Csurka, G. and Faugeras, O. 1995. Computing three-dimensional projective invariants from a pair of images using the Grassmann-Cayley algebra. In Proceedings of Europe-China Workshop on *Geometric Modeling and Invariants for Computer Vision*, Ed. Roger Mohr and Wu Chengke, Xi'an, China, April, pp. 150-157.

12. Daniilidis K. and Bayro-Corrochano E. The dual quaternion approach to hand-eye calibration. *IEEE Proceedings of ICPR'96 Viena, Austria*, Vol. I, pages 318-322, August, 1996.

13. Grassmann, H. 1877. Der Ort der Hamilton'schen Quaternionen in der Ausdehnungslehre. *Math. Ann.*, 12: 375.

14. Hestenes, D. 1986. New Foundations for Classical Mechanics *D. Reidel*, Dordrecht.

15. Hestenes, D. 1991. The design of linear algebra and geometry. *Acta Applicandae Mathematicae*, 23: 65–93.

16. Hestenes, D. and Sobczyk, G. 1984. Clifford Algebra to Geometric Calculus: A Unified Language for Mathematics and Physics. *D. Reidel*, Dordrecht.

17. Hestenes, D. and Ziegler, R. 1991. Projective geometry with Clifford algebra. *Acta Applicandae Mathematicae*, 23: 25–63.

18. Horaud R. and Dornaika F. Hand-eye calibration. *Intern. Journal of Robotics Research*, 14:195–210, 1995.

19. Lasenby, J., Fitzgerald, W.J., Lasenby, A.N. and Doran, C.J.L. 1997. New geometric methods for computer vision. To appear in *International Journal of Computer Vision*.

20. Lasenby, J., Bayro-Corrochano, E., Lasenby, A. and Sommer, G. 1996. A New Methodology for the Computation of Invariants in Computer Vision. Cambridge University Engineering Department Technical Report, CUED/F-INENG/TR.244.

21. Lasenby, J. , Bayro-Corrochano E.J., Lasenby, A. and Sommer G. 1996. A new methodology for computing invariants in computer vision. IEEE Proceedings of ICPR'96, Viena, Austria, Vol. I, pages 393-397, August, 1996.

22. Lasenby, J. , Bayro-Corrochano E.J. . 1997. Computing 3D projective invariants from points and lines. To appear in *Int. Conference on Analysis of Images and Patterns CAIP'97*, Kiel, Germany, September, 1997.

23. Longuet-Higgins, H.C. 1981. A computer algorithm for reconstructing a scene from two projections. *Nature*, 293: 133–138.

24. Luong, Q-T. and Faugeras, O.D. 1996. The fundamental matrix: theory, algorithms and stability analysis. *IJCV*, 17: 43–75.

25. McCarthy J.M. Dual orthogonal matrices in manipulator kinematics IJRR, Vol.5, Number 2, 1986.

26. Shashua, A. 1994. Projective structure from uncalibrated images: structure from motion and recognition PAMI, 16(8), 778:790.

27. Shiu Y.C. and Ahmad S. Calibration of wrist-mounted robotic sensors by solving homogeneous transform equations of the form $AX = XB$. *IEEE Trans. Robotics and Automation*, 5:16–27, 1989.

28. Tsai R.Y. and Lenz R.K. A new technique for fully autonomous and efficient 3D robotics hand/eye calibration. *IEEE Trans. Rob. and Autom.*, 5:345–358, 1989.

29. Sommer G., Bayro-Corrochano E. and Bülow T. 1997. Geometric algebra as a framework for the perception–action cycle. Workshop on Theoretical Foundation of Computer Vision, Dagstuhl, March 13-19, 1996, Springer Wien.

# Structure from Translational Observer Motion

Ambjörn Naeve

Computational Vision and Active Perception Laboratory
Department of Numerical Analysis and Computing Science
Royal Institute of Technology, 100 44 Stockholm, Sweden

ABSTRACT: This work presents a unified, globally based geometric frame-work, using congruence geometry, for the description and computation of structure from motion. It is based on projectively invariant tangent information in a sequence of monocular images, i.e. occluding contours under general perspective. The strength of the framework is demonstrated by applying it to the case of *translational* observer motion, a type of motion that is of great practical importance since it can be easily implemented with the help of various gyroscopic devices. Introducing a simple technique for the computation of the direction of motion ("Focus Of Expansion") as a function of time, the recovery of translational observer motion is reduced to a problem of determining its speed. From such speed information, we show how to reconstruct the observer motion - as well as a set of silhouette curves on the observed target - and illustrate with a few simulated examples. The FOE-reconstruction method is then generalized from the real to the complex domain, showing how to combine conjugate complex geometric elements in order to obtain real geometric information concerning the direction of observer motion. We conclude by applying this method to real image data.

## 1 Introduction

The problem of reconstructing geometric information in a scene from a sequence of mobile observations has been intensely studied within the computer vision community. Such information - notably target structure and observer motion - can be used for the purpose of object recognition and for various forms of geometric reasoning, e.g. motion planning and obstacle avoidance. Within this context, structure and motion from the outlines of space-curves or curved surfaces constitutes a difficult problem that has attracted increased attention. Among the early contributions to this field, Koenderink [(1), (2)] has related the curvature of an apparent contour to the Gaussian curvature of the observed surface; Giblin and Weiss [(3)] have worked on the positioning of objects from their occluding contours, assuming planar camera motion - a restriction that was removed by Vaillant in [(4)]. Among the later work we mention Cipolla [(5)] and Cipolla and Blake [(6)], who showed how to recover the geometry of the target surface from the deformation of its apparent contours. They assumed that the observer motion is known and used the so called epipolar parametrization, which is based on the classical conjugacy condition between a contour generator (rim) of a surface and the corresponding tangential ray of sight. Related work has been done by Vaillant and Faugeras [(7)], who used three closely related images of the surface rim - taken under known observer motion - in order to recover the local curvature properties of the surface up to second order. Also, for the case of rigid curve targets,

Faugeras and Papadopoulo [(8)] have shown how to use the spatio-temporal image of a space curve in order to derive a constraint on the viewer motion from second order spatio-temporal derivatives. Giblin and Weiss [(9)] have given criteria for the breakdown of the epipolar parametrization and Cipolla, Åström and Giblin [(10)] have used such points (which they call frontier points) in order to reconstruct the observer's motion from the occluding surface contours. Kutulakos and Dyer [(11)] have formulated rules to control observer motion so as to connect local surface patch reconstructions in a global reconstruction process of the surface. Zhao and Mohr [(12)] have used B-spline surface patches, reconstructed with the aid of the epipolar parametrization, in order to achieve global reconstruction under known observer motion.

The underlying theme of these developments is to use locally based (i.e. differential) methods in order to achieve reconstruction under more and more general forms of observer motion. In this paper we advocate a different approach. Restricting the observer motion to *translations*, we use a classical line geometric framework (notably congruence geometry) as a means to describe the geometry of the various target-observer configurations. Our framework is based on the *<Target\Observer> congruence*, which was introduced by Naeve in [(13)]. This is a type of congruence (i.e. two-parametric family of lines) that appears naturally in scene reconstruction problems in vision. It corresponds to a sequence of observations of a curve- or surface-like target by a moving, monocular observer. This congruence can be decomposed into two families of *developables* that carry information concerning the global geometry of the target surface as well as the motion of the observer. A great advantage of this approach is that it does not make any a priori assumptions about the shape of the observed target, thereby allowing a uniform treatment of various types of targets - such as lines and curves, as well as polyhedral and smoothly curved surfaces.

The restriction to translational motion is natural since it is performed by biological observers in many situations - notably by hunting predators. When e.g. a cheetah is hunting, it is easy to observe how every muscle in its body cooperates in order to keep the motion of its head translational - a fact which greatly facilitates the computation of the relative direction of motion of its prey. Moreover, such translationally balanced motion can be easily implemented in machine vision systems by the help of gyroscopic devices, as demonstrated e.g. by the so called "steady-cam" camera systems that are becoming increasingly popular within the television industry.

## 2  Background

A smoothly parametrized family of lines in 3d-space is called a *ruled surface*, a *congruence* or a *complex*, if it depends respectively on 1, 2, or 3 parameters. Examples of these three types are given e.g. by the generator lines of a *1-sheeted hyperboloid*, the *normals to an arbitrary surface* and the *tangents to an arbitrary surface*. These examples are all special in the sence that they are highly non-generic: A 1-sheeted hyperboloid can be considered as a ruled surface in *two* different ways, which is characteristic of *quadric* surfaces, the normals to a surface form a  special type of congruence, called a *normal congruence*, and the tangents to a surface constitute a special type of complex, called a *tangent complex*. The complex of lines that intersect a fixed curve is called a *secant complex* - unless the curve itself is a line. In this case, the complex of intersecting lines is known as a *singular complex*. These concepts were once well known and much used within the mathematical community. The reader is referred to Naeve [(13), (14)] or Eisenhart [(15)] for more information on this subject.

## 3    The <Target | Observer> Congruence

As mentioned above, a *congruence* is a smooth *2-parameter family of lines*. In general, these lines envelope two surface-patches, $S_1$ and $S_2$, known as the two *focal sheets* of the congruence. Taken together they constitute its *focal surface*. A congruence can be represented by its two focal sheets, in which case we denote the congruence by $<S_1|S_2>$. The lines of the congruence $<S_1|S_2>$ are precisely the lines that are tangents to both $S_1$ and $S_2$.

We will now describe the observational configuration that forms the geometric foundation for the application of congruence geometry to vision. Consider a  point like observer $O$, that is moving along an (unknown) curve $\Gamma$. While it is occupying the successive positions $O(t) = \Gamma(t)$, the observer is recording a motion picture of the silhouette $S(t)$ of a (possibly) moving target $T(t)$, using a monocular ("pin-hole") camera that is pointing in a *fixed* direction. We can think of the silhouette $S(t)$ as being recorded either on a  spherical screen with center $O(t)$ - e.g. the gaussian sphere $\mathbf{S}^2_{O(t)}$ - or on some plane screen $\pi(t)$ that is rigidly attached to $O(t)$. In the discussion below, we will take $\pi(t)$ to be the tangent plane to the gaussian sphere at its north pole, as illustrated in Figure (1).

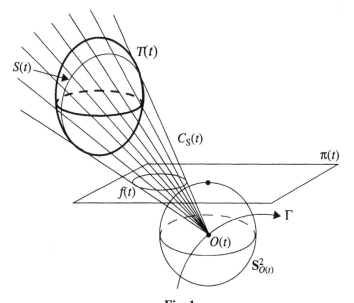

**Fig. 1.**

Keeping the direction of the camera fixed during the recording of the film is equivalent to the following geometric condition: During the motion of $O$, both $\mathbf{S}^2$ and $\pi$ are *translated* through space without any accompanying rotation, i.e. their directions remain fixed in the surrounding 3d-space. In practice, this can be achieved e.g. by the aid of an inertial navigational system based on the kind of gyroscopic compass employed by modern airplanes.

Unless we explicitly state otherwise, we will always assume that each target $T(t)$ is either a line, a curve, or a surface patch. In this case, the observed silhouette $S(t)$ is assumed to be a piecewise smooth curve, corresponding to an *observed silhouette cone* $C_S(t)$. However, what the observer actually *records* is only the *direction*s of these lines, forming the *recorded silhouette cone* $C_R(t)$. Hence, to each observation sequence there is associated a *2-parameter family of lines*, i.e. a congruence, forming a *1-parameter family of observed silhouette cones*,

**Def. 1:**     The 1-parameter family of observed silhouette cones $C_S(t)$ is called the *target-observer congruence* and will be denoted by $\langle T|O \rangle$.

# 4   The Global Structure of the *<T|O>* Configurations

Let us introduce the notation $\langle T \rangle$ for the geometric locus of lines the are tangent to the target $T(t)$ for some moment $t$ during the time interval of observation, that is $\langle T \rangle = \bigcup_t \langle T(t) \rangle$.

In the table below we present a summary of the global line-geometric structure of the various forms of *<T>* and *<T|O>* configurations that can appear in the process of observing a point-, line-, curve- or surface target, separating between the case of a stationary and a moving target.

| Target | ⟨Target⟩ | ⟨Target|Observer⟩ |
|---|---|---|
| *FixedPoint* | *StarCongruence* | *Cone* |
| *MovingPoint* | $\bigcup_t StarCongruence\,(t)$ | *RuledSurface* |
| *FixedLine* | *SingularComplex* | *DevelopedDirectedCongr* |
| *MovingLine* | $\bigcup_t SingularComplex\,(t)$ | *DevelopedDirectedCongr* |
| *FixedCurve* | *SecantComplex* | *DoublyDirectedCongr* |
| *MovingCurve* | $\bigcup_t SecantComplex\,(t)$ | *DirectedCongruence* |
| *FixedSurface* | *TangentComplex* | *DirectedCongruence* |
| *MovingSurface* | $\bigcup_t TangentComplex\,(t)$ | *DirectedCongruence* |

The deduction of these configurations is straight-forward but are omitted here for lack of space. The reader is referred to Naeve [(14)] for details.

## 5   Reconstructing the <*T*|*O*> Congruence

We will now consider the fundamental problem of reconstructing the <*T*|*O*> congruence. Before proceeding further we introduce some useful terminology:

**Def. 2:**      At a given moment $t$, the relative motion of target and observer will be called *transversal*, if the recorded silhouette curves at times $t$ and $t+\Delta$ have at least two common (real) tangents for all small enough positive $\Delta$, i.e. if the envelope of the recorded silhouette curves has at least two real tangents at time $t$.

Such a situation, for a stationary target, is depicted in Figure (2). The notation used in this figure corresponds to an infinitesimal time difference, $\Delta = \delta$, but - for the moment - let us consider the two observed silhouette cones $C_S(t)$ and $C_S(t+\delta)$ as being finitely separated in space. Now, imagine that we take a plane $\psi$ through the line that joins the points $O(t)$ and $O(t+\delta)$ and turn the plane $\psi$ around this line until it touches the target $T(t_0)$ at the point $P_1$. Denoting the corresponding plane by $\psi_1$, we obtain $\psi_1 = O(t) \vee O(t+\delta) \vee P_1$.

Since the line $O(t) \vee P_1$ is tangent to $T(t_0)$ and passes through $O(t)$, this line must belong to $C_S(t)$, and for the analogous reason the line $O(t+\delta) \vee P_1$ must belong to $C_S(t+\delta)$. It follows that the plane $\psi_1$ must be tangent to both of the cones $C_S(t)$ and $C_S(t+\delta)$. Moreover, this construction can be repeated by turning the plane $\psi$ in the opposite direction until it touches the target in another point $P_2$. For the same reason, the corresponding plane $\psi_2$ must also be tangent to both of the cones $C_S(t)$ and $C_S(t+\delta)$. Hence we can summarize in the following:

**Fact 1:** If the target and observer are subject to transversal relative motion, then for small enough positive $\Delta$, the observed silhouette cones $C_S(t)$ and $C_S(t+\Delta)$ have at least two common tangent planes.

We are now in a position to prove the following[1]:

**Fact 2:** Under transversal observer motion relative to a stationary target, the direction of the observer's motion (the so called *Focus Of Expansion*) at time $t$ is identical to the limiting position, as $\Delta \to 0$, of the point of intersection between the common tangent lines to the two recorded silhouette curves at time $t$ and time $t+\Delta$.

**Proof:**
The situation is illustrated in Figure (2). Let the recording in the plane $\pi(t)$ of the silhouette curve $S(t)$ be denoted by $f(t)$, and let $\psi_1$ and $\psi_2$ denote the two common tangent planes to the observed silhouette cones $C_S(t)$ and $C_S(t+\delta)$ [see Fact (1)]. Moreover, let $C_S^\delta(t)$ denote the translation of $C_S(t)$ along $\Gamma$ from $O(t)$ to $O(t+\delta)$, and let $f_\delta(t) = C_S^\delta(t) \cap \pi(t+\delta)$.

Now, since $C_S^\delta(t)$ is the translation of $C_S(t)$ along the observer's orbit curve, it must be identical to the recorded silhouette cone at time $t$, i.e. $C_S^\delta(t) = C_R(t)$. Therefore, the two curves $f_\delta(t)$ and $f(t+\delta)$ are identical to the two recorded silhouette curves at times $t$ respectively $t+\delta$. Moreover, since the translation takes place along the line $l_\Gamma(t) = \psi_1 \wedge \psi_2$, it follows that the two planes $\psi_1$ and $\psi_2$ must remain tangential to the translated cone $C_S^\delta(t)$. Hence, $\psi_1$ and $\psi_2$ are tangent planes to both $C_S(t)$ and $C_S^\delta(t)$, and, by construction, also to $C_S(t+\delta)$.

Consider now the two lines $l_1 = \psi_1 \wedge \pi(t+\delta)$ and $l_2 = \psi_2 \wedge \pi(t+\delta)$. Since they are located in $\psi_1$ respectively $\psi_2$, it follows that they must intersect on the line $l_\Gamma(t)$, and also that they must be tangent to the cones $C_S(t+\delta)$ and $C_S^\delta(t)$. Fur-

---

1. The author has recently discovered that Fact 2 is implicit in a paper by Cipolla , Åström and Giblin [(10), Property 1].

thermore, since $l_1$ and $l_2$ are located in the plane $\pi\,(t+\delta)$, it follows that they must be tangent to both of the recorded silhouette curves $f\,(t+\delta)$ and $f_\delta\,(t)$. Therefore, we can conclude that *the directions of the lines $l_1$ and $l_2$ are constructible from the recorded image data alone, as common tangents to two successively recorded silhouette curves.* Since the direction of motion - i.e. the Focus Of Expansion - is given by $FOE\,(t)\ =\ l_\Gamma\,(t)\wedge\pi\,(t+\delta)\ =\ l_1\wedge l_2$, we have completed the proof of Fact (2).

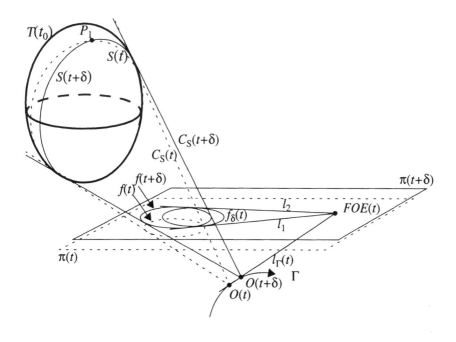

**Fig. 2.**

Note that for surface targets with a "simple enough" geometry, such as e.g. convex surfaces, . transversal motion corresponds to non-collision, while non-transversal motion corresponds to collision. For more complicated surface shapes the correspondence between transversality- and collision conditions is an interesting problem that merits further study.

From Fact (2) we can draw an interesting conclusion: Any hypothesis regarding the speed of the observer's motion and the location of the observation start point $O\,(t_0)$ constitutes the necessary input to a reconstruction process of the observer's position $O\,(t)$ as a function of time. Moreover, from the fundamental theorem of calculus, we know that variation of the constant vector $O\,(t_0)$ gives rise to congruent observer orbit curves, and therefore to congruent reconstructions of the $<\!\Pi O\!>$ congruence. Hence we can state the following

**Fact 3:**    Under transversal motion of an observer relative to a stationary target, the shape of the observer's orbit curve - as well as the shape of a net of silhouette curves and their conjugates - are reconstructible, knowing only the speed of the observer as a function of time. Knowing also the position of the starting point of observation, makes it possible to reconstruct the position of these curves in space.

## 6    Simulated Examples of *<T|O>* Congruence Reconstruction

Following the theoretical discussion presented above, we will now take a look at some examples of *<T|O>* congruence reconstruction. In this section, the examples are computer simulated in *Mathematica*®, while in the next section we will work with real image data. For reasons of simplicity, the target will always be chosen as a stationary quadric surface, where the intersection of tangent cones can be handled analytically.

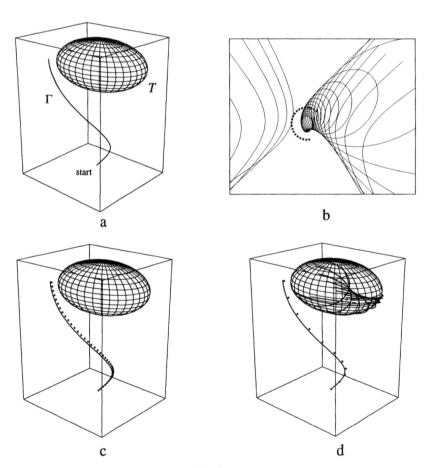

**Fig. 3.**

## 6.1 Transversal Observer Orbit Curve

In the first example, shown in Figure (3a), the stationary quadric surface target is observed transversally by an observer moving with constant speed along a circular helix orbit curve $\Gamma$. Figure (3b) shows a finite sample of the recorded silhouette curves, where the dots represent the intersections of common real tangents to neighboring curves. Hence, by Fact (2), the dotted circle represents the direction of motion as a (discretely sampled) function of time. Starting from a correct "guess" of the observer's speed and starting position $O(t_0)$, the result of reconstructing 40 points of the observer's orbit curve is shown as the dotted curve of Figure (3c). A factor 4 subsample, as well as the corresponding reconstructed silhouettes are shown in Figure (3d).

## 6.2 Non-transversal Observer Orbit Curve

Figure (4) illustrates the complications that are introduced by a non-transversal observer motion. Here the observer is moving along a circular helix, which is located so that the initial motion is transversal. Then the transversality condition is violated, and the observer starts moving towards the target. Therefore, the common tangent planes $\psi_1$ and $\psi_2$ in the proof of Fact (2) are no longer real, and the algorithm for reconstructing the direction of motion breaks down at this point. This illustrates the serious limitations of the transversality condition to the reconstruction algorithm as presented above. In order to be practically useful, we should be able to allow *generic* observer motion, and extend the algorithm to reconstruct also the non-transversal part of the observer's orbit curve. In the case of *quadric* surface targets, a method for achieving this will be presented below. This method can be naturally extended to *polynomial* surface targets, but the details of this extension will not be discussed here.

Our method is based on the fact that, for a quadric surface target, the tangent planes $\psi_1$ and $\psi_2$ will be complex-conjugates of each other, and so will the lines $l_1$ and $l_2$. Hence the line of intersection of $\psi_1$ and $\psi_2$ will still be real, and since we have made no special reality assumptions in the proof of Fact (2), the conclusion is still valid, provided only that we modify it to account for this complex-conjugacy. We formulate this as

**Fact 4:**      Under translational observer motion relative to a *quadric* target, the direction of the observer's motion at time $t$ is identical to the limiting position, as $\Delta \to 0$, of the point of intersection between one of the two pairs of complex-conjugate common tangent lines to the two recorded silhouette *conics* at time $t$ and time $t+\Delta$. If there is a *real* pair of such common tangents, then their point of intersection represents the correct direction, and in this case the corresponding observer motion is *transversal*.

Fact (4) illustrates the additional problem. Due to the fact that the motion is non-transversal, the recorded silhouette curves will be conics with no real points of intersection, and since two such conics have (in general) four common tangent lines that are complex-conjugates in pairs, we no longer have a unique reconstruction point for the direction of observer motion. Instead, we have two candidates, one of which represents the correct direction, whereas the other is a "phantom solution" introduced by the complexity of the situation. However, as will be seen below, by considering the global configuration of recorded silhouette curves, we can exclude the false solution for the following reason: The direction of motion must be closest to the part of the contours where the curves are most densely packed (See Figure (4)).

Assume that the observer motion has been started in such a way that it is inititally transversal. When the motion changes to non-transversal, the unique FOE-point will move across the border of both conics, from the outside to the inside. Hence, by keeping track of the FOE-point in the previous reconstruction step, we can discriminate the true solution from the false by a simple continuity argument. Clearly, if the time step is small enough, the correct FOE-point in the next step is the one that is closest to the previous one. Hence, always keeping track of the previous FOE-point, this continuity argument can be applied in the same way during the entire non-transversal part of the motion. If ever the motion becomes transversal again, we do not need to remember the previous FOE-point anymore.

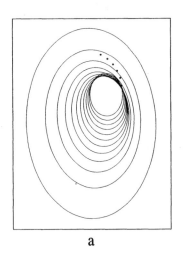 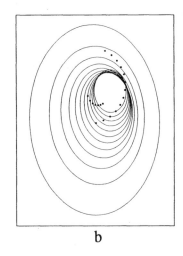

a                       b

**Fig. 4.**

Figure (4) shows the result of applying this modified type of reconstruction algorithm. The FOE-points corresponding to the transversal region of motion, with real silhouette envelope, are shown in Figure (4a), and the added FOE-points, corresponding to the intersection of conjugate-complex tangents to the silhouette envelope are shown in Figure (4b). This clearly demonstrates how the continuity condition of the direction of motion helps to discriminate between the two direction candidates.

# 7   Application to Real Image Data

We will now apply the FOE-reconstruction technique to a sequence of real images. Since we do not presently have access to equipment for controlled general 3d-motion, we have limited the experiments to detecting the motion direction under linear motion. A sequence of images of a plastic ellipsoid target[1] was recorded under linear motion of the observer. The direction of motion was chosen to be in the "colliding" region , thus assuring that the envelope of the recorded silhouette curves would be conjugate complex. For practical reasons we chose to translate the target along a straight line instead of doing the same thing with the camera. These two configurations are of course equivalent.

To arrive at an estimation of the direction of motion, we now proceed as follows: First, we perform a least square fit of a quadratic polynomial to each extracted edge-curve (i.e. recorded silhouette curve). Since we have been using a quadratic target, these fitted conics can be expected to give a good analytic approximation of the recorded silhouette curves. Second, when we have determined these approximating conics, we use them just as in the previous chapter, and determine the points of intersection to double tangents of neighboring conics in the family. This corresponds to using the time frames $t$ and $t + \Delta t$, where $\Delta t$ is the time difference between successive images. The result is shown in Figure (5a).

As can be seen, the presence of noise in our data introduces a problem. Computing instead the points of intersection to double tangents of the conics corresponding to times $t$ and $t + 2\Delta t$, we get the result of Figure (5b). For the sake of simplicity, we will refer to such curves as being related by step-2. Figures (5c) and (5d) show the corresponding computation for conics related by step-3 respectively step-4. As might be expected, we infer from Figures 5a, b, c, d that increasing the step greatly reduces the noise. If we have reason to trust one of the silhouette curves significantly more than the others, we can use it repeatedly and compute the points of intersection to double tangents of it and the other curves in the family[2]. The result of using the initial silhouette curve (smallest of the conics) in this way is shown in Figure (5e). Finally, to get an idea of the "correct" result, an idealized configuration is shown in Figure (5f). The conics here are formed from a dualized linear combination of the first and the last conic in the family, which means that all pairs of conics from this new family have the same common tangents.

---

1.  This target was manufactured by T. Henderson at the University of Utah and  made available by the courtesy of O.D. Faugeras at INRIA.

2.  We might, for instance, have kept the camera at rest for a long time, and therefore have been able to determine the corresponding silhouette curve with a significantly better precision than the others.

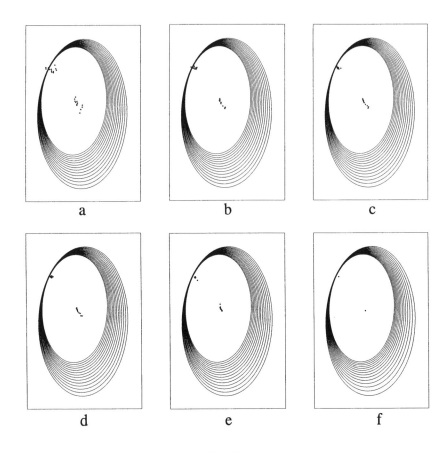

**Fig. 5.**

Note that we can exclude the false direction candidate on account of the global structure of the family of curves in Figure (5). Since the recorded silhouettes are much closer to each other at the upper left corner of the picture, the true direction of motion must be the point closest to this area. If the curves would have all touched each other in one point, then the camera would in fact have been moving towards that point - tangentially to the target object.

## 8 Conclusions and Future Work

In this paper we have discussed the scene reconstruction problem under translational observer motion. We have introduced the <Target|Observer> congruence as a framework for a globally based line-geometric approach to such problems. Moreover, we have presented a direct way to compute the direction of motion (FOE) as the intersection of two double tangents to two consecutively recorded silhouette curves. Also, we have extended this FOE reconstruction technique to include conjugate complex double tangents [Fact (4)], thereby widening its scope to include observer motion "towards" and "away from" the target. Knowing only the observer's speed as a function of time, we have shown how to recover the observer´s motion as well as a set of silhouettes on the target and illustrated this technique on synthetic images [Figure (3), Figure (4)] as well as on real image data [Figure (5)]. We stress that any hypothesis regarding observer speed provides the necessary input to a reconstruction process of the scene, the results of which can be compared with any independently available scene information, in order to verify or falsify the initial speed hypothesis.

Although we have presented some experimental results, the emphasis of this paper is on the theoretical framework and its ability to provide a unifying description of structure from motion problems that emphasizes the global geometry of the configuration. The practical applicability of this framework in recovering scene structure from real images is therefore still an open question. In the future we plan to investigate this field, aiming to demonstrate how the congruence-geometric framework can enable the construction of globally based methods for determining target shape, verifying and stabilizing observer motion, discriminating between stationary targets and targets moving relative to the background, grouping targets with the same motion relative to the moving observer and discriminating beteen curve edges and occluding surface contours.

## 9   Acknowledgements

The author is grateful to Mengxiang Li for his help with the practical experiments, and to Jan-Olof Eklundh for many interesting discussions related to these subjects, as well as for creating and maintaining a highly stimulating research environment at CVAP. This work has been performed within the project CVAP93, funded by The Swedish Research Counsil for Engineering Sciences, whose support is gratefully acknowledged.

## 10   References

[1]   J. J. Koenderink, *What does the occluding contour tell us about surface shape?*, Perception 13, 1984, pp. 321-330.

[2]   J.J. Koenderink, *Solid Shape*, MIT Press, Cambridge, Massachussetts, 1990.

[3]   P.J. Giblin and R. Weiss, *Reconstruction of Surfaces from Profiles*, Proceedings of the 1st International Conference of Computer Vision, London 1987, pp. 136-144.

[4]   R. Vaillant, *Using Occluding Contours for 3D Object Modeling*, O. Faugeras, ed., Computer Vision-ECCV '90 (First European Conference on Computer Vision, Antibes, France, April 23-27, 1990), Proceedings, Springer, Berlin, 1990, pp 454-464.

[5]   R. Cipolla, *Active Visual Inference of Surface Shape*, PhD thesis, Univ. of Oxford, 1991.

[6]   R. Cipolla and A. Blake, *Surface Shape from the Deformation of Apparent Contours*, International Journal of Computer Vision, Vol 9:2, 1992, pp 83-112

[7]     R. Valliant and O.D. Faugeras, *Using extremal boundaries for 3D object modeling*, IEEE Trans. Pattern Analysis and Machine Intelligence, 14(2), 1992, pp. 157-173.

[8]     O.D. Faugeras and T. Papadopoulo, *A Theory of the Motion Fields of Curves*, International Journal of Computer Vision, Vol 10:2, 1993, pp. 125-156.

[9]     P.J. Giblin and R. Weiss, *Epipolar Fields on Surfaces*, J.O. Eklundh, ed., Computer Vision-ECCV '94 (Third European Conference on Computer Vision, Stockholm, Sweden, May 2-6, 1994), Volume A, Springer, Berlin, 1994, pp 14-23.

[10]    R. Cipolla, K.E. Åström and P.J Giblin, *Motion from the frontier of curved surfaces*, Fifth International Conference on Computer Vision, MIT, Cambridge, Massachussetts, June 20-23, 1995, pp. 269-275.

[11]    K.N. Kutulakos and C.R. Dyer, *Global surface reconstruction by purposive control of observer motion*, Artificial Intelligence, vol. 78, no. 1-2, 1995, 147-177. surfaces, contours

[12]    ChangSheng Zhao and Roger Mohr, *Global 3-D surface reconstruction from occluding contours*, Computer Vision and Image Understanding, vol. 64, no.1, 1996, 62-96.

[13]    A. Naeve, *Focal Shape Geometry of Surfaces in Euclidean Space*, Dissertation, TRITA-NA-P9319, Computational Vision and Active Perception Laboratory, KTH, 1993.

[14]    A. Naeve, *Structure from Translational Observer Motion*, TRITA-NA-P97/02, Computational Vision and Active Perception Laboratory, KTH, 1997.

[15]    L. Eisenhart, *A Treatise on the Differential Geometry of Curves and Surfaces*, Ginn & Co, The Athaenum Press, Boston, Massachussetts, 1909.

# The Geometry of Visual Space Distortion[*]

Cornelia Fermüller[1], LoongFah Cheong[2], and Yiannis Aloimonos[1]

[1] Computer Vision Laboratory, Center for Automation Research, University of Maryland, College Park, MD 20742-3275
[2] Department of Electrical Engineering, National University of Singapore, 10 Kent Ridge Crescent, Singapore 119260

**Abstract.** The encounter of perception and action happens at the intermediate representations of space-time. In many of the computational models employed in the past, it has been assumed that a metric representation of physical space can be derived by visual means. Psychophysical experiments, as well as computational considerations, can convince us that the perception of space and shape has a much more complicated nature, and that only a distorted version of actual, physical space can be computed. This paper develops a computational geometric model that explains why such distortion might take place. The basic idea is that, both in stereo and motion, we perceive the world from multiple views. Given the rigid transformation between the views and the properties of the image correspondence, the depth of the scene can be obtained. Even a slight error in the rigid transformation parameters causes distortion of the computed depth of the scene. The unified framework introduced here describes this distortion in computational terms. We characterize the space of distortions by its level sets, that is, we characterize the systematic distortion via a family of iso-distortion surfaces which describes the locus over which depths are distorted by some multiplicative factor. Clearly, functions of the distorted space exhibiting some sort of invariance, produce desirable representations for biological and artificial systems [13]. Given that humans' estimation of egomotion or estimation of the extrinsic parameters of the stereo apparatus is likely to be imprecise, the framework is used to explain a number of psychophysical experiments on the perception of depth from motion or stereo.

## 1 Introduction

The nature of the representation of the world inside our heads as acquired by visual perception has persisted as a topic of investigation for thousands of years, from the works of Aristotle to the present [28]. In our day, answers to this question have several practical consequences in the field of robotics and automation.

---

[*] The support of the Office of Naval Research under Grant N00014-96-1-0587 is gratefully acknowledged. The second author was also supported in part by the Tan Kah Khee Postgraduate Scholarships. Special thanks to Sara Larson for her editorial and graphics assistance.

An artificial system equipped with visual sensors needs to develop representations of its environment in order to interact successfully with it. At the same time, understanding the way space is represented in the brains of biological systems is key to unraveling the mysteries of perception. We refer later to space represented inside a biological or artificial system as *perceptual space*, as opposed to *physical*, extra-personal *space*.

Interesting non-computational theories of perceptual space have appeared over the years in the fields of philosophy and cognitive science [24]. Computational theories, on the other hand, developed during the past thirty years in the area of computer vision, have followed a brute-force approach, equating physical space with perceptual space. Euclidean geometry involving metric properties has been used very successfully in modeling physical space. Thus, early attempts at modeling perceptual space concentrated on developing metric three-dimensional descriptions of space, as if it were the same as physical space. In other words, perceptual space was modeled by encoding the exact distances of features in three dimensions. The apparent ease with which humans perform a plethora of vision-guided tasks creates the impression that humans, at least, compute representations of space that have a high degree of generality; thus, the conventional wisdom that these descriptions are of a Euclidean metric nature was born and has persisted until now [1,19,28].

Computational considerations, however, can convince us that for a monocular or a binocular system moving in the world it is not possible to estimate an accurate description of three-dimensional metric structure, i.e., the exact distances of points in the environment from the nodal point of the eye or camera. This paper explains this in computational terms for the case of perceiving the world from multiple views. This includes the cases of both motion and stereo. Given two views of the world, whether these are the left and right views of a stereo system or successive views acquired by a moving system, the depth of the scene in view depends on two factors: (a) the three-dimensional rigid transformation between the views, hereafter called the *3D transformation*, and (b) the identification of image features in the two views that correspond to the same feature in the 3D world, hereafter called *visual correspondence*.

If there were no errors in the 3D transformation or the visual correspondence, then clearly the depth of the scene in view could be accurately recovered and thus a metric description could be obtained for perceptual space. Unfortunately, this is never the case. In the case of stereo, the 3D transformation amounting to the extrinsic calibration parameters of the stereo rig cannot be accurately estimated, only approximated [8]. In the case of motion, the three-dimensional motion parameters describing rotation and translation are estimated within error bounds [5–7,9,29,35]. Finally, visual correspondence itself cannot be obtained perfectly; errors are always present. Thus, because of errors in both visual correspondence and 3D transformation, the recovered depth of the scene is always a *distorted* version of the scene structure. The fundamental contribution of this paper is the development of a computational framework showing the geometric laws under which the recovered scene shape is distorted. In other words, there is a systematic way in which visual space is distorted; the transformation from physical to perceptual space belongs to the family of Cremona transformations [3,32].[1]

The power of the computational framework we introduce is demonstrated by using it to explain recent results in psychophysics. A number of recent psychophysical experiments have shown that humans make incorrect judgments of depth using either stereo [15,20] or motion [17,33]. Our computational theory explains these psychophysical results and demonstrates that, in general, perceived space is not describable using a well-established geometry such as hyperbolic, elliptic, affine or projective. Understanding the invariances of distorted perceived space will contribute to the understanding of robust representations of shape and space, with many consequences for the problem of recognition. This work was motivated by our recent work on direct perception and qualitative shape representation [11,12] and was inspired by the work of Koenderink and van Doorn on pictorial relief [23].

In the psychophysical literature it has been argued before for the interpretation of stereo data that an incorrect estimation of the viewing geometry causes incorrect estimation of the depth of the scene. This was first hypothesized but not further elaborated on by Helmholtz [36] and was explained by means of a number of tasks involving depth judgments from stereo by Foley [15]. In this paper we provide a general framework of space distortion on the basis of incorrect estimation of viewing geometry which can be used to explain estimation from motion as well as stereo. In our exposition we concentrate primarily on the experiments described in [33], which are concerned with both motion and stereo, and we use these experiments to explain in detail the utilization of the iso-distortion framework. In these experiments Tittle et al. tested how orientations of objects in space and absolute distance influence the judgment of depth, and they found very different results from the motion and stereo cues. The experiments were cleverly designed so that the underlying geometries of the motion and stereo configurations are qualitatively similar. Thus they are of great comparative interest. We also discuss an additional motion experiment [17] and some well known stereo experiments.

The computational arguments presented here are based on two ideas. First, the 2D image representation derived for stereo perception is of a different nature than the one derived for motion perception. Second, the only thing assumed about the scene is that it lies in front of the image plane, and thus all depth estimates have to be positive; therefore, the perceptual system, when estimating 3D motion, minimizes the number of image points whose corresponding scene points have negative depth values due to errors in the estimate of the motion. In [10] an error analysis has been performed to study the optimal relationship between translational and rotational errors which leads to this minimization. It has been found that for a general motion imaged on a plane the projection of the trans-

---

[1] In the projective plane, a transformation $(x, y, z) \rightarrow (x', y', z')$ with $\rho x' = \phi_1(x, y, z)$, $\rho y' = \phi_2(x, y, z)$, $\rho z' = \phi_3(x, y, z)$ where $\phi_1, \phi_2, \phi_3$ are homogeneous polynomials and $\rho$ any scalar, is called a rational transformation. A rational transformation whose inverse exists and is also rational is called a Cremona transformation.

lational error motion vector and the projection of the rotational error motion vector must have a particular relationship. Furthermore, the relative amount of translational and rotational error can be evaluated as a function of scene structure. These findings are utilized in the explanation of the psychophysical experiments.

The organization of this paper is as follows. Section 2.1 introduces the concept of iso-distortion surfaces. Considering two close views, arising from a system in general rigid motion, we relate image motion measurements to the parameters of the 3D rigid motion and the depth of the scene. Then, assuming that there is an error in the rigid motion parameters, we find the computed depth as a function of the actual depth and the parameters of the system. Considering the points in space that are distorted by the same amount, we find them to lie on surfaces that in general are hyperboloids. These are the iso-distortion surfaces that form the core of our approach. In Sect. 2.2 we further describe the iso-distortion surfaces in both 3D and visual space and we introduce the concept of the holistic or H-surfaces. These are surfaces that describe all iso-distortion surfaces distorted by the same amount, irrespective of the direction $(n_x, n_y)$ in the image in which measurements of visual correspondence are made. The H-surfaces are important in our analysis of the case of motion since measurements of local image motion can be in any direction and not just along the horizontal direction which is dominant in the case of stereo. Section 3 describes psychophysical experiments from the recent literature using motion and stereo, and Sect. 4 explains their results using the iso-distortion framework. Section 4.1 describes in detail the coordinate systems and the underlying rigid transformations for the specific experiments. Sections 4.2 and 4.3 explain the experimental results of [33] for motion and stereo respectively, and Sect. 4.4 discusses the experimental results of the additional purely motion or stereo experiments using the framework introduced here. Section 5 concludes the paper and discusses the relationship of this work to other attempts in the literature to capture the essence of perceptual space.

## 2 Distortion of Visual Space

### 2.1 Iso-distortion Surfaces

As an image formation model, we use the standard model of perspective projection on the plane, with the image plane at a distance $f$ from the nodal point parallel to the $XY$ plane, and the viewing direction along the positive $Z$ axis as illustrated in Fig. 1. We want a model that can be used both for motion and stereo. Thus, we consider a differential model of rigid motion. This model is valid for stereo, which constitutes a special constrained motion, when making the small baseline approximation that is used widely in the literature [23].

Specifically, we model the change of viewing geometry differentially through a rigid motion with translational velocity $(U, V, W)$ and rotational velocity $(\alpha, \beta, \gamma)$ of the observer in the coordinate system $OXYZ$. Stereo can be approximated as a constrained rigid motion with translation $(U, 0, W)$ and rotation $(0, \beta, 0)$, as explained in detail in Sect. 4.1. In the case of stereo the measurements obtained

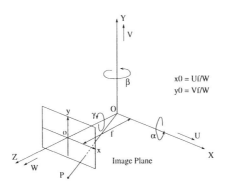

**Fig. 1.** The image formation model. $OXYZ$ is a coordinate system fixed to the camera. $O$ is the optical center and the positive $Z$ axis is the direction of view. The image plane is located at a focal length $f$ pixels from $O$ along the $Z$ axis. A point $P$ at $(X, Y, Z)$ in the world produces an image point $p$ at $(x, y)$ on the image plane where $(x, y)$ is given by $\left(\frac{fX}{Z}, \frac{fY}{Z}\right)$. The instantaneous motion of the camera is given by the translational vector $(U, V, W)$ and the rotational vector $(\alpha, \beta, \gamma)$.

on the image are the so-called disparities which we approximate here through a continuous flow field. As, due to the stereo viewing geometry, the disparities are close to horizontal, in the forthcoming analysis we only employ horizontal image flow measurements. On the other hand, in the case of continuous motion from local image information only the component of the flow perpendicular to edges (along image gradients) can, in general, be obtained. Thus, in the case of motion, we consider as input the field resulting from this component, which is known as the normal flow field. In summary, for both cases of stereo and motion we use the projection of the flow field on orientations $(n_x, n_y)$, which in the case of motion represent image gradients while in the case of stereo represent horizontal directions.

As is well known, from the 2D image measurements alone as a consequence of the scaling ambiguity, only the direction of translation $(x_0, y_0) = \left(\frac{U}{W}f, \frac{V}{W}f\right)$ represented in the image plane by the epipole (also called the FOE (focus of expansion) or FOC (focus of contraction) depending on whether $W$ is positive or negative), the scaled depth $Z/W$ and the rotational parameters can possibly be obtained from flow measurements. Using this notation the equations relating the 2D velocity $\mathbf{u} = (u, v) = (u_{\text{trans}} + u_{\text{rot}}, v_{\text{trans}} + v_{\text{rot}})$ of an image point to the 3D velocity and the depth of the corresponding scene point are

$$u = u_{\text{trans}} + u_{\text{rot}} = (x - x_0)\frac{W}{Z} + \alpha xy - \beta\left(\frac{x^2}{f} + f\right) + \gamma y$$

$$v = v_{\text{trans}} + v_{\text{rot}} = (y - y_0)\frac{W}{Z} + \alpha\left(\frac{y^2}{f} + f\right) - \frac{\beta xy}{f} - \gamma x \tag{1}$$

where $u_{\text{trans}}, v_{\text{trans}}$ are the horizontal and vertical components of the flow due to translation, and $u_{\text{rot}}, v_{\text{rot}}$ the horizontal and vertical components of the flow due to rotation, respectively.

The velocity component $\mathbf{u}_n$ of the flow in any direction $\mathbf{n} = (n_x, n_y)$ has value

$$u_n = un_x + vn_y.$$  (2)

Knowing the parameters of the viewing geometry exactly, the scaled depth can be derived from (2). Since the depth can only be derived up to a scale factor, we set $W = 1$ and obtain

$$Z = \frac{(x - x_0)n_x + (y - y_0)n_y}{u_n - u_{\text{rot}}n_x - v_{\text{rot}}n_y}$$

If there is an error in the estimation of the viewing geometry, this will in turn cause errors in the estimation of the scaled depth, and thus a distorted version of space will be computed. In order to capture the distortion of the estimated space, we describe it through surfaces in space which are distorted by the same multiplicative factor, the so-called iso-distortion surfaces. To distinguish between the various estimates, we use the hat sign " $\hat{\ }$ " to represent estimated quantities, the unmarked letters to denote the actual quantities, and the subscript "$\epsilon$" to represent errors, where the estimates are related as follows:

$$(\hat{x}_0, \hat{y}_0) = (x_0 - x_{0_\epsilon}, y_0 - y_{0_\epsilon})$$
$$(\hat{\alpha}, \hat{\beta}, \hat{\gamma}) = (\alpha - \alpha_\epsilon, \beta - \beta_\epsilon, \gamma - \gamma_\epsilon)$$
$$\hat{\mathbf{u}}_{\text{rot}} = (\hat{u}_{\text{rot}}, \hat{v}_{\text{rot}}) = \mathbf{u}_{\text{rot}} - \mathbf{u}_{\text{rot}_\epsilon} = (u_{\text{rot}} - u_{\text{rot}_\epsilon}, v_{\text{rot}} - v_{\text{rot}_\epsilon})$$

If we also allow for a noise term $N$ in the estimate $\hat{u}_n$ of the component flow $u_n$, we have $\hat{u}_n = u_n + N$. The estimated depth becomes

$$\hat{Z} = \frac{(x - \hat{x}_0)n_x + (y - \hat{y}_0)n_y}{\hat{u}_n - (\hat{u}_{\text{rot}}n_x + \hat{v}_{\text{rot}}n_y)} \qquad \text{or}$$

$$\hat{Z} = Z \cdot \left( \frac{(x - \hat{x}_0)n_x + (y - \hat{y}_0)n_y}{(x - x_0)n_x + (y - y_0)n_y + Z(u_{\text{rot}_\epsilon}n_x + v_{\text{rot}_\epsilon}n_y) + NZ} \right) \qquad (3)$$

From (3) we can see that $\hat{Z}$ is obtained from $Z$ through multiplication by a factor given by the term inside the brackets, which we denote by $D$ and call the distortion factor. In the forthcoming analysis we do not attempt to model the statistics of the noise and we will therefore ignore the noise term. Thus, the distortion factor takes the form

$$D = \frac{(x - \hat{x}_0)n_x + (y - \hat{y}_0)n_y}{\begin{array}{l}(x - x_0)n_x + (y - y_0)n_y \\ + Z\left[\left(\frac{\alpha_\epsilon xy}{f} - \beta_\epsilon\left(\frac{x^2}{f} + f\right) + \gamma_\epsilon y\right)n_x + \left(\alpha_\epsilon\left(\frac{y^2}{f} + f\right) - \beta_\epsilon\frac{xy}{f} - \gamma_\epsilon x\right)n_y\right]\end{array}}$$  (4)

or, in a more compact form

$$D = \frac{(x - \hat{x}_0)\,n_x + (y - \hat{y}_0)\,n_y}{(x - x_0 + Zu_{\text{rot}_\epsilon})\,n_x + (y - y_0 + Zv_{\text{rot}_\epsilon})\,n_y}$$

Equation (4) describes, for any fixed direction $(n_x, n_y)$ and any fixed distortion factor $D$, a surface $f(x, y, Z) = 0$ in space, which we call an iso-distortion surface. For specific values of the parameters $x_0, y_0, \hat{x}_0, \hat{y}_0, \alpha_\epsilon, \beta_\epsilon, \gamma_\epsilon$ and $(n_x, n_y)$, this iso-distortion surface has the obvious property that points lying on it are distorted in depth by the same multiplicative factor $D$. Also, from (3) it follows that the transformation from perceptual to physical space is a Cremona transformation.

It is important to realize that, on the basis of the preceding analysis, the distortion of depth also depends upon the direction $(n_x, n_y)$ and is therefore different for different directions of flow in the image plane. This means simply that if one estimates depth from optical flow in the presence of errors, the results can be very different depending on whether the horizontal, vertical, or any other component is used; depending on the direction, any value between $-\infty$ and $+\infty$ can be obtained! It is therefore imperative that a good understanding of the distortion function be obtained, before visual correspondences are used to recover the depth or structure of the scene.

In order to derive the iso-distortion surfaces in 3D space we substitute $x = \frac{fX}{Z}$ and $y = \frac{fY}{Z}$ in (4), which gives the following equation:

$$
\begin{aligned}
& D\left(\left(\alpha_\epsilon XY - \beta_\epsilon \left(X^2 + Z^2\right) + \gamma_\epsilon YZ\right) n_x \right. \\
& \left. + \left(\alpha_\epsilon \left(Y^2 + Z^2\right) - \beta_\epsilon XY - \gamma_\epsilon XZ\right) n_y\right) \\
& - \left(X - \frac{\hat{x}_0 Z}{f} - D\left(X - \frac{x_0 Z}{f}\right)\right) n_x - \left(Y - \frac{\hat{y}_0 Z}{f} - D\left(Y - \frac{y_0 Z}{f}\right)\right) n_y = 0
\end{aligned}
$$

describing the iso-distortion surfaces as quadratic surfaces—in the general case, as hyperboloids. One such surface is depicted in Fig. 2. Throughout the paper we will need access to the iso-distortion surfaces from two points of view. On the one hand we want to compare surfaces corresponding to the same $D$, but different gradient directions; thus we are interested in the families of $D$ iso-distortion surfaces (see Fig. 3a). On the other hand we want to look at surfaces corresponding to the same gradient direction $\mathbf{n}$, but different $D$'s, the families of $\mathbf{n}$ iso-distortion surfaces (see Fig. 3b). We will also be interested in the intersections of the surfaces with planes parallel to the $XZ, YZ$, and $XY$ planes. These intersections give rise to families of iso-distortion contours; for an example see Fig. 4.

## 2.2 Visualization of Iso-distortion Surfaces

The iso-distortion surfaces presented in the previous section were developed for the general case, i.e., when the 3D transformation between the views is a general rigid motion. However, the psychophysical experiments that we will explain in the sequel considered constrained motion: rotation only around the $Y$ axis and translation only in the $XZ$ plane. The only motion parameters to be considered are therefore $\beta_\epsilon, x_0$ and $\hat{x}_0$, and the iso-distortion surfaces become

$$
\begin{aligned}
& D\beta_\epsilon X^2 n_x + D\beta_\epsilon Z^2 n_x + D\beta_\epsilon XY n_y \\
& - (D - 1) X n_x - (D - 1) Y n_y - (\hat{x}_0 - Dx_0) \frac{n_x}{f} Z = 0
\end{aligned}
$$

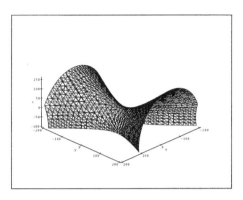

**Fig. 2.** Iso-distortion surface in $XYZ$ space. The parameters are: $x_0 = 10$, $x_{0_\epsilon} = -1$, $y_0 = -25$, $y_{0_\epsilon} = -5$, $\alpha_\epsilon = -0.05$, $\beta_\epsilon = -0.1$, $\gamma_\epsilon = -0.005$, $f = 1$, $D = 1.5$, $n_x = 0.7$.

which in general constitute hyperboloids. For horizontal flow vectors ($n_x = 1, n_y = 0$) they become elliptic cylinders and for vertical flow vectors they become hyperbolic cylinders.

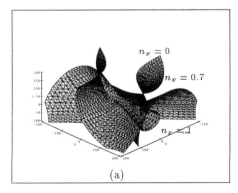

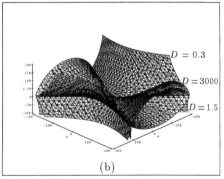

(a)                                   (b)

**Fig. 3.** (a) Family of $D$ iso-distortion surfaces for $n_x = 1, 0.7, 0$. (b) Family of $n$ iso-distortion surfaces for $D = 0.3, 3000, 1.5$. The other parameters are as in Fig. 2.

Figure 5 provides an illustration of an iso-distortion surface for a general flow direction (here $n_x = 0.7$, $n_y = 0.714$). For our purposes, only the parts of the iso-distortion surfaces within the range visible from the observer are of interest. Since in the motion considered later the FOE has a large value, these parts show very little curvature and appear to be close to planar, as can be seen from Fig. 5b.

In order to make it easier to grasp the geometrical organization of the iso-distortion surfaces we next perform a simplification and use in addition to 3D space also visual space (that is, $xyZ$ space): Within a limited field of view, terms quadratic in the image coordinates are small relative to linear and constant terms; thus we ignore them for the moment, which simplifies the rotational term for the motions considered to $(u_{\text{rot}}, v_{\text{rot}}) = (-\beta_\epsilon f, 0)$.

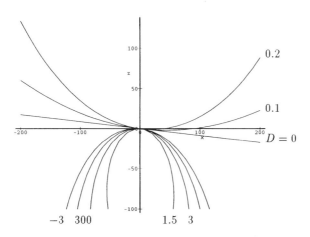

**Fig. 4.** Intersection of a family of $n$ iso-distortion surfaces (as shown in Fig. 3b) with the $XZ$ plane gives rise to a family of iso-distortion contours.

In visual space, i.e., $xyZ$ space, that is the space perceived under perspective projection, where the fronto-parallel dimensions are measured according to their size on the image plane, the iso-distortion surfaces take the following form:

$$[x(D - 1) + (\hat{x}_0 - Dx_0)]\, n_x + y(D - 1)n_y - D\beta_\epsilon f Z n_x = 0$$

That is, they become planes with surface normal vectors $((D - 1)n_x, (D - 1)n_y, -D\beta_\epsilon f n_x)$. For a fixed $D$, the family of $D$ iso-distortion surfaces obtained by varying the direction $(n_x, n_y)$ is a family of planes intersecting on a line $l$. If we slice these iso-distortion planes with a plane parallel to the $xy$ (or image) plane, we obtain a pencil of lines with center lying on the $x$ axis (the point through which line $l$ passes) (see Fig. 6a).

In our forthcoming analysis we will need to consider the family of iso-distortion surfaces for a given distortion $D$, that is, the $D$ iso-distortion surfaces for all directions $(n_x, n_y)$. Thus, we will need a compact representation for the family of $D$ iso-distortion surfaces in 3D space. The purpose of this representation is to visualize the high-dimensional family of $D$ iso-distortion surfaces in $(x, y, Z, \mathbf{n})$ space through a surface in $(x, y, Z)$ space in a way that captures the essential aspects of the parameters describing the family and thus the underlying distortion. As such a representation we choose the following surfaces, hereafter called

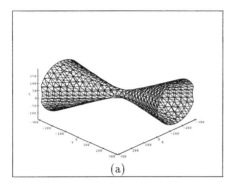
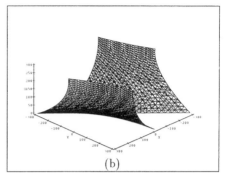

(a)        (b)

**Fig. 5.** (a) A general iso-motion surface in 3D space. The $Z$ axis corresponds to the optical axis. (b) Section of an iso-motion surface for a limited field of view in front of the image plane for large values of $x_0$.

holistic or H-surfaces, which are most easily understood through their cross sections parallel to the $xy$ plane: Considering a planar slice of the family of $D$ iso-distortion surfaces, as in Fig. 6a, we obtain a pencil of lines. As a representation for these lines we choose the circle with diameter extending from the origin to the center of the pencil (Fig. 6b). This circle clearly represents all orientations of the lines of the pencil (or the iso-distortion planes in the slicing plane). Any point $P$ of the circle represents the slice of the iso-distortion plane which is perpendicular to a line through the center $(O)$ and $P$.

If we now move the slicing plane parallel to itself, the straight lines of the pencil will trace the iso-distortion planes and the circle will change its radius and trace a circular cone with the $Z$ axis as one ruling (Fig. 6c).

The circular cones are described by the following equation:

$$x^2(D-1) + (\hat{x}_0 - Dx_0)x + y^2(D-1) - D\beta_\epsilon fZx = 0$$

$$\text{or}\quad \left(x - \frac{(Dx_0 - \hat{x}_0 + D\beta_\epsilon fZ)}{2(D-1)}\right)^2 + y^2 = \left[\frac{D(x_0 + \beta_\epsilon fZ) - \hat{x}_0}{2(D-1)}\right]^2$$

Thus their axes are given by

$$Dx_0 - \hat{x}_0 + D\beta_\epsilon fZ - 2(D-1)x = 0, \quad y = 0$$

Slicing the cones and the simplified iso-distortion surfaces with planes parallel to the $xy$ plane as in Fig. 6b, the circles we obtain have center $(x, y, Z) = \left(\frac{Dx_0 - \hat{x}_0 + D\beta_\epsilon fZ}{2(D-1)}, 0, Z\right)$ and radius $\frac{D(x_0 + \beta_\epsilon fZ) - \hat{x}_0}{2(D-1)}$. The circular cones serve as a holistic representation for the family of iso-distortion surfaces represented by the same $D$, therefore the name holistic or H-surface. It should be noted here that

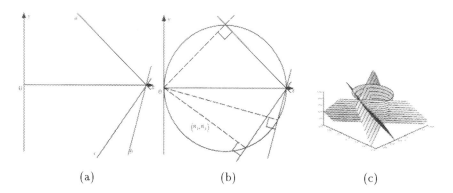

(a)                              (b)                              (c)

**Fig. 6.** Simplified iso-distortion surfaces in visual space. (a) Intersection of the family of the simplified $D$ iso-distortion surfaces (planes) for different directions $(n_x, n_y)$ with a plane parallel to the image plane. (b) A circle represents the intersections of the family of the $D$ iso-distortion surfaces with planes parallel to the image plane. (c) In visual space a family of $D$ iso-distortion surfaces is characterized by a cone (the holistic surface).

the holistic surfaces become cones only in the case of the constrained 3D motion considered in this paper. In the general case they are hyperboloids.

It must be stressed at this point that the iso-distortion surfaces should not be confused with the H-surfaces. Whereas a $D$ iso-distortion surface for a direction **n** represents all points in space distorted by the same multiplicative factor $D$ for image measurements in direction **n**, the holistic surfaces do not represent any actually existing physical quantity; they serve merely as a tool for visualizing the family of $D$ iso-distortion surfaces as **n** varies, and will be needed in explaining the distortion of space due to motion.

The H-surfaces for the families of iso-distortion surfaces vary continuously as we vary $D$. For $D = 0$ we obtain a cylinder with the $Z$ axis and the line $x = \hat{x}_0$ as diametrically opposite rulings. For $D = 1$ we obtain a plane parallel to the $xy$ plane given by $Z = \frac{-x_0 \epsilon}{\beta_\epsilon f}$; the cone for $D = \infty$ and the cone for $D = -\infty$ coincide. Thus we can divide the space into three areas: the areas between the $D = 0$ cylinder and the $D = -\infty$ cone, which only contain cones of negative distortion factor; the area between the $D = \infty$ cone and the $D = 1$ plane, with cones of decreasing distortion factor; and the area between the $D = 0$ cylinder and the $D = 1$ plane, with cones of increasing distortion factor. All the holistic surfaces intersect in the same circle, which is the intersection of the $D = 0$ cylinder and the $D = 1$ plane (see Fig. 7a). Since the holistic surfaces intersect in one plane, any family of **n** iso-distortion surfaces intersects in a line in that plane.

To go back from visual to actual space, we have to compensate for the perspective scaling. In actual 3D space the iso-distortion surfaces are given by the equation

$$D\beta_\epsilon Z^2 n_x + (1 - D)X n_x + (1 - D)Y n_x + (Dx_0 - \hat{x}_0)\frac{Z n_x}{f} = 0$$

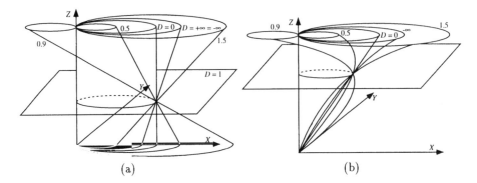

**Fig. 7.** (a) Holistic surfaces (cones) in visual space, labeled with their respective distortion factors. (b) Holistic surfaces (third-order surfaces) in 3D space.

describing parabolic cylinders curved in the $Z$ dimension. Also the circular cones have an additional curvature in the $Z$ dimension, and thus the H-surfaces in 3D space are surfaces of the form

$$X^2(D-1)f + Y^2(D-1)f + (\hat{x}_0 - Dx_0)XZ - D\beta_\epsilon XZ^2 f = 0$$

An illustration is given in Fig. 7b.

## 3  Psychophysical Experiments on Depth Perception

In the psychophysical literature a number of experiments has been reported that document a perception of depth which does not coincide with the actual situation. Most of the experiments were devoted to stereoscopic depth perception, using tasks that involved the judgment of depth at different distances. The conclusion usually obtained was that there is an expansion in the perception of depth of near distances and a contraction of depth at far distances. However, most of the studies did not explicitly measure perceived viewing distance, but asked for relative distance judgments instead. Recently a few experiments have been conducted by Tittle et al. [33] comparing aspects of depth judgment due to stereoscopic and monocular motion perception. The experiments were designed to test how the orientations of objects in space and their absolute distances influence the perceptual judgment. It was found that the stereoscopic cue and the motion cue give very different results.

The literature has presented a variety of explanations and proposed a number of models explaining different aspects of depth perception. Recently, great interest has arisen in attempts to explain the perception of visual space using well-defined geometries, such as similarity, conformal, affine, or projective transformations mapping physical space into perceived space, and it has been debated whether perceptual space is Euclidean, hyperbolic, or elliptic [37]. Our analysis shows that these models do not provide a general explanation for depth perception, and proposes that much of the data can be explained by the fact that the underlying 3D transformation is estimated incorrectly. Thus the transformation between physical and perceptual space is more complicated than previously thought. For the case of motion or stereo it is rational and belongs to the family of Cremona transformations [32].

We next describe a number of experiments and show that their results can be explained on the basis of imprecise estimation of the 3D transformation and thus can be predicted by the iso-distortion framework introduced here. Our primary focus in Sect. 3.1 is on the experiments testing the difference between motion and stereo performed by Tittle et al. [33]. In addition, in Sect. 3.2 we describe two well-known stereoscopic experiments, and in Sect. 3.3 a motion experiment.

## 3.1 Distance Judgment from Motion and Binocular Stereopsis

In the first experiment [33] that we discuss, observers were required to adjust the eccentricity of a cylindrical surface until its cross-section in depth appeared to be circular. The observers could manipulate the cylindrical surface (shown in Fig. 8) by rescaling it along its depth extent $b$ (which was aligned with the $Z$ axis of the viewing geometry when the cylinder was in a fronto-parallel orientation) with the workstation mouse. Such a task requires judgment of relative distance. In order for the cross-section to appear circular, the vertical extent and the extent in depth of the cylinder, $a$ and $b$, have to appear equal.

The experiments were performed for static binocular stereoscopic perception, for monocular motion, and for combined motion and stereopsis. The stereoscopic stimuli consisted of stereograms, and the monocular ones were created by images of cylinders rotating about a vertical axis (see Fig. 8). In all the experiments the observers had to fixate on the front of the surface where it intersected the axis of rotation, and the cylindrical surfaces were composed of bright dots.

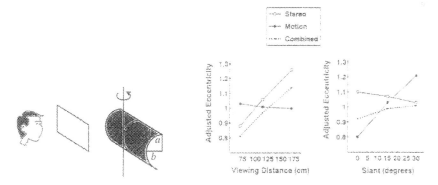

**Fig. 8.** From [33]: a schematic view of the cylinder stimulus used in Experiment 1.

**Fig. 9.** From [33]: Average adjusted cylinder eccentricity for the stereo, motion, and combined conditions as a function of simulated viewing distance and surface slant. An adjusted eccentricity of 1.0 indicates veridical performance.

The effect of the slant and distance of the cylinder on the subjective depth judgment was tested. In particular, the cylinder had a slant in the range 0° to 30°, with 0° corresponding to a fronto-parallel cylinder as shown in Fig. 8, and the distance ranged from 70 to 170 cm. Figure 9 displays the experimental results in the form of two graphs, with the $x$ axis showing either the slant or distance and the $y$ axis the adjusted eccentricity. An adjusted eccentricity of 1.0 corresponds to a veridical judgment, values less than this indicate an overestimate of $b$ relative to $a$, and values greater than 1.0 indicate an underestimate. As can be seen from the graphs, whereas the perception of depth from motion only does not depend on the viewing distance, the extent $b$ is overestimated for near distances and underestimated for far distances under stereoscopic perception. On the other hand, the slant of the surface has a significant influence on the perception of motion—at 0° $b$ is overestimated and at 30° underestimated—and has hardly any influence on perception from stereo. The results obtained from the combined stereo and motion displays showed an overall pattern similar to those of the purely stereoscopic experiments.

For stereoscopic perception only, a very similar experiment, known as apparently circular cylinder (ACC) judgment, was performed in [15,20], and the same pattern of results was reported there.

In a second experiment performed by Tittle et al. [33], the task was to adjust the angle between two connected planes until they appeared to be perpendicular to one another (see Fig. 10).

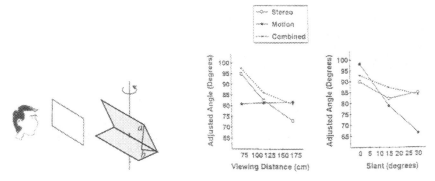

**Fig. 10.** From [33]: a schematic view of the dihedral angle stimulus used in Experiment 2.

**Fig. 11.** From [33]: Adjusted dihedral angle as a function of surface slant and simulated viewing distance. An adjusted angle of 90° indicates veridical performance.

Again the surfaces were covered with dots and the fixation point was at the intersection of the two planes and the rotation axis. As in the first experiment the influences of the cue (stereo, motion, or combined motion and stereo), the slant and the viewing distance on the depth judgment were evaluated. This task again requires a judgment of relative distance, that is, the depth extent $b$ relative to the vertical extent $a$ (as shown in Fig. 10). The results displayed in Fig. 11 are qualitatively similar to those obtained from the first experiment. An adjusted angle greater than the standard 90° corresponds to an overestimation of the extent in depth, and one less than 90° represents underestimation.

## 3.2 Stereoscopic Experiments: Apparent Fronto-parallel Plane/Apparent Distance Bisection

A classic test of depth perception for stereoscopic vision is the apparent fronto-parallel plane (AFPP) experiment [15,31]. In this experiment, an observer views a horizontal array of targets. One target is fixed, usually in the median plane ($YZ$ plane). The other targets are fixed in direction but are variable in radial distance under control of the subject. The subject sets these targets so that all of the targets appear to lie in a fronto-parallel plane. Care is taken so that fixation is maintained at one point. The results are illustrated in Fig. 12.

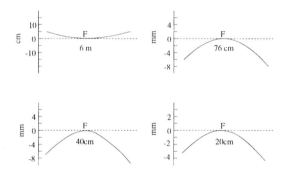

**Fig. 12.** Data for the apparent fronto-parallel plane for different observation distances. In each case, F is the point of fixation. The visual field of the target extends from $-16°$ to $16°$. From [31].

The AFPP corresponds to a physical plane only at one distance, usually between 1 m and 4 m [15]. At far distances, the targets are set on a surface convex to the observer; at near distances they are set on a surface increasingly concave to the observer. Generally, the AFPP locus is skewed somewhat, that is, one side is farther away than the other.

In another classic experiment, instead of instructing a subject to set targets in an apparent fronto-parallel plane, the subjects are asked to set one target at half of the perceived distance of another target, placed in the same direction. This is known as the apparent distance bisection task or the ADB task [14]. In practice the targets would interfere with each other if they were in exactly the same direction, so they are displaced a few degrees. The task and the results are illustrated in Fig. 13. These results were obtained with free eye movements, but the author claimed that the effect has also been replicated with fixation on one point.

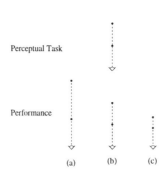

Perceptual Task

Performance

(a)    (b)    (c)

**Fig. 13.** Apparent distance bisection task: (a) Far fixation point. (b) Correct distance judgment at intermediate fixation point. (c) Near fixation point.

### 3.3 Motion Experiments

In [17], Gogel tested the distance perceived under monocular motion when fixating on points at different distances. In one of his experiments he relates motion to depth in a highly original way. The resulting task can be performed on the basis of scaled depth. The experimental set-up is shown in Fig. 14. The subjects sitting in the observation booth moved their heads horizontally while fixating on a point on either a near or far object. Between the two fixation points was a bright, moving point. Imagine the point to be moving vertically. If the distance to the point is estimated correctly the observer experiences a vertical motion. If it is estimated incorrectly the point is perceived as moving diagonally, with a horizontal component either in the direction of the head movement if there is an underestimation of depth, or in the opposite direction if there is an overestimation. In the experiment the subjects controlled the point's movement and were asked to move it in such a way that they experienced a purely vertical movement. To compensate for the additional motion component perceived, subjects moved the point diagonally with a horizontal component in the direction opposite. From the amount of horizontal movement, the estimated depth could be reconstructed. The exact dimensions of the set-up are described in Fig. 14. The results of the experiments are displayed in Table 1. As can be seen, overestimation of depth occurs with both fixation points, and it is larger for the far fixation point than for the near one.

## 4 Explanation of Psychophysical Results

### 4.1 The Viewing Geometry

*(a) Stereo* The geometry of binocular projection for an observer fixating on an environmental point is illustrated in Fig. 15. We fix a coordinate system $(LXYZ)$ on the left eye with the $Z$ axis aligned with the optical axis and the $Y$ axis perpendicular to the fixation plane. In this system the transformation relating the right eye to the left eye is a rotation around the $Y$ axis and a translation in

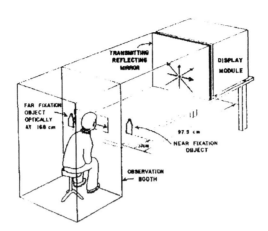

**Fig. 14.** A schematic drawing of the observation booth (from [17]). The observation booth was 50 cm wide and extending optically 194 cm in front of the observer (actually a mirror was used in the display as can be seen). The near fixation object was 15.3 cm from the right edge of the visual alley and 37 cm from the observer. The far fixation object was 2 cm from the left edge of the alley and optically 168 cm from the observer. The moving dot was between the two walls at a distance of 97.5 cm and the observer could move horizontally left and right, in one movement, 17.5 cm. The floor and the two sides of the booth were covered with white dots. All other surfaces were black.

the $XZ$ plane. If we make the small baseline assumption, we can approximate the disparity measurements through a continuous flow field. The translational and rotational velocities are $(U, 0, W)$ and $(0, \beta, 0)$ respectively, and therefore the horizontal $h$ and vertical $v$ disparities are given by

$$h = \frac{W}{Z}(x - x_0) - \beta\left(\frac{x^2}{f} + f\right) \qquad v = \frac{W}{Z}y - \frac{\beta x y}{f}$$

In the coordinate system thus defined (Fig. 15), $\beta$ is negative and $x_0$ is positive, and for a typical viewing situation very large. Therefore the epipole is far outside the image plane, which causes the disparity to be close to horizontal.

*(b) Motion* In the experiments described in Sect. 3.1 the motion of the object consists of a rotation around a vertical axis in space.

We fix a coordinate system to a point $S = (X_s, Y_s, Z_s)$ on the object in the $YZ$ plane through which the rotation axis passes. At the time of observation it is parallel to the reference coordinate system $(OXYZ)$ on the eye of the observer (see Fig. 16). In the new coordinate system on the object, the motion is purely rotational, and is given by the velocity $(0, w_y, 0)$. If we express this motion in the reference system as a motion of the observer we obtain a rotation around the $Y$ axis and an additional translation in the $XZ$ plane given by the velocity $(w_y Z_s, 0, -w_y X_s)$. Thus in the notation used before, there is a rotation with velocity $\beta = -w_y$, and a translation with epipole $(x_0, 0) = \left(-\frac{Z_s f}{X_s}, 0\right)$ or $(\infty, 0)$

**Table 1.** Results in centimeters from the experiment shown in Fig. 14. $W$ is the physical horizontal motion required for the point of light physically moving with a vertical component to appear to move vertically and $D'$ is the perceived distance of the point of light as derived using the measurement $W$.

|  | Fixation Near | | Fixation Far | |
|  | $W$ | $D'$ | $W$ | $D'$ |
| --- | --- | --- | --- | --- |
| Mean | 1.51 | 117 | 7.08 | 164 |
| Geometric Mean | — | 112 | 7.04 | 164 |
| Median | 2.42 | 113 | 7.13 | 165 |
| SD | 5.10 | 34 | 0.70 | 11 |

if $X_s = 0$. The value $u_n$ of the flow component $\mathbf{u}_n$ along a direction $\mathbf{n} = (n_x, n_y)$ is given by

$$u_n = w_y \left( \frac{X_s}{Z} \left( x + \frac{Z_s}{X_s} f \right) + \left( f + \frac{x^2}{f} \right) \right) n_x + w_y \left( \frac{y X_s}{Z} + \frac{xy}{f} \right) n_y$$

Since $X_s$ is close to zero, $x_0$ again takes on very large values. In our coordinate system (see Fig. 16) $\beta$ is positive and $x_0$ is positive, since the circular cross-section is to the right of the $YZ$ plane, and thus the locus of the fixation point most probably is biased toward the cross-section.

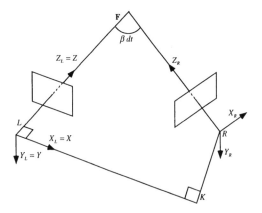

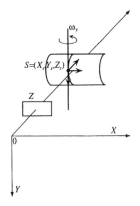

**Fig. 15.** Binocular viewing geometry. $LK = U\,dt$ (translation along the $X$ axis), $KR = W\,dt$ (translation along the $Z$ axis), $LFR = \beta\,dt$ = convergence angle (resulting from rotation around the $Y$ axis). $L$, $K$, $R$, $F$ are in the fixation plane and $dt$ is a hypothetical small time interval during which the motion bringing $X_L Y_L Z_L$ to $X_R Y_R Z_R$ takes place.

**Fig. 16.**

Although the motion in the stereo and motion configurations is qualitatively similar, the psychophysical experimental results show that the system's perception of depth is not. This demonstrates that the two mechanisms of shape perception from motion and stereo work differently. We account for this by the fact that the 2D disparity representation used in stereo is of a different nature than the 2D velocity representation computed for further motion processing.

It is widely accepted that horizontal disparities are the primary input in stereoscopic depth perception although there have been many debates as to whether vertical disparities play a role in the understanding of shape [21,30]. The fact is that for any human stereo configuration, even with fixation at nearby points, the horizontal disparities are much larger than the vertical ones. Thus, for the purpose of the forthcoming analysis, in the case of stereo we only consider horizontal disparities, although a small amount of vertical disparity would not influence the results.

On the other hand, for a general motion situation the actual 2D image displacements are in many directions. Due to computational considerations from local image measurements, only the component of flow perpendicular to edges can be computed reliably. This is the so-called aperture problem. In order to derive the optical flow, further processing based on smoothing and optimization procedures has to be performed, which implicitly requires some assumptions about the smoothness of the scene. For this reason we expect the 2D image velocity measurements used by the system to be distributed in many directions, although the optical flow in the experimental motion is mostly horizontal.

Based on these assumptions about the velocity representations used, in the next two sections the experimental data—first the data from motion perception, then the data from stereo perception—are explained through the iso-distortion framework.

## 4.2  Motion

To visualize this and later explanations let us look at the possible distortions of space for the motion and stereo configurations considered here. Figure 17a gives a sketch of the holistic surfaces (third-order surfaces) for negative rotational errors $(\beta_\epsilon)$ and Fig. 17b shows the surfaces for positive rotational errors. In both cases $x_0$ is positive. A change of the error in translation leaves the structure qualitatively the same; it only affects the sizes of the surfaces. In the overall pattern we observe a shift in the location of the intersection of the holistic surface. Since the intersection is in the $D = 1$ plane given by the equation $Z = -\frac{x_{0_\epsilon}}{\beta_\epsilon f}$, an increase in $x_{0_\epsilon}$ causes the intersection to have a smaller $Z$ coordinate in Fig. 17a and a larger one in Fig. 17b. For both the motion and the stereo experiments, the FOE lies far outside the image plane. Therefore only a small part of the illustrated iso-distortion space actually lies in the observer's field of view. This part is centered around the $Z$ axis as schematically illustrated in Fig. 17.

The guiding principle in our explanation of the motion experiments lies in the minimization of negative depth estimates. We do not assume any scene interpretation; the only thing we know about the scene is that it lies in front of the image

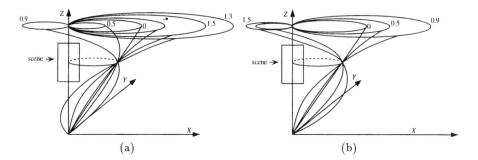

**Fig. 17.** Holistic third-order surfaces for the geometric configurations described in the experiments. (a) Positive $\beta_\epsilon$. (b) Negative $\beta_\epsilon$.

plane, and thus all depth estimates have to be positive. Therefore, we want to keep the number of image points, whose corresponding scene points would yield negative depth values due to erroneous estimation of the 3D transformation, as small as possible.

To represent the negative depth values we use a geometric statistical model: The scene in view lies within a certain range of depths between $Z_{min}$ and $Z_{max}$. The flow measurement vectors on the image are distributed in many directions; we assume that they follow some distribution. We are interested in the points in space for which we would estimate negative depth values.

For every direction **n** the points with negative depths lie between the $D = 0$ and $D = -\infty$ distortion surfaces within the range of depths covered by the scene. Thus, for every gradient direction we obtain a 3D subspace, covering a certain volume. The sum of all volumes for all gradient directions, normalized by the flow distribution considered here, represents a measure of the likelihood of negative depth estimates being derived from the image flow on the basis of some motion estimate. We call this sum the *negative depth volume*.

Let us assume there is some error in the estimate of the rotation, $\beta_\epsilon$. We are interested in the translation error $x_{0_\epsilon}$ that will minimize the negative depth volume. Under the assumption that the distribution of flow directions is uniform (that is, the flow directions are uniformly distributed in every direction and at every depth within the range between $Z_{min}$ and $Z_{max}$), and that the simplified model is used (i.e., quadratic terms are ignored) and the computations are performed in visual space, the minimum occurs when the intersection of the iso-distortion cones is at the middle of the depth range of the scene. That is, the $D = 1$ plane is given as $Z = -\frac{x_{0_\epsilon}}{\beta_\epsilon f} = \frac{Z_{min} + Z_{max}}{2}$, and $x_{0_\epsilon} = -\beta_\epsilon f \frac{Z_{min} + Z_{max}}{2}$ [4].

Of course, we do not know the exact flow distribution, or the exact scene depth distribution, nor do we expect the system to optimally solve a minimization problem. We do, however, expect that the estimation of motion is such that the negative depth volume is kept rather small and thus that $x_{0_\epsilon}$ and $\beta_\epsilon$ are of opposite sign and the $D = 1$ plane is between the smallest and largest depth of the object observed.

In the following explanation we concentrate on the first experiment, which was concerned with the judgment about the circular cylinder.

We assume that the system underestimates the value of $x_0$, because the observer is fixating at the rotation axis in the image center while judging measurements to the right of the center. As this does not correspond to a natural situation (fixation center and object of attention coinciding), the observer should perceive the fixation center closer to the object resulting in an underestimation in the value of $x_0$. Thus, $x_{0_e} > 0$ which implies $\beta_\epsilon < 0$ and the distortion space of Fig. 17b becomes applicable.

The holistic surfaces corresponding to negative iso-distortion surfaces in the field of view are very large in their circular extent, and thus the flow vectors leading to negative depth estimates are of large slope, close to the vertical direction. Figure 4.2 shows a cross-section through the negative iso-distortion surfaces and the negative holistic surfaces for a value $Z$ in front of the $D = 1$ plane.

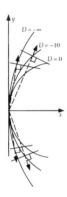

**Fig. 18.** Cross-sections through negative iso-distortion surfaces and negative holistic surfaces. The flow vectors yielding negative depth values have large slopes.

The rotating cylinder constitutes the visible scene. Its vertical cross-section along the axis of rotation lies in the space where $x$ is positive. The most frontal points of the cross-section always lie in front of the $D = 1$ plane, and as the slant of the cylinder increases, the part of the cross-section which lies in front of the $D = 1$ plane increases as well.

The minimization of the negative depth volume and thus the estimation of the motion is independent of the absolute depth of the scene. Therefore a change in viewing distance should not have any effect on the depth perceived by the observer, *which explains the first experimental observation.*

The explanation of the second result lies in a comparison of the estimated vertical extent, $\hat{a}$, and the extent in depth, $\hat{b}$.

Figures 19a–c illustrate the position of the circular cross-section in the distortion space for the fronto-parallel position of the cylinder. Section $a = (AC)$ lies at one depth and intersects the cross section of the holistic surface as shown in Fig. 19b. Section $b = (BC)$ lies within a depth interval between depth values $Z_B$ and $Z_C$. The cross-sections of the holistic surfaces are illustrated in Fig. 19c. To make quantitative statements about the distortion $D$ at any depth value, we assume that at any point $P$, $D$ is the average value of all the iso-distortion surfaces passing through $P$. With this model we derive $\hat{a}$ and $\hat{b}$ as follows:

$$\hat{a} = Da \qquad (5)$$

where $D$ is the average distortion at the depth of section $AC$. The estimate $\hat{b}$ is derived as the difference of the depth estimate at points $B$ and $C$. We denote by $\delta$ the difference between the average distortion factor of extent $a$ and the distortion at point $C$, and we use $\epsilon$ to describe the change in the distortion factor from point $C$ to point $B$. Thus

$$\hat{b} = \hat{Z}_C - \hat{Z}_B = (D+\delta)Z_C - (D+\delta+\epsilon)(Z_C - b) = (D+\delta)b - \epsilon(Z_C - b) \quad (6)$$

$Z_C$ is much larger than $b$ and thus $(Z_C - b)$ is always positive. Comparing (5) and (6) we see that for $a = b$ the factor determining the relative perceived length of $a$ and $b$ depends primarily on $\delta$ and $\epsilon$.

For the case of a fronto-parallel cylinder, where extent $a$ appears behind the $D = 1$ plane, $\delta$ is positive (see Fig. 19b) and $\epsilon$ is negative (see Fig. 19c), which means that $b$ will be perceived to be greater than $a$.

As the cylinder is slanted (see Figs. 19d–f), the circular cross-section also becomes slanted. As a consequence the cylinder covers a larger depth range and extent $a$ appears closer to or even in front of the $D = 1$ plane (see Fig. 19e). Points on section $b$ have increasing $X$-coordinates as $Z$ increases (see Fig. 19f). As the slant becomes large enough $\delta$ reaches a negative value, $\epsilon$ reaches a positive value and $b$ is perceived to be smaller than $a$. Therefore the *results for the experiments involving the cylindrical surface* for the case of motion *can be explained* in terms of the iso-distortion diagrams with $D$ that decreases or increases with $Z$.

The second experiment, concerned with the judgment of right angles, can be explained by the same principle. The estimate is again based on judgment of the vertical extent $a$ relative to the extent in depth $b$ (see Fig. 10). *Either* we encounter the situation where the sign of $x_0$ is positive, so that $a$ and $b$ are measured mostly to the right of the $YZ$ plane, and Fig. 17b explains the iso-distortion space; *or* $x_0$ is negative, so that $a$ and $b$ are mostly to the left of the $YZ$ plane, and the iso-distortion space is obtained by reflecting the space of Fig. 17b in the $YZ$ plane. In both cases the explanation given for the first experiment still applies. Due to the changes of position of the two planes in iso-distortion space with a change in slant, the extent in depth will be overestimated for the fronto-parallel position and underestimated for larger slants.

## 4.3   Stereo

In the case of stereoscopic perception the primary 2D image input is horizontal disparity. Due to the far-off location of the epipole the negative part of the distortion space for horizontal vectors does not lie within the field of view, as can be seen from Fig. 17.

Since depth estimation in stereo vision has long been of concern to researchers in psychophysics, a large amount of experimental data has been published, and

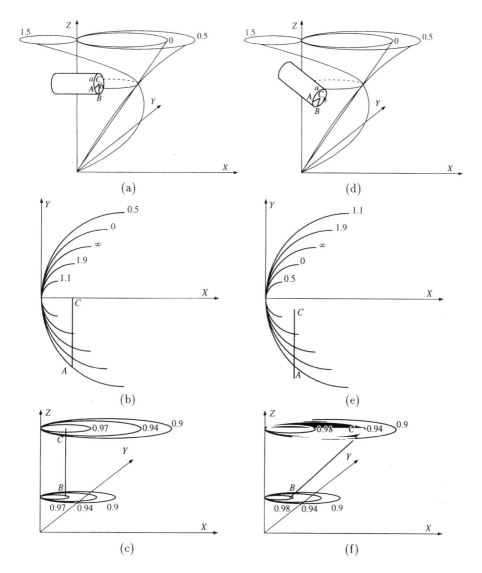

**Fig. 19.** (a–c) Position of fronto-parallel cylinder in iso-distortion space. (d–f) Position of slanted cylinder in iso-distortion space. The figure shows that extent $a$ appears behind the $D = 1$ plane in (b–c) and in front of the $D = 1$ plane in (e–f).

the parameters of the human viewing geometry are well documented. In [15] Foley studied the relationship between viewing distance and error in the estimation of convergence angle ($\beta$ in our notation). From experimental data he obtained the relationship between perceived convergence angle and actual convergence angle shown in Fig. 20.

According to his data, the convergence angle is overestimated at far distances and underestimated at near distances. Foley expressed the data through the

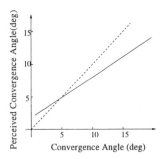

**Fig. 20.** Perceived convergence angle as a function of convergence angle.

following relationship:

$$-\hat{\beta} = E + G(-\beta)$$

with $E$ and $G$ in the vicinity of 0.5; in the figures displayed here the following parameters based on data of Ogle [31] have been chosen: $E = 0.91°$ and $G = 0.66°$.

On the basis of these data, models have been proposed [14,15,31] that explain the perception of concavity and convexity for objects in a fronto-parallel plane. To account for the skewing described in the AFPP task an additional model has been employed which assumes the ocular images are of different sizes.

In our explanation we use the experimental data of Fig. 20 to explain $\beta_\epsilon$. As will be shown, the iso-distortion framework alone allows us to explain all aspects of the experimental findings. For far fixation points $\beta_\epsilon$ is positive (since $\beta < 0$) and the iso-distortion space of Fig. 17a applies. If we also take into account the quadratic term in the horizontal disparity formula of Sect. 4.1(a) (that is, the rotational part $\beta_\epsilon(\frac{x^2}{f} + f)$), we obtain an iso-distortion configuration for horizontal vectors as shown in Fig. 21. In particular Fig. 21a shows the contours obtained by intersecting the iso-distortion surfaces with planes parallel to the $xZ$ plane in visual space, and Fig. 21b shows the same contours in actual 3D space. Irrespective of $x_{0_\epsilon}$ the iso-distortion factor decreases with depth $Z$. The sign of $x_{0_\epsilon}$ determines whether the $D = 1$ contour (the intersection of the $D = 1$ surface with the $xZ$ plane) is in front of or behind the image plane, and the exact position of the object with regard to the $D = 1$ contour determines whether the object's overall size is over- or underestimated.

For near fixation points, $\beta_\epsilon$ is negative and the iso-distortion space appears as in Fig. 17b. The corresponding iso-distortion contours derived by including the quadratic term are illustrated in Fig. 21c and d.

The perceived estimates $\hat{a}$ and $\hat{b}$ are modeled as before. However, this time it is not necessary to refer to an average distortion $D$, since only one flow direction is considered. Section $a$ lies in the $yZ$ plane and $\hat{a}$ is estimated as $aD$, with $D$ the distortion factor at point $C$. The estimate for $b$ is

$$\hat{b} = Db - \epsilon(Z_C - b)$$

As can be seen from Figs. 21a and c, $\epsilon$ is increasing if the fixation point is distant and decreasing if the fixation point is close, and we thus obtain the under- and overestimation of $\hat{b}$ as experimentally observed. A slanting of the object has very little effect on the distortion pattern because the fixation point is not affected by it. As long as the slant is not too large, causing $\epsilon$ to change sign, the qualitative estimation of depth should not be affected by a change in slant. The slant might, however, influence the amount of over- and underestimation. There should be a decrease in the estimation error as the slant increases, since section $b$ covers a smaller range of the distortion space. This can actually be observed from the experimental data in Fig. 9.

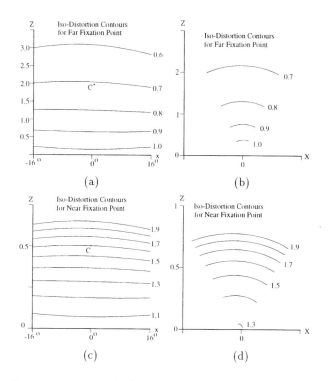

**Fig. 21.** Iso-distortion contours for horizontal disparities: (a, b) for far fixation point in visual space (a) and actual space (b); (c, d) for near fixation point in visual and actual space.

The same explanation covers the second experiment related to the judgment of angles.

## 4.4 Explanation of Purely Stereoscopic and Purely Motion Experiments

The iso-distortion patterns outlined in Sect. 4.3 also explain the purely stereoscopic experiments. With regard to the AFPP task it can be readily verified that the iso-distortion diagram of Fig. 21a (far fixation point) causes a fronto-parallel plane to appear on a concave surface, and thus influences the observer to set them at a convex AFPP locus, whereas the diagram of Fig. 21c (near fixation point) influences the observer to set them on a concave AFPP locus. In addition, the skewing of the AFPP loci is also predicted by the iso-distortion framework.

With regard to the ADB task, the iso-distortion patterns predict that the target will be set at a distance closer than half-way to the fixation point if the latter is far, and at a distance further than half-way to the fixation point if the latter is near, which is in agreement with the results of the task.

In the motion experiment of [17] the 3D motion parameters are as follows. When fixating on one of the two objects the $z$ axis passes through the fixation point, the translation of the observer's head is along the $x$ axis, the rotation of the observer's head (due to fixation) is around the $y$ axis, and the vertical motion component of the point is parallel to the $y$ axis. The scene in view is the observation booth covered with dots.

When the observer fixates on the near point $x_0 > 0$. As in the experiments in Sect. 4.2, it is assumed that the value of $x_0$ is underestimated, that is, $\hat{x}_0 < x_0$, and $\beta_\epsilon < 0$. The resulting distortion space corresponds to the one sketched in Fig. 17b. The moving point appears to the left of the $YZ$ plane, and since the observer fixates on a point in the front part of the scene, it should be behind the $D = 1$ plane. The flow vectors originating from the movement of the point are in a diagonal direction. As can be seen, in the area of the moving point the distortion for that direction is greater than one, and thus the distance of the point is overestimated. When the observer fixates on the far point $x_0 < 0$. If again the absolute value of $x_0$ is underestimated, $\hat{x}_0 > x_0$ and $\beta_\epsilon > 0$. The distortion space is the one we obtain by reflecting the space of Fig. 17a in the $YZ$ plane. In this reflected space the moving point appears to the right of the $YZ$ plane and since the observer fixates on a point in the back of the scene, it should be in front of the $D = 1$ plane. In this area too there occurs an overestimation of distances. This explains the general overestimation found in [17].

In order to assess the exact amount of overestimation we would need to know a number of parameters exactly. The estimated motion, the exact position of the point in the distortion space, and the estimated flow directions determine the distortion factor. Our intuitive argument is as follows. It can be seen that the negative distortion space in Fig. 17b behind the $D = 1$ plane increases very quickly with the distance from the plane. It is therefore assumed that the moving point lies closer to the $D = 1$ plane for the near fixation than for the far fixation, and thus the distortion should be smaller for the near fixation than for the far one, as observed in the experiment.

# 5 Conclusions

The geometric structure of the visual space perceived by humans has been a subject of great interest in philosophy and perceptual psychology for a long time [2,25,26,31]. With the advent of digital computers and the possibility of constructing anthropomorphic robotic devices that perceive the world in a way similar to the way humans and animals perceive it, computational studies are beginning to be devoted to this problem [22].

Many synthetic models have been proposed over the years in an attempt to account for the systematic distortion between physical and perceptual space. These range from Euclidean geometry [16] to hyperbolic [26] and affine [34,37] geometry. Many other interesting approaches have also been proposed, such as the Lie group theoretical studies of Hoffman [18] and the work of Koenderink and van Doorn [23], that are characterized by a deep geometric analysis concerned with the invariant quantities of the distorted perceptual space. It is generally believed in the biological sciences that a large number of shape representations are computed in our heads and different cues are processed with different algorithms. For the case of motion and/or stereo, there might exist more than one process performing local analysis of motion or stereo disparity, that might work at several levels of resolution [27]. The analysis proposed here has concentrated on a global examination of motion or disparity fields to explain a number of psychological results about the distortion of visual space that takes place over an extended field of view.

In contrast to the synthetic approaches in the literature, we have offered an analytic account of a number of properties of perceptual space. Our starting point was the fact that when we have multiple views of a scene (motion or stereo), then the 3D rigid transformation relating the views, and functions of local image correspondence, determine the perceived depth of the scene. However, even slight miscalculations of the parameters of the 3D transformation result in computing a distorted version of the actual physical space. In this paper, we studied geometric properties of the computed distorted space. We have concentrated on analyzing the distortions from first principles, through an understanding of iso-distortion loci, and introduced an analytic, geometric framework as a tool for modeling the distortion of depth.

It was found that, in general, the transformation between physical and perceptual space (i.e., actual and computed space) is a Cremona transformation; however, it turns out that this transformation can be approximated locally quite closely by an affine transformation in the inverse depth or a hyperbolic transformation in the depth. This can be easily understood from the equations. For the case of stereo, where $n_x = 1$, if we ignore the quadratic terms in the image coordinates which are very small with regard to the linear and constant terms, we obtain from (3)

$$\hat{Z} = Z \cdot \frac{x - \hat{x}_0}{x - x_0 + \beta_\epsilon f Z} \qquad \text{or} \qquad \frac{1}{\hat{Z}} = \frac{\frac{x - x_0}{Z} + \beta_\epsilon f}{x - \hat{x}_0} \qquad (7)$$

If we consider $x$ to be locally constant (7) describes $\hat{Z}$ locally as a hyperbolic function of $Z$ or $1/\hat{Z}$ as an affine function of $1/Z$.

Thus our model is in accordance with some of the models previously employed in the psychophysical literature. For all practical purposes the locally described approximations are so close to the real Cremona transformation that they cannot possibly be distinguished from experimental data. [3] contains a detailed analysis of the properties of the distortion for the case of stereo.

Finally, in the light of the misperceptions arising from stereopsis and motion, the question of how much information we should expect from these modules must be raised. The iso-distortion framework can be used as an avenue for discovering other properties of perceived space. Such properties may lead to new representations of space that can be examined through further psychophysical studies. We are interested ultimately in the invariances of the perceived distorted space, since these invariances will reveal the nature of the representations of shape that flexible vision systems, biological or artificial, extract from images.

# References

1. J. Y. Aloimonos and D. Shulman. Learning early-vision computations. *Journal of the Optical Society of America A*, 6:908–919, 1989.
2. A. Ames, K. N. Ogle, and G. H. Glidden. Corresponding retinal points, the horopter and the size and shape of ocular images. *Journal of the Optical Society of America A*, 22:538–631, 1932.
3. G. Baratoff. *Qualitative Space Representations Extracted from Stereo*. PhD thesis, Department of Computer Science, University of Maryland, 1997.
4. L. Cheong, C. Fermüller, and Y. Aloimonos. Interaction between 3D shape and motion: Theory and applications. Technical Report CAR-TR-773, Center for Automation Research, University of Maryland, June 1996.
5. K. Daniilidis. *On the Error Sensitivity in the Recovery of Object Descriptions*. PhD thesis, Department of Informatics, University of Karlsruhe, Germany, 1992. In German.
6. K. Daniilidis and H. H. Nagel. Analytical results on error sensitivity of motion estimation from two views. *Image and Vision Computing*, 8:297–303, 1990.
7. K. Daniilidis and M. E. Spetsakis. Understanding noise sensitivity in structure from motion. In Y. Aloimonos, editor, *Visual Navigation: From Biological Systems to Unmanned Ground Vehicles*, chapter 4. Lawrence Erlbaum Associates, Hillsdale, NJ, 1997.
8. O. D. Faugeras. *Three-Dimensional Computer Vision*. MIT Press, Cambridge, MA, 1992.
9. C. Fermüller and Y. Aloimonos. Direct perception of three-dimensional motion from patterns of visual motion. *Science*, 270:1973–1976, 1995.
10. C. Fermüller and Y. Aloimonos. Algorithm independent stabiity analysis of structure from motion. Technical Report CAR-TR-840, Center for Automation Research, University of Maryland, 1996.
11. C. Fermüller and Y. Aloimonos. Ordinal representations of visual space. In *Proc. ARPA Image Understanding Workshop*, pages 897–903, February 1996.
12. C. Fermüller and Y. Aloimonos. Towards a theory of direct perception. In *Proc. ARPA Image Understanding Workshop*, pages 1287–1295, February 1996.

13. C. Fermüller and Y. Aloimonos. Ambiguity in structure from motion: Sphere versus plane. *International Journal of Computer Vision*, 1997. In press.
14. J. M. Foley. Effects of voluntary eye movement and convergence on the binocular appreciation of depth. *Perception and Psychophysics*, 11:423–427, 1967.
15. J. M. Foley. Binocular distance perception. *Psychological Review*, 87:411–434, 1980.
16. J. J. Gibson. *The Perception of the Visual World*. Houghton Mifflin, Boston, 1950.
17. W. C. Gogel. The common occurrence of errors of perceived distance. *Perception & Psychophysics*, 25(1):2–11, 1979.
18. W. C. Hoffman. The Lie algebra of visual perception. *Journal of Mathematical Psychology*, 3:65–98, 1966.
19. B. K. P. Horn. *Robot Vision*. McGraw Hill, New York, 1986.
20. E. B. Johnston. Systematic distortions of shape from stereopsis. *Vision Research*, 31:1351–1360, 1991.
21. R. Julesz. *Foundations of Cyclopean Perception*. University of Chicago Press, Chicago, IL, 1971.
22. J. J. Koenderink and A. J. van Doorn. Two-plus-one-dimensional differential geometry. *Pattern Recognition Letters*, 15:439–443, 1994.
23. J. J. Koenderink and A. J. van Doorn. Relief: Pictorial and otherwise. *Image and Vision Computing*, 13:321–334, 1995.
24. S. Kosslyn. *Image and Brain*. MIT Press, Cambridge, MA, 1993.
25. J. M. Loomis, J. A. D. Silva, N. Fujita, and S. S. Fukusima. Visual space perception and visually directed action. *Journal of Experimental Psychology*, 18(4):906–921, 1992.
26. R. K. Luneburg. *Mathematical Analysis of Binocular Vision*. Princeton University Press, Princeton, NJ, 1947.
27. H. A. Mallot, S. Gillner, and P. A. Arndt. Is correspondence search in human stereo vision a coarse-to-fine process? *Biological Cybernetics*, 74:95–106, 1996.
28. D. Marr. *Vision*. W.H. Freeman, San Francisco, CA, 1982.
29. S. J. Maybank. *Theory of Reconstruction from Image Motion*. Springer, Berlin, 1993.
30. J. E. W. Mayhew and H. C. Longuet-Higgins. A computational model of binocular depth perception. *Nature*, 297:376–378, 1982.
31. K. N. Ogle. *Researches in Binocular Vision*. Hafner, New York, 1964.
32. J. G. Semple and L. Roth. *Inroduction to Algebraic Geometry*. Oxford University Press, Oxford, United Kingdom, 1949.
33. J. S. Tittle, J. T. Todd, V. J. Perotti, and J. F. Norman. Systematic distortion of perceived three-dimensional structure from motion and binocular stereopsis. *Journal of Experimental Psychology: Human Perception and Performance*, 21:663–678, 1995.
34. J. T. Todd and P. Bressan. The perception of three-dimensional affine structure from minimal apparent motion sequences. *Perception and Psychophysics*, 48:419–430, 1990.
35. R. Y. Tsai and T. S. Huang. Uniqueness and estimation of three-dimensional motion parameters of rigid objects with curved surfaces. *IEEE Transactions on Pattern Analysis and Machine Intelligence*, 6:13–27, 1984.
36. H. L. F. von Helmholtz. *Treatise on Physiological Optics*, volume 3. Dover, 1962. J. P. C. Southhall, trans. Originally published in 1910.
37. M. Wagner. The metric of visual space. *Perception and Psychophysics*, 38:483–495, 1985.

# The Cascaded Hough Transform as Support for Grouping and Finding Vanishing Points and Lines

Tinne Tuytelaars*, Marc Proesmans** and Luc Van Gool

Katholieke Universiteit Leuven, ESAT-MI2
Kardinaal Mercierlaan 94, B-3001 Leuven, BELGIUM

**Abstract.** In the companion paper [7] a grouping strategy with a firm geometrical underpinning and without the problem of combinatorics is proposed. It is based on the exploitation of structures that remain fixed under the transformations that relate corresponding contour segments in regular patterns. In this paper we present a solution for the complementary task of extracting these fixed structures in an efficient and non-combinatorial way, based on the iterated application of the Hough transform. Apart from grouping, this 'Cascaded Hough Transform' or CHT for short can also be used for the detection of straight lines, vanishing points and vanishing lines.

## 1  Introduction

Grouping is the process of combining bits and pieces of visual information into perceptually salient structures. It is an important stepping stone towards a deeper understanding of observed shapes structure and scene organisation.

Recently, a more systematic geometrical basis for grouping has been propounded on the basis of so-called fixed structures [7, 8]. These are the structures in the image that remain fixed under the grouping transformation. As an example, under the transformation that maps one half of a symmetric pattern onto the other the symmetry axis and all its points remain fixed . This also holds when looking at such pattern from an oblique viewpoint.

A problem that has been left open is that exploitating the grouping-specific invariants is only possible when the fixed structures are known. Therefore, methods to find the fixed structures are called for. This paper proposes a way of doing just that, again avoiding combinatorics.

Therefore, a modified version of the Hough transform is introduced — coined the 'Cascaded Hough Transform' or CHT for short — that finds structures at the different hierarchical levels by iterating one single kind of Hough transform. It directly supports the quest for fixed points, fixed lines, lines of fixed points, and pencils of fixed lines.

---

\* Research Assistant of the Fund for Scientific Research – Flanders (FWO–Vlaanderen)
\*\* Postdoctoral Fellow of the Flemish Institute for the advancement of Science and Technology in Industry (IWT)

Apart from this framework of efficient grouping, the CHT can also serve to find vanishing points and lines, which are special types of fixed structures that play an increasingly important role in automatic navigation and three-dimensional scene reconstruction. Previously, Hough transforms were already used for line extraction or vanishing point detection, but these steps were not combined into a single scheme. Moreover, the Hough transforms involved were of a different nature. This is discussed in more detail later. Having a single scheme for the extraction of these different features greatly simplifies the interaction between the different levels at which they are found.

The paper is organised as follows. Section 2 introduces the Cascaded Hough Transform, the central tool introduced in the paper to find good fixed structure candidates. Section 3 discusses some implementation issues of the CHT. The CHT is illustrated in section 4 and section 5 concludes the paper.

## 2 The Cascaded Hough Transform

The Hough transform's history stretches a long way back, as P.V.C. Hough ([2]) patented it in 1962. Yet, now it probably has more applications than ever before and it still is the subject of research and a favourite approach to feature extraction for many [3, 4]. Attempts to replace it by alternative schemes such as SLIDE have not been very successful [6].

The Hough transform is a global, robust technique for the detection of pre-defined shapes in images, esp. straight lines. It is based on the transformation of the edge points to a parameter space. In fact, in its most efficient forms, the Hough transform exploits the symmetry of simple parametric shapes (translational for lines, rotational for circles) in order to reduce the dimensionality of the search [5].

Here, only the extraction of straight lines is considered. In the original version of the Hough transform, straight lines were detected using a slope-intercept parametric representation. In order to detect edge pixels $(x, y)$ that lie on a straight line $ax + b + y = 0$, a voting mechanism in the $(a, b)$ parameter space is used. The important thing is that the Hough transform finds structures in a non-combinatorial fashion.

Every point $(x_0, y_0)$ of an edge in the image is transformed into a line $x_0 a + y_0 + b = 0$ in parameter space. Following this parameterisation, there is a degree of symmetry between image space and parameter space. Indeed, the parameters $a$ and $b$ are to image space $(x, y)$ what $x$ and $y$ are to parameter space $(a, b)$. Lines in one space can be detected as points in the other space and, vice versa, for every point there is also a corresponding line.

This high symmetry not only emphasizes the duality between lines and points, but also is the very basis of the proposed 'Cascaded Hough Transform', that amounts to iterating the same procedure. Indeed, we can apply a second Hough transform to the (filtered) result of the first one, transforming parameter space back into image space. This is reminiscent of the Fourier transform. There also one goes to the other domain (i.c. the frequency domain) to perform operations that can be carried out more efficiently there, in order then to go back to the original (spatial) domain. In contrast to the Fourier transform, the 'forward' and 'backward' Hough transforms are identical.

The first Hough transform can be used to detect lines in the image, where it is usual to only keep the dominant peaks in the resulting $(a, b)$-parameter space. A second Hough transform could then detect lines of collinear peaks in parameter space. This second Hough goes back to image space where the positions of the peaks that it generates correspond to vertices where several straight lines in the original image intersect. Vanishing points are an important case of such vertices. Similarly, a third Hough transform can be applied to the peaks of the second one to detect collinear vertices. An important case of this kind of feature is given by vanishing lines (e.g. the horizon line), containing several of the vanishing points detected by the second Hough transform. These observations are summarised in the flowchart of Fig. 1.

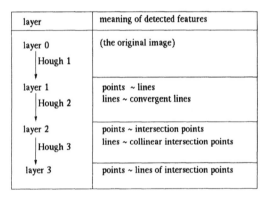

**Fig. 1.** Flowchart summarising the meaning of the features detected at the different layers of the CHT.

The structures found at the second and third layers of the CHT correspond well to the fixed structures we are looking for. Certainly structures that receive many votes are strong candidates to consider as possible fixed structures. In a sense the non-accidentalness principle is applied, but to the interpretation of fixed structures rather than to the interpretation of image contours.

Having this selection process run on straight lines in the original image exclusively, would lower the chance of finding fixed structures except for indoor and building type of scenes, where straight lines abound. Therefore, it is useful to add well-chosen structures at the different levels, as auxiliary input for the next Hough transform. One could for instance add points that correspond to tangent lines at inflections at the level of the first $(a, b)$-space. Such lines are bound to reemerge at the next $(a, b)$-level, but also structures like symmetry axes can be expected to emerge. The original lines can easily be suppressed.

# 3 Implementation of the CHT

## 3.1 Parameterisation of the Spaces

The main drawback of the slope-intercept parameterisation of the lines is the unbounded parameter space. Indeed, both $a$ and $b$ can take on infinite values. Therefore, another parameterisation was introduced by Duda and Hart [1]. It is based on the orthogonal distance $\rho$ of the line to the origin and the direction $\theta$ of the normal to the line. This parameterisation is the one most often used. It yields a bounded parameter space. However, it also breaks the symmetry between the image and parameter spaces. A point in the image is transformed to a trigonometric function in parameter space, and a line to a point. So the basis for the CHT is gone.

In order to obtain a bounded parameter space, while keeping the symmetry intact, the original, unbounded $(a, b)$-space is split into three bounded subspaces. This is shown in Fig. 2. The first subspace also has coordinates $a$ and $b$, but only applies to lines with $|a| \leq 1$ and $|b| \leq 1$. If $|a| > 1$ and $|b| \leq |a|$, the point $(a, b)$ turns up in the second subspace, with coordinates $1/a$ and $b/a$. If, finally, $|b| > 1$ and $|a| < |b|$, we use a third subspace with coordinates $1/b$ and $a/b$. In this way, the unbounded $(a, b)$-space is split into three bounded subspaces. Their coordinates are restricted to the interval $[-1, 1]$, while a point $(x, y)$ in the original space is still transformed into a line in each of the three subspaces. As can be seen at the left side of Fig. 2, this subdivision of the parameter space can also be interpreted as an inhomogenous discretisation of the unbounded parameter space, with cells growing larger as they get further away from the origin.

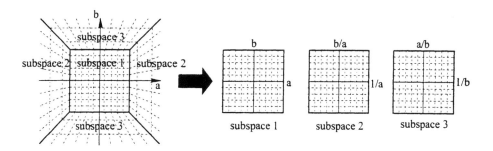

**Fig. 2.** The original, unbounded space is split into three, bounded subspaces

In order to keep both spaces as symmetrical as possible, the same structure is given to $(x, y)$-image space, subdividing it into three subspaces, with coordinates

$(x, y)$, $(1/x, y/x)$ and $(1/y, x/y)$. But symmetry is not the only reason for this subdivision. $(x, y)$-space is in fact also an unbounded space, not restricted to the size of the image itself. Vanishing points for instance can lie outside the image boundaries, or even at infinity. With the proposed parameterisation, the points lying at infinity are included in a natural way. The fact that cells grow larger when going out of the field of view reflects the fact that points lying further away are normally determined less accurately anyhow.

In order to give the $(x, y)$-parameterisation an additional, practical meaning, the original image is rescaled, such that it fits entirely within the first subspace and is discretized in a homogeneous way. Hence the first subspace actually corresponds to the image itself. Subspace 2 captures information left and right from the field of view, and subspace 3 does the same for information from above or below.

It might seem that setting up and clustering three two-dimensional spaces at each layer causes an enormous computational burden. However, it is not the number of spaces that is important, but their total size. For instance, the computational cost of the clustering stage mainly depends on the number of clusters, which is the same for any kind of Hough transform, whatever parameterisation has been used. On the other hand, our approach does require a lot of memory, especially if a high resolution is needed for the Hough spaces, since each layer needs its own three spaces.

## 3.2 Filtering Stages

Before applying a new Hough transform to the result of a previous one, appropriate data handling may be called for. In general, this will include deciding on

1. which data to eliminate from further consideration by subsequent layers,
2. which data to read out as important information emerging at each layer,
3. which data to add to the input of the next Hough Transform.

A first aspect is that only peaks, i.e. relatively high concentrations of votes, are considered for further processing. In order to better emphasise the initial peaks (at the output of the first Hough transform), locally around edge pixels an estimate of the tangential line is made, and only a smalll blob around the corresponding $(a, b)$ values is introduced, rather than a complete line. In this way, the strength of the peak will still reflect the length of the straight line in the image.

Only truly 'non-accidental' structures are read out at the different layers. What that means is layer dependent: for the subsequent layers these are straight lines of a minimum length, points where at least three straight lines intersect, and lines that contain at least three line intersections. Note that in the latter case, these can be three intersections of each time two lines.

There also are a number of additional possibilities, that have not yet been implemented completely, but are under study. These include:

- As was mentioned before, one can add structures at the different levels, such as $(a, b)$ peaks that correspond to tangent lines at inflections. This generalises the approach to scenes and shapes containing few straight lines.
- It can help to suppress structures found at earlier layers. For instance, a line will intersect with all other lines. As a consequence, it is bound to reappear as a line of collinear line intersections. Such lines clearly are not very promising candidates, unless more lines intersect it in the same point.
- Structures found can be fed back into earlier layers. This allows additional structures to be found. As an example, a vanishing line added to the image may yield additional vanishing points as non-accidental read-outs.

# 4    Examples of the CHT

As a first illustration of the CHT, consider Fig. 3. It shows a historic building. In such scenes, straight lines abound, so this example was run without the addition of any extra information (e.g. without adding tangents at inflections). It is illustrated how the CHT yields the vanishing points and lines for the principal directions in the scene. Figure 4 shows the different layers of the CHT in its different rows, with the result of the first Hough transform at the first row, and so on. The different subspaces correspond to the columns, following the order of Fig. 2.

One can clearly see that the peaks in the first layer of the CHT, representing the lines in the original image, are arranged in lines themselves. These lines of peaks correspond to lines in the image that are convergent. The point where they intersect can be detected by applying a second Hough-transform. After filtering, three main peaks can be found in this second layer. They correspond to the three main vanishing points, as can be seen in the bottom part of Fig. 3, where all lines contributing to the detected intersection points are shown.

By applying a third Hough transform we can detect lines of vanishing points, i.e. the vanishing lines in the image. Since we only have three vanishing points, we can find three vanishing lines, all going through two points only. One of them is the horizon line, also shown in Fig. 3.

A similar experiment was carried out for the aerial image shown in Fig. 5. Again, this aerial image contains enough straight lines, allowing to apply the CHT directly, without the addition of any extra features such as tangent lines. Figure 6 shows the different subspaces and layers of the CHT. Peaks in the first row correspond to straight lines in the original image. Some collinear configurations are already salient and correspond to lines intersecting in a single point. These intersections are picked up by the second Hough and hence show up as peaks in the second row. There are two dominant peaks, that correspond to the vanishing points of the two dominant line orientations in Fig. 5. Finally, the third row has local maxima at lines that contain at least two intersections. The two dominant peaks of the second row show up as two dominant lines here and their intersection yields the strongest peak of all. This is the horizon line.

**Fig. 3. Upper left:** the original image. **Upper right**: idem, with the detected horizon line superimposed. **Bottom left to right** : lines belonging to one of the three detected vanishing points.

Note that this aerial image is a nadir rather than an oblique view. Perspective effects have previously been used almost exclusively for oblique, reconnaissance type of views. Aerial imagery for civil applications has the camera oriented almost vertically, however. Nevertheless, even if the perspective effects are much weaker in such case, the results show them to be of sufficient importance to warrant their explicit detection. Thanks to the special parameterisation, vanishing points and lines 'approaching' infinity can be handled adequately.

Especially for the orientation going from top-left to bottom-right, there is a perspective effect that cannot be discarded. As an example, it is interesting to focus on the pedestrian walkway with this orientation, that is visible at the left side of the image. Figure 7a shows the line that is obtained from a single application of the Hough transform. Although it is positioned rather well, this line does not follow the walkway precisely. Part of the problem are the cars and their shadows that have pulled the edge to the right.

One can expect a better result, by pooling information on the corresponding, dominant orientation. Therefore, the average orientation of the different Hough-based lines belonging to this group was calculated. Only sufficiently long straight edge segments were used, as indicated in Fig. 9a. They were found as segments

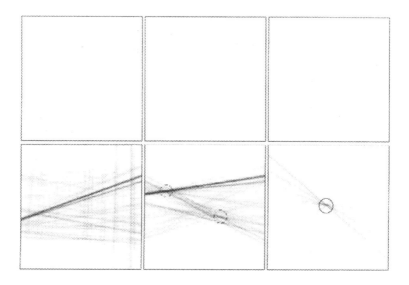

**Fig. 4.** The different layers of the CHT of the image of Fig. 3. The result of the first Hough transform is shown at the first row, and so on. The different subspaces correspond to the columns, following the order of Fig. 2.

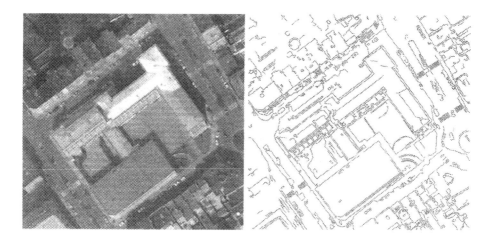

**Fig. 5.** Aerial image of high resolution with a school building and surrounding streets (left), and the corresponding edges (right).

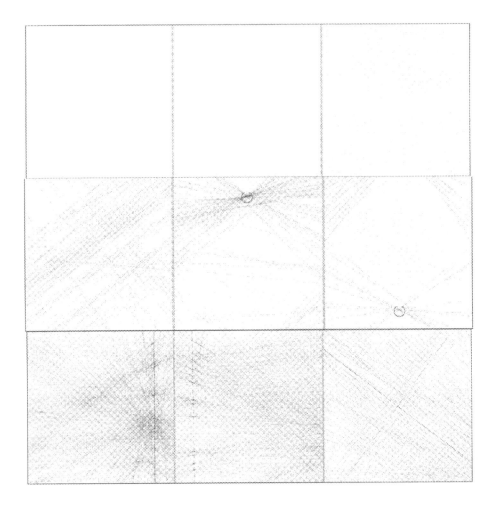

**Fig. 6.** Nine CHT subspaces for the image of Fig. 5. The first, second, and third rows give the results for the first, second, and third Hough, resp. The columns give the first, second, and third subspaces, ordered from left to right according to Fig. 2.

along the lines corresponding to peaks in the first Hough space where contrast over the lines was sufficiently high. This threshold was set quite conservatively. Figure 7b shows a second estimate for the walkway edge. It has the average orientation and runs through the center of gravity of the walkway segments in Fig. 9a. This line follows the walkway edge more closely. A deviation from the real edge is still visible, however. The edge should be rotated slightly clockwise. This is a deviation that is in agreement with what we would expect from the perspective distortion in the image. Finally, Fig. 7c shows another line, through the same center of gravity and the vanishing point corresponding to this dominant orientation. The vanishing point was found after the second step of the CHT. As the system keeps track of the lines that contribute to the different vanishing points, they are easy to match. As can be seen in the figure, the line now follows the walkway edge more closely.

The relative quality of the three edge estimations is better illustrated on a detail near the bottom of the image, because there the differences show up most clearly. Figure 8 shows that part of the walkway. The three lines are shown together with, from top to bottom, the line found on the basis of a single Hough, based on the averaged orientation, and based on the vanishing point. Only the latter line remains within a 1-pixel distance from the visual edge throughout the image. A similar correction can be applied to the other lines of the same direction. When looking again for support along these lines, we now find longer edge segments (Fig. 9b).

## 5 Conclusion

A non-combinatorial approach for finding the fixed structures in an image was propounded based on the Cascaded Hough Transform or CHT. Applying several Hough transforms in a cascaded way was shown to yield an efficient method for finding fixed point and line candidates. Such fixed structures are the key element for the geometric grouping proposed in [7].

Of course, the structures found at the different layers of the CHT can also be useful as such. For instance, they can be used as cues for three dimensional reconstructions, or they can serve as a starting point towards robot navigation by e.g. detecting the horizon and the vanishing point of the street which a robot is on.

## References

1. Duda, R.O., Hart, P.E.: Use of the Hough transform to detect lines and curves in pictures. Commun. ACM **15** (1972) 11–15
2. Hough, P.V.C.: Method and Means for Recognising Complex Patterns. U.S. Pattern No. 3069654 (1962)
3. Illingworth, J., Kittler, J.: A survey of the Hough transform. CVGIP **44** (1988) 87–116
4. Leavers, V.F.: Which Hough transform ? CVGIP **58** (1993) 250–264
5. Pintsov, D.: Invariant pattern recognition, symmetry, and the Radon transform. J. of the Optical Soc. of America A, **6** (1989) 1544–1554
6. Sheinvald, J., Kiryati, N.: On the magic of SLIDE. Machine Vision and Applications **9** (1997) 251–261
7. Van Gool, L.: From Fixed Structures to Geometrical Grouping. Proc. Int. workshop Algebraic Frames for the Perception-Action Cycle (1997)
8. L. Van Gool, M. Proesmans, T. Moons, Groups for Grouping, SPIE Int. Symp. on Optical Science, Appl. of Digital Image Processing XVIII, Vol.2564, pp.402-413, 1995.

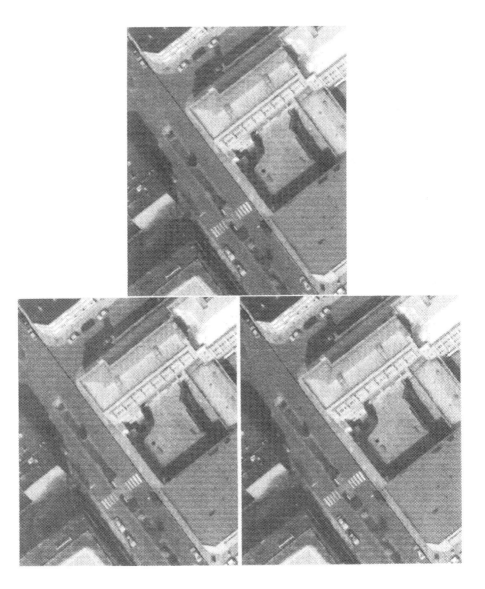

**Fig. 7.** The line detected by the first Hough-transform (top) clearly is not very accurate. One can improve the orientation of this line by giving it the average orientation of that direction (bottom left). However, a better result is obtained if the line is adapted using the vanishing point of that direction (bottom right).

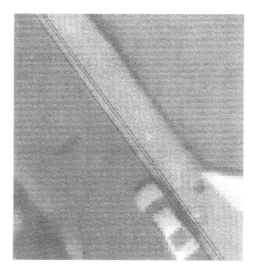

**Fig. 8.** A detail of Fig. 5 with the three lines fitted to the walkway: the Hough line at the right, the average line in the middle and the line through the vanishing point at the left.

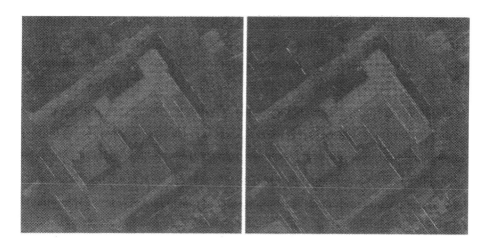

**Fig. 9.** a/ Line segments belonging to the same direction as the walkway (top) b/ idem, after adapting the orientation of the lines such that they pass through the vanishing point (bottom).

# Visual Perception Strategies for 3D Reconstruction

Éric Marchand and François Chaumette

IRISA - INRIA Rennes
Campus de Beaulieu
35 042 Rennes cedex, France

**Abstract.** We propose in this paper an active vision approach for performing the 3D reconstruction of static scenes. The perception-action cycles are handled at various levels: from the definition of perception strategies for scene exploration downto the automatic generation of camera motions using visual servoing. To perform the reconstruction we use a structure from controlled motion method which allows a robust estimation of primitive parameters. As this method is based on particular camera motions, perceptual strategies able to appropriately perform a succession of such individual primitive reconstructions are proposed in order to recover the complete spatial structure of complex scenes. Two algorithms are proposed to ensure the exploration of the scene. The former is an incremental reconstruction algorithm based on the use of a prediction/verification scheme managed using decision theory and Bayes Nets. It allows the visual system to get a complete high level description of the observed part of the scene. The latter, based on the computation of new viewpointsC ensures the complete reconstruction of the scene.

## 1 Active Vision to handle the perception action cycles

Most of the approaches proposed to solve vision problems are inspired from the Marr paradigm which considers a sensor, static or mobile but not controlled. Unfortunately, this approach appears to be inadequate to solve many problems where appropriate modifications of intrinsic and/or extrinsic parameters of the sensor are necessary. This is why Aloimonos, Bajcsy, or Ballard (among others) have proposed to modify the Marr concept. They proposed a new paradigm named **active vision**. Since the major shortcomings which limit the performance of vision systems are their sensitivity to noise, their low accuracy, and their lack of reactivity, the aim of active vision is generally to elaborate strategies for adaptively setting camera parameters (position, velocity,... ) in order to improve the perception task. Thus, function of the specified task and of the data extracted from the acquired image, an active vision system might be induced to modify its parameters (position, orientation, ocular parameters such as focus or aperture), but also the way data are processed (region of interest, peculiar image processing, etc). It controls either the sensor parameters, either the processing resources allocated to the system [20].

However, this is a general definition of active vision and the different authors who had introduced this concept had different motivations. What is usually called active vision can be divided into four main classes : the *active vision* introduced by Aloimonos [3] is a mathematical analysis of complex problems such as stability, linearity and uniqueness of solutions ; the goal of *active perception* as defined by Bajcsy [4] is to elaborate control strategies for setting sensor parameters in order to improve the knowledge of the environment. Thus, active vision is defined as an intelligent data acquisition process ; the *animate vision* [5] is based on the analysis of human perception. Animate vision mainly uses binocular camera heads. Its goal is on one hand to solve the *gaze control* problem, and on a second hand to facilitate the computational process ; dealing with *purposive vision* [2], the goal is to acquire and extract from the environment only the information needed to ensure the realization of a given task. Actions irrelevant to the specified problem will not be executed.

Despite these differences, the goal of the active vision community is to show that an active system is more *relevant* to the application (usually because it is goal driven), more robust (because they can handle either uncertainty and/or dynamic environment) and more accurate (because they are able to modify their own configuration). From our point of view we think that these different approaches are closely related. The methodology used in this paper to define efficient exploration and reconstruction strategies is based on the three following relations:

- the perception-action cycle. The main point of the proposed approach is the relation between the motion of the camera and the information acquired during this motion. Visual information is used to control the camera motion, and camera motions are used to acquire information. We see this feedback loop as a fundamental characteristic of an active vision system. At this level, real time implementation (*i.e.*, images handled at video rate) is a fundamental issue to allow an efficient feedback between perception and action.
- the relation between *global* and *local*. A task is usually defined in a global way (by the goal). However, data available to ensure this goal are usually local. The relation between the global modeling of the goal and this set of local sub-model (closely related to the parameters and the location of the camera) must be studied in order to ensure the execution of the nominal task. Describing a task as a scheduling of elementary tasks is a fundamental step to describe and implement such systems. However efficient techniques are necessary to link the local and global models.
- the relation between *continuous* and *discrete*. This aspect of the problem is closely related to the previous one. In one hand, the local elementary tasks can be handled in real time using continuous schemes. In that case, information must be seen as an infinite flow of data acquired by the sensor. In an other hand, the scheduling of these different tasks may require sensor planning strategies and therefore discrete camera motions. In that case, we manipulate discrete information (logic, temporal, etc.).

Let us now consider how this methodology has been applied to the scene reconstruction and exploration problem. Our concern is to deal with the problem of recovering the 3D spatial structure of a whole scene without any knowledge on the localization and the dimension of the different geometrical primitives of the scene (assumed to be composed of polygons, cylinders and segments). The autonomous system we propose deals with various issues from the automatic generation of camera motion using image-based visual servoing to sensor planning to ensure a reconstruction as complete as possible of the scene. The whole system is described using a hierarchical parallel automata (see Fig. 1). It has three main **perception-action cycles**. The main one is the exploration-reconstruction cycle which ends only when the reconstruction is complete. This cycle deals with global information and the resulting camera motions are discrete. However, when an object is observed, the system enters in the second cycle which is the incremental reconstruction loop. The main goal of this level is to bridge the gap between a local modeling of the scene and a global one. The latter cycle deals with the active reconstruction itself which is here intrinsically based on a local/continuous approach. There, for each observed segment, a recognition task is performed in order to determine the nature of the primitive (cylinder or segment). Then, if a cylinder has been recognized, an estimation of its parameters based on its two limbs is performed in order to get a more robust reconstruction. Finally in both cases, we have to compute the length of the primitive. In parallel to the reconstruction tasks, due to the camera motion, occlusions and manipulator joint limits avoidance tasks are realized. Let us now examine the various issues raised by this reconstruction problem.

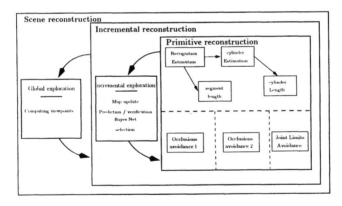

**Fig. 1.** Hierarchical parallel automata describing the whole reconstruction process

**Exploration - Complete reconstruction.** The first issue deals with the exploration. The goal is to determine where the objects are and to ensure the completeness of the reconstruction (for all the most a reconstruction as complete as possible). Previous works have been done in order to answer the "*where to look next*" question [9,21,8,22,23]. As far as we are concerned, in the high level perception strategies of our reconstruction scheme, active vision is used to determine the camera position which provides the maximum of new information (Section 4). The resulting gaze planning strategy proposes a solution to the next best view problem that mainly uses a representation of known and unknown areas as a basis for selecting viewpoints. We have chosen to handle this problem as a function minimization problem. We define a function to be minimized which integrates the constraints imposed by the system and evaluates the quality of the viewpoint. When an object is observed, the exploration process ends and its reconstruction is realized.

**Primitive reconstruction and camera motion generation.** The approach we have chosen to get an accurate three-dimensional geometric description of a scene is based on a continuous structure from motion approach [7]. Very noticeable improvements can be obtained in the 3D reconstruction if the camera viewpoint is properly selected and if adequate camera motions are generated (Section 2). These motions are generated using the visual servoing approach [11,13].

This approach has many advantages. First of all, visual servoing allows to generate automatically the camera motions defined for an optimal estimation of the primitives. To this purpose, we define a secondary task such as a trajectory tracking in "parallel" with a priority task (*e.g.*, gaze control). Second, as the camera motions are dedicated to the estimation of one primitive, we have only one features to track in the images sequence. Therefore, we are able to perform a real time estimation of the primitives parameters. Furthermore, as we used a continuous structure from motion approach, the motion of the primitive in the image is very small during the estimation since the primitive must remain at a constant position in the image. Therefore, the spatio-temporal matching process is quite straightforward and can be handled in real time. This real time computation of the camera motion allows us to deal on-line with some other constraints such as occlusions and kinematics problems specific to the manipulator.

**Incremental reconstruction.** However, since the camera motion is controlled for the estimation of one primitive at a time, this implies to successively focus on each primitive of the scene, using a **local exploration** algorithm. The proposed method is based on a prediction/verification scheme (Section 3). Bayes nets [18,19,6,10] seem to be well adapted to manage this process. They allow us to model "expert" reasoning. Furthermore, they are adapted to the automatic generation of action while performing this reasoning. Thus we can directly introduce perception strategies within the scene interpretation process. This algorithm proposes a partial solution to the occlusion problem and allows us to obtain a high level description of the scene. This way, we can bridge the gap between a set of local sub-models (obtained using a continuous method) and a global model of the scene (thus obtained using a discrete method).

The remainder of this paper is organized as follows. Section 2 is devoted to the local aspect of our reconstruction scheme and describes the structure from motion framework based on an active vision paradigm. Section 3 describes the Bayes Nets-based prediction / verification scheme used to get a complete description of the observed part of the scene. The last cycle is described in Section 4 where the computing viewpoint issue used to ensure a reconstruction as complete as possible of the scene is proposed. Finally, Section 5 presents experiments carried out on a robotic cell which have demonstrated the validity of our approach.

## 2  3-D structure estimation using active dynamic vision

The measure of the camera motion, which is necessary for the 3D structure estimation, characterizes a domain of research called dynamic vision. The method used here is a continuous approach [1,12] which stems on the measure of the camera velocity and of the motion of the considered primitive in the image. More precisely, we use a "*structure from controlled motion*" method which consists in constraining the camera motion in order to obtain a precise and robust estimation [7].

For most of the geometrical primitives, it is possible to determine the interaction matrix $L_{\underline{P}}^T$ defined by the classical equation [11]:

$$\dot{P} = L_P^T(P, p_l)T_c \tag{1}$$

where $\dot{P}$ is the time variation of $P$ due to the camera motion $T_c$. The parameters $P$ describe the position of the object in the image while the parameters $p_l$ describe the position of the object limb surface (*i.e.*, for a volumetric primitive, it defines the 3D surface in which the limbs lie).

From the resolution of a linear system derived from (1), it is possible to obtain the parameters $p_l$ [7]. Then, using geometrical constraints related to the considered primitive, we can estimate the parameters $\underline{p}$ which fully define its 3D configuration.

When no particular strategy concerning camera motion is defined, important errors on the 3D structure estimation can be observed. This is due to the fact that the quality of the estimation is very sensitive to the nature of the successive camera motions. An active vision approach is thus necessary to improve the accuracy of the results by generating adequate camera motions. In fact, two main results dealing with this problem have been achieved [7]:

1. A sufficient and general condition that suppresses the discretization error is to constrain the camera motions such that the projection of the primitive must be kept constant in the image and no variation must occur on the limb surface parameters during the camera motion (*i.e.*, $\dot{P} = \dot{p}_l = 0, \forall t$).
2. A more robust estimation with respect to measurement errors is obtained if the location between the camera and the primitive is considered. Indeed, particular positions of the primitive in the image do minimize the influence of the measurements errors. Thus, in order to obtain an **optimal estimation**, a gaze control task which constrains the camera motion so that the object remains fixed at its specified position in the image is realized (see Fig. 2).

For example, in the case of a cylinder, it has been shown that the optimal camera motion is such that the cylinder constantly appears as two static centered, vertical or horizontal, straight lines in the image sequence. Visual servoing [11,13], which is the main point of low level perception-action cycles, is very well qualified to control camera motions in order to satisfy these constraints.

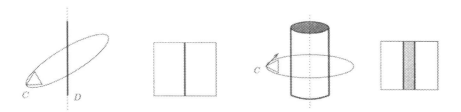

**Fig. 2.** Optimal camera motion and resulting image in the cases of a straight line and a cylinder

# 3 A Bayes-Nets Based Prediction / Verification Scheme

The next level of our reconstruction scheme is the incremental reconstruction of the objects observed from a computed viewpoint. In order to obtain as accurate results as possible, we have chosen to perform the reconstruction in sequence. The resulting algorithm [16] allows us to perform an estimation of all the primitives which appear in the camera field of view. However:

– The description of the scene is a low level and local description which only contains a list of 3D segments and cylinders. It might be more interesting to get high level global information such as junctions, polygons, and faces.
– The scene reconstruction is incomplete for two main reasons. First, the projection in the image of some segments have a too small length to make their reconstruction possible. Second, as this algorithm only deals with the observed objects, it as a local perception of the scene. According to this, some objects may not appear in the camera field of view (because of occlusions or because they are located in an unknown and unobserved area).

To cope with these problems, we propose a Bayes Nets based prediction/verification scheme. A Bayes Net [18] is a directed acyclic graph where nodes represent the discrete random variables and where links between nodes represent the causality between the variables. Such a net can be used to represent the knowledge available on a particular domain. The graph structure and the *a priori* knowledge introduced in the graph (as conditional probability tables) must be defined by the conceptor of the application. The advantages of Bayes Nets lies in the ability to reflect the *a priori* knowledge available on the application. This knowledge is reflected in the structure of the net through the nature and the number of nodes (variables), the different states of these variables and the relations (links) between these variables. The knowledge is also present in the conditional probability tables associated with the variables of the net. These tables reflect the expert reasoning as well as the uncertainty associated with the observations. Finally, the propagation allows to take each new observation into account. The influence of an observation is propagated to the other variables of the net according to the causality relations.

The goal of our prediction/verification scheme is to determine the relations between reconstructed 3D segments and to infer either the presence of new segments, either the existence of more complex objects. As our reconstruction scheme is incremental, we determine the consequence of the introduction of a new segment $S_t$ in the 3D map of the scene as soon as the structure of $S_t$ is known. Our approach can be decomposed into three steps. For each couple of segments $(S_{t'}, S_t)$, we propose hypotheses on the relation between these two segments. Then, we verify if these hypotheses match the observations. Finally, the system proposes a new model of the scene resulting from the integration of the new segment.

**Prediction** Dealing with two segments $S_{t'}$ and $S_t$, the possible actions are the followings: fuse the segments, create a junction, or add a link (a new segment) between $S_{t'}$ and $S_t$. Therefore the aim of the prediction step is to create some hypotheses leading to the realization of one (or more) of these actions. The hypotheses are directly linked to the actions:

- $H_1$: there is a junction between $S_{t'}$ and $S_t$ ;
- $H_2$: there are one or two segments between $S_{t'}$ and $S_t$.
- $H_3$: $S_{t'}$ and $S_t$ are identical ;
- $H_4$: there are no (or some other) relation between $S_{t'}$ and $S_t$.

We have a multi-step strategy. First, we compute the belief we have in simple topological relations (proximity $(p(N))$, coplanarity $(p(C))$, collinearity $(p(P))$) between $S_{t'}$ and $S_t$. Then, according to these beliefs, it is possible to classify the pair of segments into five classes (see the first raw of the Table in the Fig. 4). Classes are $C_1$: $CNP$ (coplanar, neighbor and parallel) , $C_2$: $CN\neg P$, $C_3$: $C\neg NP$, $C_4$: $C\neg N\neg P$, and $C_5$: $\neg C\neg N\neg P$.

Using the belief we have in the belonging of the couple of segments to each class, the system can infer the belief in each possible hypothesis. We have defined decision strategies which are able to determine the best hypothesis according to the available knowledge. These strategies are coded in conditional probability tables $P(\mathcal{H}|C)$ where $\mathcal{H}$ is the hypothesis and $C$ the class (see Fig. 4). These tables are defined in an empirical way from a set of elementary considerations about topological relationship that we usually find in a group of segments. These considerations often reflect the truth, though they provide no guarantee. However, extreme precision is not required. Rather, they must reflect the knowledge we want to transmit to the system.

The prediction step reasoning can be encoded in a simple Bayes Net (see Fig. 3.a). It is composed of six nodes. Links between these nodes depict the causality relations between the different steps of reasoning and thus its progression. One node is associated to each topological relation, another to the class, and one node is associated to each set of hypotheses. Indeed, two sets of hypotheses are emitted. The first concerns the relation between the closest extremities

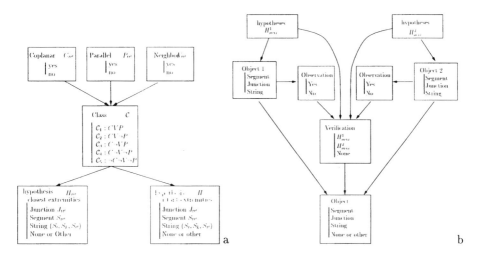

**Fig. 3.** (a) Prediction net and (b) Verification net

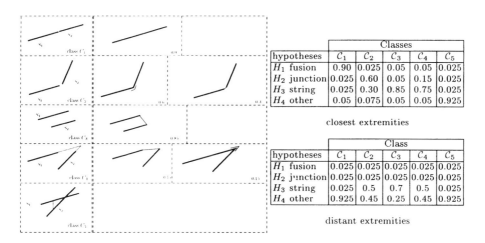

|  | Classes | | | | |
|---|---|---|---|---|---|
| hypotheses | $C_1$ | $C_2$ | $C_3$ | $C_4$ | $C_5$ |
| $H_1$ fusion | 0.90 | 0.025 | 0.05 | 0.05 | 0.025 |
| $H_2$ junction | 0.025 | 0.60 | 0.05 | 0.15 | 0.025 |
| $H_3$ string | 0.025 | 0.30 | 0.85 | 0.75 | 0.025 |
| $H_4$ other | 0.05 | 0.075 | 0.05 | 0.05 | 0.925 |

closest extremities

|  | Class | | | | |
|---|---|---|---|---|---|
| hypotheses | $C_1$ | $C_2$ | $C_3$ | $C_4$ | $C_5$ |
| $H_1$ fusion | 0.025 | 0.025 | 0.025 | 0.025 | 0.025 |
| $H_2$ junction | 0.025 | 0.025 | 0.025 | 0.025 | 0.025 |
| $H_3$ string | 0.025 | 0.5 | 0.7 | 0.5 | 0.025 |
| $H_4$ other | 0.925 | 0.45 | 0.25 | 0.45 | 0.925 |

distant extremities

**Fig. 4.** Elementary classes and associated hypothesis (closest extremities) and Conditional probabilities table $P$(hypotheses | classes) for the closest and distant extremities

of the segments (see Fig. 4) and the second concerns the relation between their distant extremities. In both cases, the same hypotheses can be emitted, though the associated conditional probabilities can be very different. As already stated, the hypothesis with the higher belief is not always the correct one, and this is the reason why we will always consider for each case (closest and distant extremities) the two hypotheses with the highest belief ($H^1_{max}$ and $H^2_{max}$). These two hypotheses are then verified or invalidated.

**Verification** In order to verify the two selected hypotheses, we use the reasoning encoded in the Bayes Net depicted in Fig. 3.b. We use two similar nets, each associated with one of the two sets of hypotheses (*i.e* close and distant extremities). Considering the two hypotheses, we first define the nature (segment, junction, string) and the position of the created object associated with each hypothesis. Then, we compute the belief in the existence of this object using the observation node. Finally, knowing the belief in each hypothesis and the belief in the related observation, it is possible to determine the most probable hypothesis (or to reject both).

The most important node in the verification net is the observation node. Sometimes, the hypotheses can be verified (or invalidated) using direct observation in the images previously acquired. In such cases, the validation is performed using the 3D information associated with the hypotheses and the 2D observation. We perform a back-projection of the 3D objects in each image previously acquired by the camera and we try to associate this projection to the observed data in more than one image (to avoid false matching). For each possible matching, we compute the belief granted to this matching. The case of a single segment or of a junction is simple. If this junction exists, it has already been observed (because the presence of the two segments has been already verified). Thus, the verification is performed as described above. In the case of a string, with three segments, the presence of two of them is certain (they have been used to predict the presence of the third). However, the last one has not been yet reconstructed (most of the time), and its presence is not validated. When no matching is found in the images previously acquired, it is necessary to know why. The first possibility is that the segment under consideration does not exist, the second is that it is occluded by another object. In the latter case, it necessary to move the camera to a new viewpoint from which the segment can be observed. Rather than computing explicitly a viewpoint (*e.g.* [9,21]) and researching *off-line* the considered segment, we prefer to turn the camera around a segment which belongs either to the occluding polygon or to a plane to which the considered segment belongs. During this motion, automatically generated by visual servoing [11], an image processing is performed *on-line* to detect the appearance of the researched segment.

**Modeling.** At this step of the reconstruction process, we have a model of the scene composed of 3D segments, 3D junctions, or even a coplanar string of segments. It is finally quite easy to use this information in order to get 3D polygons. To this end, we use the junction information and the coplanarity information already used in the hypotheses generation (see [14] for further details).

This three-step approach allows us to get a high level and more complete representation of the scene. Section 5 will present experimental results which illustrate the different key points of this algorithm.

## 4  Global exploration - complete scene reconstruction

Since it is not possible to ensure that the model of the scene issued from the local exploration process is complete, we present now the last perception-action cycle which includes the two previous ones and ensures a reconstruction as complete as possible. We have to determine viewpoints able to bring more information about the scene. By *information*, we mean either a new object, either the certainty that a given area is object-free.

Knowing the set of viewpoints since the beginning of the reconstruction process, it is possible to maintain a map of the observed and unexplored areas using a ray tracing scheme. The knowledge is composed by: the objects already reconstructed $\mathcal{O}$, the known free space $\mathcal{V}$, and the unknown area $\mathcal{U}$. Using this knowledge, we have defined a gaze planning strategy which proposes a solution to the next best view problem that mainly uses a representation of known and unknown areas as a basis for selecting viewpoints. We have chosen to handle the "where to look next" problem as a function minimization problem. Such a function $\mathcal{F}(\phi)$ has to integrate the constraints imposed by the system and to evaluate the quality of a viewpoint in order to select the next camera viewpoint $\phi_{t+1}$ which corresponds to its minimal value. The cost function is minimized using a fast deterministic relaxation scheme corresponding to a modified version of the ICM algorithm. The camera viewpoints are constrained inside an hemisphere located around the scene, but only in the region already observed and object-free (in order to avoid collision). At the beginning of the exploration process, as the observed area is null, the camera motion is limited to the surface of the sphere.

The function $\mathcal{F}$ is taken as a weighted sum of the following measures:

*Quality of a new position.* The quality of a new position $\phi_{t+1}$ is function of the volume of the unknown area which appears in the camera field of view. The new observed area $\mathcal{G}(\phi_{t+1})$ is given by:

$$\mathcal{G}(\phi_{t+1}) = \mathcal{V}(\phi_{t+1}) - \mathcal{V}(\phi_{t+1}) \cap \mathcal{V} \qquad (2)$$

where $\mathcal{V}(\phi_{t+1})$ defines the part of the scene observed from the position $\phi_{t+1}$ and $\mathcal{V}(\phi_{t+1}) \cap \mathcal{V}$ defines the sub-part of $\mathcal{V}(\phi_{t+1})$ which has been already observed (see Fig. 5).

*Displacement Cost.* A term reflecting the cost of the camera displacement between two viewpoints $\phi_t$ and $\phi_{t+1}$ is introduced in the cost function $\mathcal{F}$, in order to reduce the total camera displacement (see [16]).

*Reachability Constraints.* To avoid unreachable viewpoints, we use a binary test which returns an infinite value when the position is unreachable. A position is unreachable if it is not in the operational space of the manipulator, or if this position is located in an unknown area (leading to a collision risk).

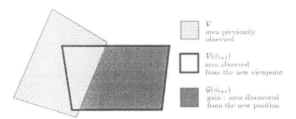

**Fig. 5.** Quality of a new position (2D projection).

If a new object is observed from the selected viewpoint, its reconstruction is performed. On the other hand, if the observed space is free of new objects, a next viewpoint has to be computed. This process is iterated until the end of the exploration. In theory, it must end when all the space has been observed, *i.e.*, if $\mathcal{U} = \emptyset$. However, this condition is usually unreachable. Ensuring the completeness of the reconstruction is not always possible. Some areas may remain observed only from a set of viewpoints unreachable by the camera. Furthermore, due to the objects topology, some areas may be unobserved whatever the camera position. Thus the exploration process is said to be as complete as possible if, for all reachable viewpoints, the camera looks at a known part of the scene. We thus can be sure that, at the end of the exploration process, all the areas of the scene are either free-space, either an object which has been reconstructed, either an unobservable area.

## 5    Experimental results

The whole application presented in this paper has been implemented on an experimental testbed composed of a CCD calibrated camera mounted on the end effector of a six degrees of freedom cartesian robot. Describing the complete implementation of our system is not the goal of this paper ; however we want to underline the fact that, if it is important to bridge the gap between continuous/local and discrete/global in the vision/control part of an active vision system, it is also important to consider this gap from a software engineering point of view in order to obtain a safety implementation of such system. As classical asynchronous languages are not really adapted to specify and program either the continuous and the discrete part of our algorithm, we have implemented the control/estimation algorithm and the task controller using SIGNAL [17]. SIGNAL is a real-time synchronous data-flow language adapted to implementation of vision-based tasks such as visual servoing and estimation. Dealing with the high level PAC, we have used SIGNAL *GTi*, an extension that introduces intervals of time. SIGNAL *GTi* provides constructs for the specification of hierarchical preemptive tasks executed on these intervals. It allows to consider in an unified framework the various aspects of the perception action cycle: from data-flow task (estimation, visual servoing) to multi-tasking and hierarchical task preemption (perception strategies).

**Structure from controlled motion.** As already stated, we are interested in the reconstruction of cylinders and segments. We here presents the results obtained for the structure estimation of a cylinder. Fig. 6.a represents the initial image acquired by the camera and the selected cylinder. Fig. 6.b contains the image acquired by the camera after the convergence of the visual gazing task.

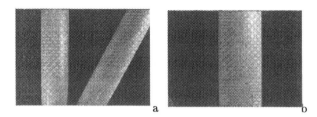

**Fig. 6.** Position of the cylinder in the image before (a) and after (b) the focusing task

Fig. 7 describes the evolution of the estimation of the parameters of the cylinder displayed in Fig. 6. Fig. 7.a shows its radius $r$ and the coordinates $x_0, y_0, z_0$ of a point of its axis. Let us note that the cylinder radius is determined with an accuracy less than 0.5 mm whereas the camera is one meter away from the cylinder (and even less than 0.1 mm with good lighting conditions). Fig. 7.b reports the error between the estimated value of the radius and its true value (*i.e.*, $r_i - r*$) using the two limbs-based estimation. As far as depth is concerned, its standard deviation is less than 2.5 mm (that is 0.25%).

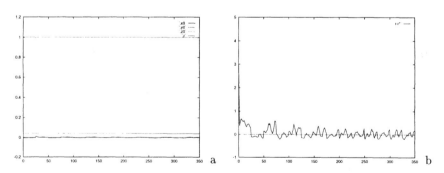

**Fig. 7.** Estimation of the parameters of a cylinder in the camera frame (a) estimated position of a point on the axis ($x_0$, $y_0$, $z_0$) and radius ($r$) (in $mm$) (b) error between the real and estimated radius of the cylinder (in $mm$)

The experiment of the cylinder structure estimation has been carried out fifty times from different initial camera locations. For each one of the 50 experiments, we have computed the estimated radius $\hat{r}$, and the estimated depth $\hat{z}_0$. Each time, the measured error $\hat{r} - r^*$ is less than 0.5 mm and the standard deviation of all the estimations (*i.e.*, $\sigma_{\hat{r}}$) is around 0.02 mm (resp. $\sigma_{\hat{z}_0} = 0.23$ mm). These results underline the fact that our estimation algorithm is particulary robust, stable and accurate.

**From a local to a global description of the scene.** We present in this section the reconstruction results obtained for a polyhedral object (see Fig. 8.a). This scene allows us to illustrate the interests of the proposed method. The result of the scene reconstruction using the simple incremental reconstruction process is depicted on Fig. 8.b. As already stated, as they are too small, some of the vertices of the polyhedron have not been reconstructed. Furthermore, due to the local approach used in that process, others remain occluded and thus non reconstructed. We now focus on two aspects of the Bayes Nets prediction verification scheme.

**Fig. 8.** Polyhedral scene: (a) view of the scene (b) model of the "polyhedron" scene acquired using the incremental algorithm (c) model of the same scene acquired using the prediction/verification scheme and numbering of the reconstructed segments in the order of their introduction in the map of the scene

**Fig. 9.** Polyhedral scene : arrows point at the next primitive to be estimated

Consider that segments $S_0$ and $S_1$ have been already estimated and that $S_2$ has just been reconstructed (see Fig. 9.abc), the system considers the relation between $S_2$ and $S_0$ and between $S_2$ and $S_1$. Dealing with $S_2$ and $S_0$, the system concludes easily to the presence of a junction between them. Dealing with the couple $(S_1,S_2)$, there is around 1cm between their closest extremities. The belief for $S_2$ and $S_1$ to be neighbor is 61% and to be coplanar is 99% ; thus they are likely to belong to the class $\mathcal{C}_2$. According to the strategies encoded in the Hypotheses Bayes Net, it is likely that there exists a junction with a 46% belief and a a segment between them with a 41% belief. The remaining 13% are shared between the two other hypotheses. After the verification process, and according to the observations, the former hypothesis (junction) is verified with a 60% belief. This high value (even if this hypothesis is false, see Fig. 8.a) results from the fact that these two segments are very close in the different images (around 5 pixels). Thus the observations reinforce this hypothesis. However, the latter hypothesis is verified with a 95% belief. Indeed, a 2D segment is observed at the predicted position in many images. Finally, according to the belief in each hypothesis, to the belief in the observations, a new segment $S_3$ is added to the model of the scene (with a confidence of 53%, while the confidence in a junction creation is only 37%). This underlines the interest to consider a multi-hypotheses approach. A classical approach might have chosen the first (and wrong) hypothesis.

Let us now consider a second interesting case. When segment $S_7$ is reconstructed, within relations with other segments, the system proposes the creation of a junction with $S_4$ and the creation of a segment between their two distant extremities. Such a segment has never been observed (and could not have been observed according to the current knowledge on the scene and on the camera trajectory). Therefore, as described in the previous section, the camera focuses on $S_7$, and turns around it (see Fig. 10). During this motion, automatically generated by visual servoing, observers are looking for a moving segment located at its expected position in the images. The discovered segment is then reconstructed and introduced in the scene model (see Fig. 10.c).

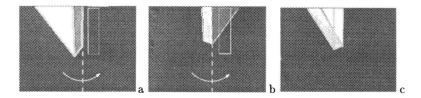

**Fig. 10.** Verification of a hypothesis: (a) rotation around $S_7$ (b) $S_{10}$ is discovered and (c) reconstructed

**Scene exploration - computing viewpoints.** The example reported here (see Fig. 11.a) deals with a scene composed of a cylinder and four polygons which lie in different planes. In Fig. 11.b is displayed the initial image acquired by the camera. Only the cylinder and a polygon are reconstructed during the first local incremental reconstruction process (see Fig. 11.c). Fig. 12 presents the different steps of the global exploration of the scene. Each figure shows the obtained 3D scene, the camera trajectory and the projection on a virtual plane of the unknown areas. Fig. 12.a corresponds to the camera position $\phi_6$ obtained just after the local exploration process. The first camera displacements allows to reduce significantly the unknown areas (see Fig. 12.b). A new object is then detected. A local exploration process is performed. It ends at position $\phi_{24}$ (Fig. 12.c). At this step, the two polygons on the "top" of the scene have been reconstructed. A new global exploration is then performed. After a last exploration process, the last polygon is observed and reconstructed (Fig. 12.d). At this step, 99% of the space has been observed, which ensures that the reconstruction of the scene is complete. Fig. 11 shows the final 3D model of the scene (to be compared to Fig. 11) and the camera trajectory.

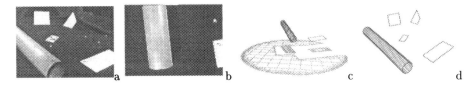

**Fig. 11.** (a) External view and (b) first view of the scene and results of the first local exploration/incremental reconstruction process : (c) reconstructed scene and projection on a virtual plane of the unknown area (d) 3D model of the reconstructed scene.

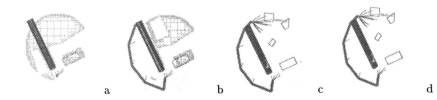

**Fig. 12.** Different steps of the global exploration process (camera trajectory, 3D model of the reconstructed scene and projection on an virtual plane of the unknown area).

# 6 Conclusion

We have proposed an active vision approach to the 3D reconstruction of static scenes composed of cylinders and polyedral objects. The perception-action cycles are handled at various levels: from the definition of perception strategies for scene exploration downto the automatic generation of camera motions using visual servoing. As the structure from controlled motion approach used to perform primitives estimation is based on particular camera motions, perceptual strategies able to appropriately perform a succession of such individual primitive reconstructions have been proposed in order to recover the complete spatial structure of complex scenes. An important feature of our approach is its ability to easily determine the next primitive to be estimated without any knowledge or assumption on the number, the localization and the spatial relation between objects. To this purpose, an algorithm has been proposed to ensure the incremental reconstruction and exploration of the scene. It is based on a computing viewpoints algorithm and the use of a prediction/verification scheme managed using decision theory and Bayes Nets. Finally, experiments have proved the validity of our approach (accurate, stable and robust results with efficient exploration algorithms).

# References

1. G. Adiv. Inherent ambiguities in recovering 3D motion and structure from a noisy flow field. *IEEE Trans. on PAMI*, 11(5):477–489, May 1989.
2. Y. Aloimonos. Purposive and qualitative active vision. In *ICPR'90*, pp. 346–360, Atlantic City, June 1990.
3. Y. Aloimonos, I. Weiss, A. Bandopadhay. Active vision. *IJCV*, 1(4):333–356, January 1987.
4. R. Bajcsy. Active perception. *Proc. of the IEEE*, 76(8):996–1005, August 1988.
5. D.H. Ballard. Animate vision. *Artificial Intelligence*, 48(1):57–86, February 1991.
6. H. Buxton, S. Gong. Visual surveillance in a dynamic and uncertain world. *Artificial Intelligence*, 78(1-2):431–459, October 1995.
7. F. Chaumette, S. Boukir, P. Bouthemy, D. Juvin. Structure from controlled motion. *IEEE Trans. on PAMI*, 18(5):492–504, May 1996.
8. C. Connoly. The determination of next best views. In *IEEE Int. Conf. on Robotics and Automation*, pp. 432–435, St Louis, March 1985.
9. C.K. Cowan, P.D. Kovesi. Automatic sensor placement from vision task requirements. *IEEE Trans. on PAMI*, 10(3):407–416, May 1988.
10. D. Djian, P. Probert, and P. Rives. Active sensing using bayes nets. In *Proc. of Int. Conf. on Advanced Robotics, ICAR'95*, pp. 895–902, Sant Feliu de Guixols, Spain, September 1995.
11. B. Espiau, F. Chaumette, P. Rives. A new approach to visual servoing in robotics. *IEEE Trans. on Robotics and Automation*, 8(3):313–326, June 1992.
12. B. Espiau, P. Rives. Closed-loop recursive estimation of 3D features for a mobile vision system. In *IEEE Int. Conf. on Robotics and Automation*, pp. 1436–1443, Raleigh, April 1987.
13. S. Hutchinson, G. Hager, P Corke. A tutorial on visual servo control. *IEEE Trans. on Robotics and Automation* 12(5):651–670, October 1996.
14. E. Marchand. *Stratégies de perception par vision active pour la reconstruction et l'exploration de scènes statiques*. PhD thesis, Université de Rennes 1, June 1996.
15. E. Marchand, F. Chaumette. Controlled camera motions for scene reconstruction and exploration. In *CVPR'96*, pp. 169–176, San Francisco, June 1996.
16. E. Marchand, F. Chaumette, A. Rizzo. Using the task function approach to avoid robot joint limits and kinematic singularities in visual servoing. In *IROS'96*, pp. 1083–1090, Osaka, Japan, November 1996.
17. E. Marchand, E. Rutten, F. Chaumette. From data-flow task to multi-tasking : Applying the synchronous approach to active vision in robotics. *IEEE Trans. on Control Systems Technology*, 5(2):200–216, Mars 1997.
18. J. Pearl. *Probabilistic reasoning in intelligent systems : Networks of plausible inference*. Morgan Kaufmann Publisher Inc., San Mateo, California, 1988.
19. R.D. Rimey, C. Brown. Control of selective perception using bayes nets and decision theory. *IJCV*, 12(2/3):173–207, April 1994.
20. M.J. Swain, M.A. Stricker. Promising direction in active vision. *International Journal of Computer Vision*, 11(2):109–127, October 1993.
21. K. Tarabanis, P.K. Allen, R. Tsai. A survey of sensor planning in computer vision. *IEEE Trans. on Robotics and Automation*, 11(1):86–104, February 1995.
22. B. Triggs, C. Laugier. Automatic camera placement for robot vision. In *IEEE Int. Conf. on Robotics and Automation*, pp. 1732–1738, Nagoya, Japon, May 1995.
23. L.E. Wixson. Viewpoint selection for visual search. In *CVPR'94*, pp. 800–805, Seattle, June 1994.

# Statistical Optimization
# and Geometric Visual Inference

Kenichi Kanatani

Department of Computer Science, Gunma University, Kiryu, Gunma 376 Japan
kanatani@cs.gunma-u.ac.jp

**Abstract.** This paper gives a mathematical formulation to the computer vision task of inferring 3-D structures of the scene based on image data and geometric constraints. Introducing a statistical model of image noise, we define a geometric model as a manifold determined by the constraints and view the problem as model fitting. We then present a general mathematical framework for proving optimality of estimation, deriving optimal schemes, and selecting appropriate models. Finally, we illustrate our theory by applying it to structure from motion.

## 1  Introduction

The goal of computer vision is to infer 3-D structures of the scene from image data. The key that makes this possible is our prior knowledge about the environment. This knowledge takes the form of *constraints*: if something is to be seen in the image, the image data should satisfy a certain relationship; if another thing is to be seen, another relationship should hold. Let us call a particular constraint a *(geometric) model*. The inference takes the following two stages:

**Model selection:** We decide which model is to be adopted from among all possibilities.

**Model fitting:** We obtain a detailed description of the selected model by optimally fitting it to the data.

In this paper, we give a mathematical formulation to these tasks in very general terms and apply it to structure from motion as a typical computer vision problem.

## 2  Model Fitting

Suppose we observe $m$-dimensional vectors $a_1$, ..., $a_N$ constrained to be in an $m'$-dimensional manifold $\mathcal{A} \in \mathcal{R}^m$, which we call the *data space*. We write

$$a_\alpha = \bar{a}_\alpha + \Delta a_\alpha, \tag{1}$$

where $\bar{a}_\alpha$ is the position supposedly observed in the absence of noise. We regard the noise term $\Delta a_\alpha$ as a Gaussian random variable of mean $\mathbf{0}$ and covariance matrix $V[\boldsymbol{a}_\alpha]$, independent for each $\alpha$. Since each $\boldsymbol{a}_\alpha$ is constrained to be in $\mathcal{A}$, its covariance matrix

$$V[\boldsymbol{a}_\alpha] = E[\Delta\boldsymbol{a}_\alpha \Delta\boldsymbol{a}_\alpha^\top] \tag{2}$$

is singular ($E[\cdot]$ denotes expectation, and $\top$ denotes transpose). We assume that it is a positive semi-definite symmetric matrix of rank $m'$ whose range coincides with the tangent space $T_{\bar{\boldsymbol{a}}_\alpha}(\mathcal{A})$ to the data space $\mathcal{A}$ at $\bar{\boldsymbol{a}}_\alpha$. We also assume that $V[\boldsymbol{a}_\alpha]$ is known *only up to scale*, i.e., we decompose it into the *noise level* $\epsilon$ and the *normalized covariance matrix* $V_0[\boldsymbol{a}_\alpha]$ in the form

$$V[\boldsymbol{a}_\alpha] = \epsilon^2 V_0[\boldsymbol{a}_\alpha], \tag{3}$$

and assume that $V_0[\boldsymbol{a}_\alpha]$ is known but $\epsilon$ is unknown.

Suppose the true values $\bar{\boldsymbol{a}}_\alpha$, $\alpha = 1, ..., N$, are known to satisfy a set of equations parameterized by an $n$-dimensional vector $\boldsymbol{u}$. We assume that the domain of the vector $\boldsymbol{u}$ is an $n'$-dimensional manifold $\mathcal{U} \subset \mathcal{R}^n$, which we call the *parameter space*. Let $F^{(k)}(\boldsymbol{a}, \boldsymbol{u}): \mathcal{R}^m \times \mathcal{R}^n \to \mathcal{R}$, $k = 1, ..., L$, be smooth functions of arguments $\boldsymbol{a} \in \mathcal{R}^m$ and $\boldsymbol{u} \in \mathcal{R}^n$, and consider the following problem:

**Problem 1** *Estimate* $\bar{\boldsymbol{a}}_\alpha \in \mathcal{A}$, $\alpha = 1, ..., N$, *and* $\boldsymbol{u} \in \mathcal{U}$ *that satisfy*

$$F^{(k)}(\bar{\boldsymbol{a}}_\alpha, \boldsymbol{u}) = 0, \quad k = 1, ..., L, \tag{4}$$

*from the noisy data* $\boldsymbol{a}_\alpha \in \mathcal{A}$, $\alpha = 1, ..., N$.

The $L$ equations $F^{(k)}(\boldsymbol{a}, \boldsymbol{u}) = 0$, $k = 1, ..., L$, need not be algebraically independent with respect to the argument $\boldsymbol{a}$; we call the number $r$ of independent equations the *rank* of the constraint. In order to avoid pathological cases, we assume that each of the $L$ equations defines a manifold of *codimension 1* in the data space $\mathcal{A}$ in such a way that the $L$ manifolds intersect each other *transversally* [6]. It follows that eq. (4) defines a manifold $\mathcal{S} \subset \mathcal{A}$ of *codimension* $r$ parameterized by $\boldsymbol{u} \in \mathcal{U}$. We call $\mathcal{S}$ the *geometric model* of eq. (4). Then, Problem 1 can be rephrased as the following *geometric model fitting*:

**Problem 2** *Estimate the model* $\mathcal{S} \subset \mathcal{A}$ *that passes through* $N$ *points* $\{\bar{\boldsymbol{a}}_\alpha\} \in \mathcal{A}$ *from the noisy data* $\{\boldsymbol{a}_\alpha\} \in \mathcal{A}$.

## 3   Theoretical Accuracy Bound

Let $\hat{\boldsymbol{u}}$ be an arbitrary *unbiased* estimator of $\hat{\boldsymbol{u}}$. The unbiasedness is usually defined by $E[\hat{\boldsymbol{u}}] = \boldsymbol{u}$. However, the parameter space $\mathcal{U}$ is generally 'curved'. Hence, although $\hat{\boldsymbol{u}} \in \mathcal{U}$, we have $E[\hat{\boldsymbol{u}}] \notin \mathcal{U}$ in general. Here, we define the unbiasedness by

$$E[\boldsymbol{P}_{\boldsymbol{u}}^{\mathcal{U}}(\hat{\boldsymbol{u}} - \boldsymbol{u})] = \boldsymbol{0}, \tag{5}$$

where $\boldsymbol{P}_{\boldsymbol{u}}^{\mathcal{U}}$ is the projection matrix onto the tangent space $T_{\boldsymbol{u}}(\mathcal{U})$ to the parameter space $\mathcal{U}$ at $\boldsymbol{u}$. Define the *a posteriori covariance matrix* of the estimator $\hat{\boldsymbol{u}}$ by

$$V[\hat{\boldsymbol{u}}] = \boldsymbol{P}_{\boldsymbol{u}}^{\mathcal{U}} E[(\hat{\boldsymbol{u}} - \boldsymbol{u})(\hat{\boldsymbol{u}} - \boldsymbol{u})^\top] \boldsymbol{P}_{\boldsymbol{u}}^{\mathcal{U}}, \tag{6}$$

which defines a positive semi-definite symmetric matrix whose range is restricted to be in $T_{\boldsymbol{u}}(\mathcal{U})$. The estimator $\hat{\boldsymbol{u}}$ is assumed to be such that $\boldsymbol{P}_{\boldsymbol{u}}^{\mathcal{U}}(\hat{\boldsymbol{u}}-\boldsymbol{u})$ distributes in all directions in $T_{\boldsymbol{u}}(\mathcal{U})$ so that the range of $V[\hat{\boldsymbol{u}}]$ coincides with $T_{\boldsymbol{u}}(\mathcal{U})$, having rank $n'$ ($=$ the dimension of $\mathcal{U}$).

Let $\nabla_{\boldsymbol{u}}(\cdot)$ denote the $n$-dimensional vector $(\partial(\cdot)/\partial u_1, ..., \partial(\cdot)/\partial u_n)^\top$, and $\nabla_{\boldsymbol{a}}(\cdot)$ the $m$-dimensional vector $(\partial(\cdot)/\partial a_1, ..., \partial(\cdot)/\partial a_m)^\top$. The following inequality can be proved [6]:

$$V[\hat{\boldsymbol{u}}] \succ \epsilon^2 \left( \sum_{\alpha=1}^{N} \sum_{k,l=1}^{L} \bar{W}_\alpha^{(kl)} (\boldsymbol{P}_{\boldsymbol{u}}^{\mathcal{U}} \nabla_{\boldsymbol{u}} \bar{F}_\alpha^{(k)})(\boldsymbol{P}_{\boldsymbol{u}}^{\mathcal{U}} \nabla_{\boldsymbol{u}} \bar{F}_\alpha^{(l)})^\top \right)^-, \tag{7}$$

$$\left( \bar{W}_\alpha^{(kl)} \right) = \left( (\nabla_{\boldsymbol{a}} \bar{F}_\alpha^{(k)}, V_0[\boldsymbol{a}_\alpha] \nabla_{\boldsymbol{a}} \bar{F}_\alpha^{(l)}) \right)^-. \tag{8}$$

Here, $(\cdot)^-$ denotes the (Moore-Penrose) generalized inverse; $\nabla_{\boldsymbol{u}} \bar{F}_\alpha^{(k)}$ and $\nabla_{\boldsymbol{a}} \bar{F}_\alpha^{(k)}$ are the abbreviations of $\nabla_{\boldsymbol{u}} F^{(k)}(\bar{\boldsymbol{a}}_\alpha, \boldsymbol{u})$ and $\nabla_{\boldsymbol{a}} F^{(k)}(\bar{\boldsymbol{a}}_\alpha, \boldsymbol{u})$, respectively. The relation $\boldsymbol{A} \succ \boldsymbol{B}$ for symmetric matrices $\boldsymbol{A}$ and $\boldsymbol{B}$ means that $\boldsymbol{A} - \boldsymbol{B}$ is a positive semi-definite matrix. Eq. (8) means that $\bar{W}_\alpha^{(kl)}$ is the $(kl)$ element of the generalized inverse of the matrix whose $(kl)$ element is defined by the expression inside $(\cdot)^-$. It can be proved that $(\bar{W}_\alpha^{(kl)})$ is a positive semi-definite symmetric matrix of rank $r$ [6].

The logic that derives eq. (7) is the same as that for what is known as the *Cramer-Rao lower bound* in statistics. So, we call eq. (7) the *Cramer-Rao inequality* and the right-hand side the *Cramer-Rao lower bound*. Eq. (7) is a special case of a more general form for a non-Gaussian noise distribution, for which the *Fisher information matrix* plays the role of the covariance matrix [6].

**Remark 1.** In statistics a (*statistical*) *model* is a mechanism that *predicts* future observation: it consists of a deterministic part which specifies the structure of the phenomenon and a random fluctuation part which accounts for all factors not modeled in the deterministic part. Mathematically, the problem is to estimate the parameters involved in a probability density by observing data sampled from it. Problem 1 does not fit this framework: we cannot predict where the 'next datum' $\boldsymbol{a}_{N+1}$ will appear; the structure is given as an implicit relationship in the form of eq. (4).

## 4 Computation Methods

The following three are the most widely used method for solving Problem 1:

**Minimal approach:** Eq. (4) gives $r$ constraints on $\boldsymbol{u}$, which has $n'$ degrees of freedom, so we observe $n' - r$ data, substitute them for their true values in eq. (4), and solve the resulting simultaneous equations:

$$F^{(k)}(\boldsymbol{a}_\alpha, \boldsymbol{u}) = 0, \quad k = 1, ..., L, \quad \alpha = 1, ..., n'. \tag{9}$$

**Least-squares approach:** We observe $N$ $(\geq n' - r)$ data and compute the value of $\boldsymbol{u}$ that minimizes

$$J_{LS} = \sum_{\alpha=1}^{N} \sum_{k=1}^{L} W_{\alpha}^{(k)} F^{(k)}(\boldsymbol{a}_{\alpha}, \boldsymbol{u})^2 \tag{10}$$

for appropriate weights $W_{\alpha}^{(k)}$ $(> 0)$.

**Maximum likelihood approach:** We compute the values $\{\bar{\boldsymbol{a}}_{\alpha}\}$ that minimize

$$J = \sum_{\alpha=1}^{N} (\boldsymbol{a}_{\alpha} - \bar{\boldsymbol{a}}_{\alpha}, V_0[\boldsymbol{a}_{\alpha}]^{-}(\boldsymbol{a}_{\alpha} - \bar{\boldsymbol{a}}_{\alpha})), \tag{11}$$

under the constraint that eq. (4) holds; the value $\boldsymbol{u}$ is chosen so as to minimize $J$ (we denote the inner product of vectors $\boldsymbol{a}$ and $\boldsymbol{b}$ by $(\boldsymbol{a}, \boldsymbol{b})$).

**Remark 2.** Minimization of eq. (11) is called *MLE* (*maximum likelihood estimation*); since the likelihood that the data $\{\boldsymbol{a}_{\alpha}\}$ are observed is constant$\times e^{-J/2\epsilon^2}$ under our assumption, minimizing $J$ is equivalent to maximizing the likelihood. The summand in eq. (11) is known as the squared *Mahalanobis distance* of $\boldsymbol{a}_{\alpha}$ from $\bar{\boldsymbol{a}}_{\alpha}$ with respect to $V_0[\boldsymbol{a}_{\alpha}]$.

**Remark 3.** Many variants exist for the first two methods. For example, we may regard $\boldsymbol{u}$ as an unconstrained variable. This usually simplifies the computation, but the resulting value of $\boldsymbol{u}$ does not necessarily satisfy the constraint $\boldsymbol{u} \in \mathcal{U}$ (the *decomposability condition*). So, we project it onto the manifold $\mathcal{U}$ by some criterion. This approach is called the *linearization technique*.

The MLE approach is evidently preferable to the other two. The minimum approach is sensitive to noise, linearized or not, because the errors in the data directly affect the solution. The least-squares approach may be robust in the sense that the solution is not so very sensitive to the errors in the individual data, but there is no guarantee that the resulting solution is close to the true value. In fact, the solution is usually statistically biased whatever weights $W_{\alpha}^{(kl)}$ are used [6]. In contrast, the MLE solution can be shown to be optimal in the sense we describe shortly.

## 5  Optimal Model Fitting

In order to minimize eq. (11), we first fix the value of $\boldsymbol{u}$ and compute the values $\{\bar{\boldsymbol{a}}_{\alpha}\}$ that minimize $J$. This is equivalent to 'projecting' each datum $\boldsymbol{a}_{\alpha}$ onto a point $\hat{\boldsymbol{a}}_{\alpha}$ in $\mathcal{S}$ optimally measured in the Mahalanobis distance (figure 1). The first order solution obtained by ignoring terms of $O(\epsilon^2)$ is given in the form

$$\hat{\boldsymbol{a}}_{\alpha} = \boldsymbol{a}_{\alpha} - V_0[\boldsymbol{a}_{\alpha}] \sum_{k,l=1}^{L} W_{\alpha}^{(kl)} F_{\alpha}^{(k)} \nabla_{\boldsymbol{a}} F_{\alpha}^{(l)}. \tag{12}$$

310

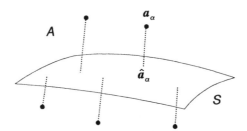

**Fig. 1.** Datum $a_\alpha$ is projected onto manifold $\mathcal{S}$.

Here, $F_\alpha^{(k)}$ and $\nabla_{\boldsymbol{a}} F_\alpha^{(k)}$ are the abbreviations of $F^{(k)}(\boldsymbol{a}_\alpha, \boldsymbol{u})$ and $\nabla_{\boldsymbol{a}} F^{(k)}(\boldsymbol{a}_\alpha, \boldsymbol{u})$, respectively; $W_\alpha^{(kl)}$ is defined by replacing in eq. (8) the true value $\bar{\boldsymbol{a}}_\alpha$ by the corresponding data value $\boldsymbol{a}_\alpha$ and the generalized inverse $(\cdots)^-$ by the rank-constrained generalized inverse $(\cdots)_r^-$, by which we mean the operation performed by replacing all the eigenvalues of $\cdots$ other than the $r$ largest ones by zero in the canonical form and computing the (Moore-Penrose) generalized inverse [6]. This operation is necessary for preventing numerical instability, because the operand approaches a singular matrix of rank $r$ in the limit $\boldsymbol{a}_\alpha \to \bar{\boldsymbol{a}}_\alpha$.

Substituting eq. (12) for $\bar{\boldsymbol{a}}_\alpha$ in eq. (11), we can express $J$ in terms of $\boldsymbol{u}$ alone in the form

$$J = \sum_{\alpha=1}^{N} \sum_{k,l=1}^{L} W_\alpha^{(kl)} F_\alpha^{(k)} F_\alpha^{(l)}. \tag{13}$$

Here, the 'weight' $W_\alpha^{(kl)}$ is not a constant but a function of $\boldsymbol{u}$. We call the value $\hat{\boldsymbol{u}}$ that minimizes eq. (13) the *MLE estimator*, and the minimum value of $J$ the *residual*, which we denote by $\hat{J}$. It can be proved that $\hat{J}/\epsilon^2$ is subject to a $\chi^2$ distribution with $rN - n'$ degrees of freedom in the first order [6]. Hence, an unbiased estimator of the squared noise level $\epsilon^2$ is obtained in the form

$$\hat{\epsilon}^2 = \frac{\hat{J}}{rN - n'}. \tag{14}$$

It can be shown that MLE in this form is optimal in the sense that the covariance matrix of the resulting solution attains the Cramer-Rao lower bound given by eq. (7) in the first order for small noise (i.e., if terms of $O(\epsilon^4)$ are ignored) [6]. This fact corresponds to the fact that MLE in traditional statistics is optimal in the sense that the resulting solution attains the (traditional) Cramer-Rao lower bound *asymptotically* for repeated observations (i.e., if terms of $O(1/n^2)$ are ignored for $n$ observations).

**Remark 4.** We should distinguish the 'number of observations' from the 'number of data'. Suppose we do $n$ observations and obtain $N$ data for each observation; each $N$ data are assumed to be a sample from a probability density of $N$ variables that involves unknown parameters which we want to estimate. In traditional statistics, MLE is optimal in the limit $n \to \infty$.

**Remark 5.** In Problem 1, the unknown parameters that characterize the probability density of the data are the true values $\{\bar{a}_\alpha\}$ of the data $\{a_\alpha\}$ and the value $\boldsymbol{u}$ that parameterizes the constraint. Hence, as the number of data increases, the number of unknowns increases at the same rate. This means that if we observe an increased number of data, they are still 'one' sample from a new probability density that involves an increased number of unknowns. In other words, the number of observation is always $n = 1$ however large the number of the data $N$ is. Thus, asymptotic analysis in the limit $N \to \infty$ does not make sense for Problem 1.

**Remark 6.** Suppose we hypothetically repeat observations of the *same* values $\{\bar{a}_\alpha\}$ $n$ times and obtain $n$ sets of data $\{a_\alpha^{(i)}\}$, $i = 1, ..., n$, $\alpha = 1, ..., N$. If we take the average $a_\alpha = \sum_{i=1}^{n} a_\alpha^{(i)}/n$, the resulting values $\{a_\alpha\}$ have errors of $O(1/\sqrt{n})$ times the original errors. In other words, increasing the number of (hypothetical) observations $n$ means *effectively reducing the noise level* $\epsilon$. Thus, it is natural that the optimality of MLE, which holds asymptotically (i.e., $n \to \infty$) in traditional statistics, should hold for small noise (i.e., $\epsilon \to 0$) for Problem 1.

# 6  Model Selection

If there are multiple candidates of geometric models $S_1$, $S_2$, ..., we should choose the 'best' one for observed data $\{a_\alpha\}$. But how can we evaluate the 'goodness' of a model $S$? First of all, a good model should explain the data $\{a_\alpha\}$ well, which implies that the residual $\hat{J}$ should be small. Since $\hat{J}/\epsilon^2$ is subject to a $\chi^2$ distribution with $rN - n'$ degrees of freedom, the residual becomes smaller as $n'$ becomes larger. In particular, the residual is zero in general if the model $S$ has more than $rN$ free parameters, meaning that we can make $S$ pass through all the data points by adjusting the parameters. Such an artificial model cannot be regarded as a good model, because it only explains the *current data* $\{a_\alpha\}$ which happen to be observed; there is no guarantee that it could explain the data if the noise occurred differently.

This observation suggests that the 'goodness' of a model can be measured by its 'predicting capability' [1]. Let $\{a_\alpha^*\}$ be the *future data* that have the same probability distribution as the current data $\{a_\alpha\}$ but are independent of $\{a_\alpha\}$. Consider the residual $J^*$ for the maximum likelihood estimators $\{\hat{a}_\alpha\} \in \hat{S}$, which are computed from the current data $\{a_\alpha\}$, *with respect to the future data* $\{a_\alpha^*\}$:

$$J^* = \sum_{\alpha=1}^{N} (a_\alpha^* - \hat{a}_\alpha, V_0[a_\alpha]^-(a_\alpha^* - \hat{a}_\alpha)). \tag{15}$$

A model $S$ is expected to have high prediction capability if $J^*$ is small. Put differently, we are 'validating' the optimal estimate $\hat{S}$ of $S$ by measuring its discrepancy from data *yet to be observed*. It can be shown that $\hat{J}$ *is smaller than*

$J^*$ by $2(pN + n')\epsilon^2$ in expectation [6]. So, we define the *geometric information criterion*, or the *geometric AIC* for short, by

$$AIC(\mathcal{S}) = \hat{J} + 2(pN + n')\epsilon^2, \tag{16}$$

and use it as a measure of the goodness of the model.

**Remark 7.** The most distinctive characteristic of the geometric AIC is the fact that the number $N$ of data, which does not appear in the usual AIC [1], explicitly appears in the expression; the traditional AIC contains, other than the residual, only the number of the model parameters. This is because, as we pointed out earlier, the number of data in traditional statistics means the number of 'observations' while the number $N$ of the data $\{a_\alpha\}$ means the number of the parameters $\{\bar{a}_\alpha\}$ (often called *nuisance parameters*) of the model; *the number of observation is always one*. This fact results in the following features of the geometric AIC:

- The degree of freedom $n'$ of the model has no significant effect for the geometric AIC if the number $N$ of data is large, whereas it plays a dominant role in the usual AIC [1].
- The dimension $p$ of the model manifold plays a dominant role in the geometric AIC, while no such geometric concepts are involved in the usual AIC.

## 7 Noise Estimation

In order to compute the geometric AIC, we need to estimate the noise level $\epsilon$ appropriately. This is obvious; distinguishing one model from another is meaningless if the noise level is high, while a small difference between the residuals gives a strong clue if the noise level is low. However, estimating the noise level a priori is in general very difficult. Here, we resolve this difficulty by focusing on the *inclusion relationship* of the models.

Let $\mathcal{S}_1$ be a model of dimension $p_1$ and codimension $r_1$ with $n_1'$ degrees of freedom, and $\mathcal{S}_2$ a model of dimension $p_2$ and codimension $r_2$ with $n_2'$ degrees of freedom. Let $\hat{J}_1$ and $\hat{J}_2$ be their respective residuals. Suppose model $\mathcal{S}_2$ is obtained by adding an additional constraint to model $\mathcal{S}_1$. We say that model $\mathcal{S}_2$ is *stronger* than model $\mathcal{S}_1$, or model $\mathcal{S}_1$ is *weaker* than model $\mathcal{S}_2$, and write

$$\mathcal{S}_2 \succ \mathcal{S}_1. \tag{17}$$

If model $\mathcal{S}_1$ is correct, the squared noise level $\epsilon^2$ is estimated by eq. (14). Substituting it into the expression for the geometric AIC, we obtain

$$AIC(\mathcal{S}_1) = \hat{J}_1 + \frac{2(p_1 N + n_1')}{r_1 N - n_1'}\hat{J}_1, \quad AIC(\mathcal{S}_2) = \hat{J}_2 + \frac{2(p_2 N + n_2')}{r_1 N - n_1'}\hat{J}_1. \tag{18}$$

Since the geometric AIC estimates the expected sum of squared Mahalanobis distances, the ratio of the deviations from the two models can be evaluated by

$$K = \sqrt{\frac{AIC(\mathcal{S}_2)}{AIC(\mathcal{S}_1)}} = \sqrt{\frac{r_1 N - n_1'}{(2p_1 + r_1)N + n_1'} \left( \frac{\hat{J}_2}{\hat{J}_1} + \frac{2(p_2 N + n_2')}{r_1 N - n_1'} \right)}. \qquad (19)$$

This quantity measures *how good* model $\mathcal{S}_2$ is *compared with model* $\mathcal{S}_1$: if $K < 1$, model $\mathcal{S}_2$ is expected to have more predicting capability.

**Remark 8.** Our approach for noise estimation is very different from that in statistics, where the noise level is estimated *model by model* in such a way that the AIC is minimized, which is equivalent to MLE. If the noise level is so estimated and substituted back into the AIC expression, the residual term is effectively replaced by its logarithm. We cannot do that in our problem for two reasons:

- In statistics, noise is a characteristic of the model, which specifies how deterministic causes are separated from random effects. Since noise characteristics are model-dependent, it makes sense to estimate them model by model. In our problem, noise is a characteristic of the devices and data processing operations involved and is *independent of the models we are comparing*. Hence, it should be estimated once for all the models.
- MLE of the noise level produces a very poor estimator for our problem. In fact, if we estimate $\epsilon^2$ by MLE, its expectation is approximately $r\epsilon^2/m'$, e.g., it is $\epsilon^2/2$ for curve fitting in two dimensions and $\epsilon^2/3$ for surface fitting in three dimensions. This does not occur in traditional statistics, because MLE estimators rapidly approach their true values as the number of observations increases, i.e., as the number of data increases. In our problem, the number of observation is always one; increasing the number of data has no effect in improving the accuracy of estimating $\epsilon^2$. In fact, the optimality of MLE can be established only in the limit $\epsilon^2 \to 0$.

## 8    3-D Motion Analysis

Define an $XYZ$ camera coordinate system in such a way that the origin $O$ is at the center of the lens and the $Z$-axis is in the direction of the optical axis. With an appropriate scaling, the image plane can be identified with the plane $Z = 1$, on which an $xy$ image coordinate system is defined in such a way that the origin $o$ is on the $Z$-axis and the $x$- and $y$-axes are parallel to the $X$- and $Y$-axes, respectively. Suppose the camera moves to a position defined by translating the first camera by vector $h$ and rotating it around the center of the lens by matrix $R$; we call $\{h, R\}$ the *motion parameters* (figure 2).

Let $(x_\alpha, y_\alpha)$, $\alpha = 1, ..., N$, be the image coordinates of $N$ feature point before the motion, and $(x_\alpha', y_\alpha')$, $\alpha = 1, ..., N$, those after the motion. We use the following three-dimensional vectors to represent them:

$$x_\alpha = (x_\alpha, y_\alpha, 1)^\top, \qquad x_\alpha' = (x_\alpha', y_\alpha', 1)^\top. \qquad (20)$$

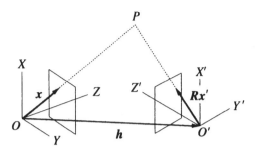

**Fig. 2.** Geometry of camera motion

We regard these as Gaussian random variables and denote their their covariance matrices by $V[\boldsymbol{x}_\alpha]$ and $V[\boldsymbol{x}'_\alpha]$. Since the third components of $\Delta\boldsymbol{x}_\alpha$ and $\Delta\boldsymbol{x}'_\alpha$ are identically 0, $V[\boldsymbol{x}_\alpha]$ and $V[\boldsymbol{x}'_\alpha]$ are singular matrices of rank 2. Assuming that the covariance matrices are known only up to scale, we decompose them into the noise level $\epsilon$ and the normalized covariance matrices $V_0[\boldsymbol{x}_\alpha]$ and $V_0[\boldsymbol{x}'_\alpha]$ in the form

$$V[\boldsymbol{x}_\alpha] = \epsilon^2 V_0[\boldsymbol{x}_\alpha], \qquad V[\boldsymbol{x}'_\alpha] = \epsilon^2 V_0[\boldsymbol{x}'_\alpha]. \tag{21}$$

### 8.1 General model

Since vectors $\{\boldsymbol{x}, \boldsymbol{x}'\}$ represent the same point in the scene if and only three vectors $\boldsymbol{x}$, $\boldsymbol{R}\boldsymbol{x}'$, and $\boldsymbol{h}$ are coplanar, we obtain the *epipolar equation* [2, 3, 18]

$$|\boldsymbol{x}, \boldsymbol{h}, \boldsymbol{R}\boldsymbol{x}'| = 0, \tag{22}$$

where $|\boldsymbol{a}, \boldsymbol{b}, \boldsymbol{c}|$ denotes the scalar triple product of vectors $\boldsymbol{a}$, $\boldsymbol{b}$, and $\boldsymbol{c}$. This equation defines a model $S$ of codimension 1 in the four-dimensional data space $\mathcal{X} = \{(x, y, z, x', y', z')|z = 1, z' = 1\} \in \mathcal{R}^6$. The scale of the translation $\boldsymbol{h}$ is indeterminate, so we normalize it into $\|\boldsymbol{h}\| = 1$. As a result, the parameter space $\mathcal{U}$ is a five-dimensional manifold in $\mathcal{R}^{12}$; it is the collection of points $(h_1, h_2, h_3, R_{11}, R_{12}, ..., R_{33})$ such that $(h_i)$ is a unit vector and $(R_{ij})$ is a rotation matrix.

**Errors in translation:** Since translations constitute an additive group, we measure the deviation of the computed value $\hat{\boldsymbol{h}}$ from its true value $\boldsymbol{h}$ by the 'difference' $\hat{\boldsymbol{h}} - \boldsymbol{h}$. Both $\hat{\boldsymbol{h}}$ and $\boldsymbol{h}$ are normalized to a unit vector, so they are both on a unit sphere. Assuming that error is small, we identify the domain of the error $\Delta\boldsymbol{h}$ with the tangent plane to the sphere at $\boldsymbol{h}$ and define the error of translation by

$$\Delta\boldsymbol{h} = \boldsymbol{P}_{\boldsymbol{h}}(\hat{\boldsymbol{h}} - \boldsymbol{h}), \tag{23}$$

where $\boldsymbol{P}_{\boldsymbol{h}} (= \boldsymbol{I} - \boldsymbol{h}\boldsymbol{h}^\top)$ is the projection matrix onto a subspace perpendicular to $\boldsymbol{h}$ ($\boldsymbol{I}$ denotes the unit matrix).

**Errors in rotation:** Since rotations constitute a multiplicative group, we measure the deviation of the computed rotation $\hat{R}$ from the true rotation $R$ by the 'quotient' $\hat{R}R^{-1}$, which is a small rotation. Let $l$ and $\Delta\Omega$ be its axis (unit vector) and angle of rotation, respectively. We define the error of rotation by

$$\Delta\boldsymbol{\Omega} = \Delta\Omega l. \tag{24}$$

From these observations, we define the covariance matrices of the motion parameters $\{\hat{h}, \hat{R}\}$ as follows [3]:

$$V[\hat{h}] = E[\Delta h \Delta h^\top], \quad V[\hat{h}, \hat{R}] = E[\Delta h \Delta\Omega^\top],$$

$$V[\hat{R}, \hat{h}] = E[\Delta\Omega \Delta h^\top], \quad V[\hat{R}] = E[\Delta\Omega \Delta\Omega^\top]. \tag{25}$$

For this representation, the Cramer-Rao inequality (7) has the following form:

$$\begin{pmatrix} V[\hat{h}] & V[\hat{h}, \hat{R}] \\ V[\hat{R}, \hat{h}] & V[\hat{R}] \end{pmatrix} \succ \epsilon^2 \begin{pmatrix} \sum_{\alpha=1}^{N} \bar{W}_\alpha \bar{p}_\alpha \bar{p}_\alpha^\top & \sum_{\alpha=1}^{N} \bar{W}_\alpha \bar{p}_\alpha \bar{q}_\alpha^\top \\ \sum_{\alpha=1}^{N} \bar{W}_\alpha \bar{q}_\alpha \bar{p}_\alpha^\top & \sum_{\alpha=1}^{N} \bar{W}_\alpha \bar{q}_\alpha \bar{q}_\alpha^\top \end{pmatrix}^{-}, \tag{26}$$

$$\bar{p}_\alpha = \bar{x}_\alpha \times R\bar{x}'_\alpha, \quad \bar{q}_\alpha = (\bar{x}_\alpha, R\bar{x}_{\alpha'})h - (h, R\bar{x}'_\alpha)\bar{x}_\alpha, \tag{27}$$

$$\bar{W}_\alpha = \frac{1}{(h \times R\bar{x}'_\alpha, V_0[x_\alpha](h \times R\bar{x}'_\alpha)) + (h \times \bar{x}_\alpha, RV_0[x'_\alpha]R^\top(h \times \bar{x}_\alpha))} \tag{28}$$

This bound is attained in the first order by MLE; we minimize the following function [5, 6]:

$$J = \sum_{\alpha=1}^{N} \frac{|x_\alpha, h, Rx'_\alpha|^2}{(h \times R\bar{x}'_\alpha, V_0[x_\alpha](h \times R\bar{x}'_\alpha)) + (h \times \bar{x}_\alpha, RV_0[x'_\alpha]R^\top(h \times \bar{x}_\alpha))}. \tag{29}$$

Direct minimization of this function requires numerical search [4, 19], but the computation can simplified by combining the *linearization* technique with renormalization and an optimal correction scheme [5].

## 8.2 Planar surface model

Consider a planar surface in the scene. Let $n$ be its unit surface normal, and $d$ its distance from the origin $O$ (positive in the direction of $n$); we call $\{n, d\}$ the *surface parameters*. A pair $\{x, x'\}$ is a projection of a feature point on that plane if and only if

$$x' \times Ax = 0, \tag{30}$$

where $A$ is a matrix that determines the projective transformation (or *homography*) of the image and has the following form [2, 3, 11, 12, 17]:

$$A = R^\top(hn^\top - dI). \tag{31}$$

Eq. (30) defines a model $S_{II}$ of codimension 2 in the four-dimensional data space $\mathcal{X} = \{(x, y, z, x', y', z')|z = 1, z' = 1\} \in \mathcal{R}^6$. Since the scale of the homography $A$

is indeterminate, we may normalize it to $\|A\| = 1$. So, the parameter space $\mathcal{U}$ is a eight-dimensional manifold in $\mathcal{R}^9$. An optimal solution of the surface parameters $\{n, d\}$ and the motion parameters $\{h, R\}$ in the sense of MLE is obtained by minimizing the following function [6, 9]:

$$J_\Pi = \sum_{\alpha=1}^{N}(x'_\alpha \times Ax_\alpha, W_\alpha x'_\alpha \times Ax_\alpha). \tag{32}$$

$$W_\alpha = \left(x'_\alpha \times AV_0[x_\alpha]A^\top \times x'_\alpha + (Ax_\alpha) \times V_0[x'_\alpha] \times (Ax_\alpha)\right)_2^-. \tag{33}$$

The product $v \times T$ of a vector $v$ and a matrix $T$ is defined to be the matrix whose columns are the vector products of $v$ and the three columns of $T$. For a vector $v$ and a matrix $T$, the symbol $v \times T \times v$ is an abbreviation of $v \times T(v \times I)^\top$. The renormalization procedure can be applied to compute the homography $A$ that minimizes eq. (32) [6, 9]. The surface parameters $\{n, d\}$ and the motion parameters $\{h, R\}$ are analytically computed from the resulting matrix $A$ [3, 11, 17].

## 8.3 Pure rotation model

No 3-D information can be obtained if the camera motion is pure rotation around the center of the lens; all we can estimate is the amount of the camera rotation $R$. A pair $\{x, x'\}$ is a projection of the same feature point if and only if

$$x \times Rx' = 0. \tag{34}$$

This equation defines a model $\mathcal{S}_R$ of codimension 2 in the four-dimensional data space $\mathcal{X} = \{(x, y, z, x', y', z')|z = 1, z' = 1\} \in \mathcal{R}^6$. The parameter space $\mathcal{U}$ is a three-dimensional manifold defined by all rotation matrices in $\mathcal{R}^9$. An optimal solution of $R$ in the sense of MLE is obtained by minimizing the following function [6]:

$$J_R = \sum_{\alpha=1}^{N}(x_\alpha \times Rx'_\alpha, W_\alpha x_\alpha \times Rx'_\alpha), \tag{35}$$

$$W_\alpha = \left((Rx'_\alpha) \times V_0[x_\alpha] \times (Rx'_\alpha) + x_\alpha \times RV_0[x'_\alpha]R^\top \times x_\alpha\right)_2^-. \tag{36}$$

## 8.4 Model selection

Eq. (30) implies the epipolar eq. (22). Eq. (33) is obtained from eq. (30) by letting $d = 1$ and $h = 0$ in eq. (31). Hence, $\mathcal{S}_R \succ \mathcal{S}_\Pi \succ \mathcal{S}$. Thus, the following tests can be done [6, 7]:

**Planarity test:** The object is judged to be planar if

$$K_\Pi = \sqrt{\frac{N-5}{7N+5}\left(\frac{\hat{J}_\Pi}{\hat{J}} + \frac{4N+16}{N-5}\right)} < 1. \tag{37}$$

**Fig. 3.** (Left) A two-dimensional manifold with three degrees of freedom can be fitted. (Middle) A two-dimensional manifold with eight degrees of freedom can be fitted. (Right) A three-dimensional manifold with five degrees of freedom can be fitted.

**Rotation test:** The camera motion is judged to be a pure rotation if

$$K_R = \sqrt{\frac{N-5}{7N+5}\left(\frac{\hat{J}_R}{\hat{j}} + \frac{4N+6}{N-5}\right)} < 1. \tag{38}$$

## 8.5 Self-evaluation

In order to do robust 3-D reconstruction, the camera must be displaced over a long distance so that the resulting disparity is sufficiently large. As the disparity increases, however, it is increasingly difficult to match feature points over the two images due to possible occlusions and illumination changes. Feature matching becomes easier as the disparity decreases; for each feature point, only a small neighborhood needs to be searched. It is therefore desirable to keep the camera displacement minimum in such a way that the resulting disparity is sufficient for reliable 3-D reconstruction. This can be done by measuring of the 'goodness' of the images by the values $(K_R, K_{II})$ defined by eqs. (37) and (38) if we note the following (figure 3):

**Small motion:** If the camera motion is small, the data points are concentrated near a two-dimensional manifold in the four-dimensional data space $\mathcal{X}$. As a result, we can robustly fit the rotation model $\mathcal{S}_R$, but we cannot robustly fit $\mathcal{S}_{II}$ or $\mathcal{S}$, so we are unable to perceive any 3-D structure of the scene.

**Intermediate motion:** As the camera motion increases, the data spread more in $\mathcal{X}$, so we can robustly fit the planar surface model $\mathcal{S}_{II}$, but we cannot robustly fit $\mathcal{S}$. As a result, the scene we can perceive is merely a planar surface.

**Large motion:** If the camera motion is sufficiently large, the distribution of the data is sufficiently three-dimensional in $\mathcal{X}$, so we can robustly fit the general model $\mathcal{S}$ and thereby perceive the full 3-D structure of the scene.

# 9    3-D Interpretation of Optical Flow

Suppose we observe flow $(\dot{x}, \dot{y})$ at point $(x, y)$. As in the case of finite motion, the flow and the position are represented by three-dimensional vectors

$$x = (x, y, 1)^\top, \quad \dot{x} = (\dot{x}, \dot{y}, 0)^\top. \tag{39}$$

Let $v$ and $\omega$ be, respectively, the instantaneous translation velocity and rotation velocity of the camera; we call $\{v, \omega\}$ the (instantaneous) *motion parameters*. Suppose we observe flow $\{\dot{x}_\alpha\}$ at points $\{x_\alpha\}$, $\alpha = 1, ..., N$, and let $\{\bar{\dot{x}}_\alpha\}$ be the (unknown) true flow.

Let $\epsilon^2 V_0[\dot{x}_\alpha]$ be the covariance matrix of flow $\dot{x}_\alpha$. The normalized covariance matrix $V_0[\dot{x}_\alpha]$ is determined up to scale in the course of optical flow detection [13, 15]. For a stationary scene, we consider the following models:

## 9.1    General motion model $\mathcal{S}_{\mathrm{gm}}$

The motion parameters $\{v, \omega\}$ are unconstrained. In the absence of noise, we have the following *epipolar equation* [14, 6]

$$|x_\alpha, \bar{\dot{x}}_\alpha, v| + (v \times x_\alpha, \omega \times x_\alpha) = 0. \tag{40}$$

**Errors in translation:** The deviation of the computed value $\hat{v}$ from its true value $v$ is represented by the difference $\hat{v} - v$. Since $\hat{v}$ and $v$ are both unit vectors, we project them onto the plane perpendicular to $v$ and define

$$\Delta v = P_v(\hat{v} - v). \tag{41}$$

**Errors in rotation:** The deviation of the computed value $\hat{\omega}$ from its true value $\omega$ is represented by the difference

$$\Delta \omega = \hat{\omega} - \omega. \tag{42}$$

The covariance matrices of the motion parameters $\{\hat{v}, \hat{\omega}\}$ are defined by

$$V[\hat{v}] = E[\Delta v \Delta v^\top], \quad V[\hat{v}, \hat{\omega}] = E[\Delta v \Delta \omega^\top],$$

$$V[\hat{\omega}, \hat{v}] = E[\Delta \omega \Delta v^\top], \quad V[\hat{\omega}] = E[\Delta \omega \Delta \omega^\top]. \tag{43}$$

The Cramer-Rao inequality (7) has the following form:

$$\begin{pmatrix} V[\hat{v}] & V[\hat{v}, \hat{\omega}] \\ V[\hat{\omega}, \hat{v}] & V[\hat{\omega}] \end{pmatrix} \succ \epsilon^2 \begin{pmatrix} \sum_{\alpha=1}^N \bar{W}_\alpha \bar{p}_\alpha \bar{p}_\alpha^\top & \sum_{\alpha=1}^N \bar{W}_\alpha \bar{p}_\alpha \bar{q}_\alpha^\top \\ \sum_{\alpha=1}^N \bar{W}_\alpha \bar{q}_\alpha \bar{p}_\alpha^\top & \sum_{\alpha=1}^N \bar{W}_\alpha \bar{q}_\alpha \bar{q}_\alpha^\top \end{pmatrix}^{-}, \tag{44}$$

$$\bar{p}_\alpha = x_\alpha \times \bar{\dot{x}}_\alpha + \|x_\alpha\|^2 \omega - (x_\alpha, \omega)x_\alpha, \quad \bar{q}_\alpha = \|x_\alpha\|^2 v - (x_\alpha, v)x_\alpha \tag{45}$$

$$\bar{W}_\alpha = \frac{1}{(v, (x_\alpha \times V_0[\dot{x}] \times x_\alpha)v)} \tag{46}$$

This bound is attained in the first order by MLE; we minimize the following function [6, 14]:

$$J_{\text{gm}}[v, \omega] = \sum_{\alpha=1}^{N} \frac{|x_\alpha, \dot{x}_\alpha, v| + (v \times x_\alpha, \omega \times x_\alpha)^2}{(v, (x_\alpha \times V_0[\dot{x}_\alpha] \times x_\alpha)v)}. \tag{47}$$

Let $\hat{J}_{\text{gm}}$ be its residual. The geometric AIC is

$$AIC(S_{\text{gm}}) = \hat{J}_{\text{gm}} + 2(N + 5)\epsilon^2. \tag{48}$$

An unbiased estimator of $\epsilon^2$ is obtained as follows:

$$\hat{\epsilon}^2 = \frac{\hat{J}_{\text{gm}}}{N - 5}. \tag{49}$$

## 9.2 General translation model $S_{\text{gt}}$

The camera translates without rotation. In the absence of noise, we have

$$|x_\alpha, \bar{\dot{x}}_\alpha, v| = 0. \tag{50}$$

The scale of the translation velocity $v$ is indeterminate, so we normalize it into $\|v\| = 1$. An optimal estimate of $v$ is obtained by minimizing

$$J_{\text{gt}}[v] = \sum_{\alpha=1}^{N} \frac{|x_\alpha, \dot{x}_\alpha, v|^2}{(v, (x_\alpha \times V_0[\dot{x}_\alpha] \times x_\alpha)v)}. \tag{51}$$

Let $\hat{J}_{\text{gt}}$ be its residual. The geometric AIC is

$$AIC(S_{\text{gt}}) = \hat{J}_{\text{gt}} + 2(N + 2)\epsilon^2. \tag{52}$$

An unbiased estimator of $\epsilon^2$ is obtained as follows:

$$\hat{\epsilon}^2 = \frac{\hat{J}_{\text{gt}}}{N - 2}. \tag{53}$$

## 9.3 Special translation model $S_{\text{st}}$

The camera undergoes a known translation $v^0$ without rotation. In the absence of noise, we have

$$|x_\alpha, \bar{\dot{x}}_\alpha, v^0| = 0, \tag{54}$$

which involves no unknowns. Let

$$\hat{J}_{\text{st}} = \sum_{\alpha=1}^{N} \frac{|x_\alpha, \dot{x}_\alpha, v^0|^2}{(v^0, (x_\alpha \times V_0[\dot{x}_\alpha] \times x_\alpha)v^0)}. \tag{55}$$

The geometric AIC is

$$AIC(S_{\text{st}}) = \hat{J}_{\text{st}} + 2N\epsilon^2. \tag{56}$$

**Remark 9.** Direct minimization of eqs. (47) and (51) requires numerical search, but the computation can simplified by combining the *linearization* technique with renormalization and an optimal correction scheme [14]. Caution is necessary, however, if the *focus of expansion* appears in the image; it becomes a singularity of MLE with an infinite weight in eqs. (47) and (51), causing numerical instability [6]. This can be avoided by removing flow components that give the largest values of a fixed percentage in the summand in eqs. (47) and (51).

## 9.4 Model comparison

The following model comparison criteria are obtained [6, 8]:

$S_{gm}$ **vs.** $S_{gt}$: The square noise level is estimated from Eq. (49). The general translation model $S_{gt}$ is preferred if

$$K_{gt} = \sqrt{\frac{N-5}{3N+5}\left(\frac{\hat{J}_{gt}}{\hat{J}_{gm}} + \frac{2(N+2)}{N-5}\right)} < 1. \tag{57}$$

$S_{gt}$ **vs.** $S_{st}$: The square noise level is estimated from Eq. (53). The special translation model $S_{st}$ is preferred if

$$K_{st} = \sqrt{\frac{N-2}{3N+2}\left(\frac{\hat{J}_{st}}{\hat{J}_{gt}} + \frac{2N}{N-2}\right)} < 1. \tag{58}$$

## 9.5 Moving object detection

In autonomous vehicle navigation by visual sensing, detecting objects that are moving independently of the background, such as people and other vehicles, is of utmost importance. This type of "intelligent processing" has been usually done by incorporating various thresholds to be adjusted empirically. They depend on the environment, the device, and the image processing techniques by which the images are obtained. In particular, they heavily depend on the magnitude of noise: the decision should be strict for a low noise level; a large deviation should be tolerated for a high noise level. Hence, values adjusted in one environment become meaningless in another environment.

This difficulty can be avoided by applying the model selection criterion stated above, allowing us to detect moving objects in such a way that in principle

– no knowledge about the magnitude of the noise in the flow is required and
– any thresholds to be adjusted empirically are involved.

Suppose a camera is fixed to a vehicle moving ahead. The camera observes a special translation flow if the scene is stationary. If an object is translating without rotating in the scene, the camera observes a general translation flow in the object region. Hence, we can detect a moving object by sliding a window over the image frame and comparing the special translation model $\mathcal{S}_{st}$ with the general translation model $\mathcal{S}_{gm}$ within the window. We judge that no object is moving if $\mathcal{S}_{st}$ is favored. In other words, we compute Eq. (58) and test if $K_{st} <$ 1. Admitting the fact that numerous sources of uncertainty that are not modeled in the theory exist, it is more realistic to regard the value $K_{st}$ as the "degree of non-existence of moving objects".

## 10    Concluding Remarks

We have given a mathematical formulation to the computer vision task of infer-ring 3-D structures of the scene based on image data and geometric constraints. Introducing a statistical model of image noise, we defined a geometric model as a manifold determined by the constraints and viewed the problem as model fitting. We then presented a general mathematical framework for proving optimality of estimation, deriving optimal schemes, and selecting appropriate models. Finally, we illustrated our theory by applying it to structure from motion. Geometric model selection based on the geometric AIC has many other applications than described here—inferring true 2-D structures from distorted shapes [8, 16] and inferring 3-D structures by stereo vision [10], for instance.

This work was in part supported by the Ministry of Education, Science, Sports and Culture, Japan under a Grant in Aid for Scientific Research C(2) (No. 09680352).

## References

1. Akaike, H., A new look at the statistical model identification, *IEEE Trans. Au-tomation Control*, **19**-6, 176–723, 1974.
2. Faugeras, O., *Three-Dimensional Computer Vision: A Geometric Viewpoint*, MIT Press, Cambridge, U.S.A., 1993,
3. Kanatani, K., *Geometric Computation for Machine Vision*, Oxford University Press, Oxford, U.K., 1993.
4. Kanatani, K., Unbiased estimation and statistical analysis of 3-D rigid motion from two views, *IEEE Trans. Patt. Anal. Mach. Intell.*, **15**-1, 37–50, 1993.
5. Kanatani, K., Renormalization for motion analysis: Statistically optimal algorithm, *IEICE Trans. Inf. & Sys.*, **E 77-D**-11, 1233–1239, 1994.
6. Kanatani, K., *Statistical Optimization for Geometric Computation: Theory and Practice*, Elsevier Science, Amsterdam, the Netherlands, 1996.
7. Kanatani, K., Automatic singularity test for motion analysis by an information criterion, *Proc. 4th European Conf. Computer Vision*, Cambridge, UK, 15–19 April 1996, pp. 697–708, 1996.
8. Kanatani, K., Comments on 'Symmetry as a continuous feature', *IEEE Trans. Patt. Anal. Mach. Intell.*, **19**-3, 246–247, 1997.
9. Kanatani K. and Takeda, S., 3-D motion analysis of a planar surface by renormal-ization, *IEICE Trans. Inf. & Syst.*, **E 78-D**-8, 1074–1079, 1995.
10. Kanazawa, Y. and Kanatani, K., Infinity and planarity test for stereo vision, *IEICE Trans. Inf. & Syst.*, **E 80-D**, 1997, to appear.

11. Longuet-Higgins, H. C., The reconstruction of a plane surface from two perspective projections. *Proc. Roy. Soc. Lond.*, **B 227**, 399–410, 1986.

12. Maybank, S., *Theory of Reconstruction from Image Motion*, Springer, Berlin, Germany, 1993.

13. Ohta, N., Image movement detection with reliability indices, *IEICE Trans. Inf. & Syst.*, **E 74-D**-10, 3379–2288, 1991.

14. Ohta, N. and Kanatani, K., Optimal structure from motion algorithm for optical flow, *IEICE Trans. Inf. & Syst.*, **E 78-D**-12, 1559–1566, 1995.

15. Ohta, N., Optical flow detection using general noise model, *IEICE Trans. Inf. & Syst.*, **E 79-D**-7, 951–957, 1996.

16. Triono, I. and Kanatani, K., Automatic recognition of regular figures by geometric AIC, *Proc. IAPR Workshop Machine Vision Applications*, Tokyo, Japan, 12–14 November 1996, pp. 393—396, 1996.

17. Weng, J., Ahuja, N. and Huang, T. S., Motion and structure from point correspondences with error estimation: Planar surfaces, *IEEE Trans. Sig. Proc.*, **39**-12, 2691–2717, 1991.

18. Weng, J., Huang, T.S. and Ahuja, N., *Motion and Structure from Image Sequences*, Springer, Berlin, Germany, 1993.

19. Weng, J., Ahuja, N. and Huang, T. S., Optimal motion and structure estimation, *IEEE Trans. Patt. Anal. Mach. Intell.*, **15**-9, 864–884, 1993.

# Path Prediction and Classification Based on Non-linear Filtering

S.J. Maybank and A.D. Worrall

Department of Computer Science, University of Reading, Whiteknights, PO Box 225, Reading, Berkshire, RG6 6AY, UK.

**Abstract.** In path based filtering the usual state space is replaced by a finite dimensional approximation to the path space of the system. The information in the system model and the measurements is summarised by a probability density function on the approximating space. Path based filters are well suited to the inference of system behaviour over time. They have the advantage that the predicted or estimated paths are always physically plausible, in that they are realisations of the system model.

The filter is applied to the model based tracking of cars. Measurements of the position and orientation of a moving car are obtained by fitting a wire frame model to a sequence of video images. The filter estimates the velocity, acceleration and steering angle of the car. Experiments show that the steering angle can be estimated after tracking for 1 s and the acceleration after 2 s.

## 1 Introduction

In conventional approaches to filtering, as described in [9], [10] for example, the state of the system is specified at each moment of time by a point in a state space $\mathbb{R}^n$. The information about the state available at time $s$ is summarised by a probability density function $\rho_s$ defined on $\mathbb{R}^n$. The true state at time $s$ is estimated by the expectation $E(\rho_s)$. In applications such as vehicle tracking the path must be estimated over an extended time, to allow a classification of the motion. A sequence of expectations can be taken, $s \mapsto E(\rho_s)$, but in the non-linear case this has the disadvantage that the estimated path is usually physically impossible, in that it is not a realisation of the random process defined by the system model.

Sections 2,3 below develop an approach to filtering in which the state space is replaced by the space of possible paths of the system. The filter estimates a probability density function on a finite dimensional approximation to the path space. The point at which the density has a global maximum is the path of maximum likelihood (PML), which is adopted here as a good estimate of the true path of the system. The advantage of using a path space is that the filter only considers paths which are physically plausible, in that they can be realisations of the system model.

The filter is applied in §4 to the tracking of road vehicles. Vehicle tracking is essential in order to obtain the information needed for optimal management of a road network. For example, many vehicles are detected moving into the same area, then signalling patterns might be varied to reduce the build up of congestion. The vision group at Reading University has developed a model based car tracker which obtains accurate estimates of the position and orientation of a car from a sequence of images taken by a static camera The development of the tracker can be followed in [7], [8], [14] and [16]. The model is a 'wire frame' which gives the relative positions of the prominent lines and edges on the car body. It stabilises the tracker and allows accurate estimates of three dimensional position and orientation even from low quality video sequences.

The path space approach has a great flexibility. The path space can be reduced to take account of prior knowledge of the vehicle's destination or to make hypotheses about the future behaviour of the vehicle. An example is developed in §5, using the first leg of a three point turn.

Other applications of the paths to parameter estimation and filtering can be found in [5], [6]. The binomial tree method for pricing stock options, as described in [5], explores a range of possible future changes in prices in order to calculate the fair price of an option. In the trellis filter of [6] the state space is partitioned into a finite set of segments $S_1, \ldots, S_r$ and the paths approximated by a Markov chain in which the $S_i$ are the states of the chain.

## 2    The Filter

A general introduction to filtering can be found in [9], [10]. The system model for the path based filter is of a standard type, specified by linked stochastic differential equations (SDEs). Background information on SDEs and the path based description of Brownian motion can be found in [11], [13].

### 2.1    System Model

The purpose of the system model is to generate a physically plausible family of motions for the vehicle. The possible motions are realisations of a stochastic process in $\mathbb{R}^5$, $M = (X, Y, \Theta, V, \Phi)^\top$, where $(X, Y)^\top$ is the position of the vehicle in a known ground plane, $\Theta$ is the orientation of the body, $V$ is the tangential velocity and $\Phi$ is the steering angle. The process $(V, \Phi)^\top$ models the actions of the driver as he or she presses on the accelerator and turns the steering wheel.

Let $w$ be the wheelbase of the car, i.e. the distance from the rear axle to the front axle. The evolution of $(X, Y, \Theta)^\top$ over times $s \geq 0$ is described by the linked SDEs

$$dX_s = V_s \cos(\Theta_s)\, ds$$
$$dY_s = V_s \sin(\Theta_s)\, ds$$
$$d\Theta_s = w^{-1} V_s \Phi_s\, ds \tag{1}$$

The quantity $V_s^{-1} d\Theta_s/ds$ is equal to the curvature of the path $s \mapsto (X_s, Y_s)$ in the ground plane. The first two equations of (1) ensure that the velocity is constrained to have direction $\Theta$. The physical interpretation of this constraint is that the vehicle does not slip sideways during the motion. The last equation of (1) links the steering angle to the rate of change of orientation of the vehicle body. It is an approximation to a more accurate equation given in [4],

$$d\Theta_s = w^{-1} V_s \tan(\Phi_s) \, ds$$

The error in the approximation is usually small. For a car with $w = 2.5\,\text{m}$ and a turning circle of radius $r = 5\,\text{m}$ the maximum value of the steering angle $\phi$ is $\tan^{-1}(w/r) \approx 0.46\,\text{rad}$. In the range $\pm 0.46\,\text{rad}$ the error on replacing $\tan(\phi)$ by $\phi$ is less than 8%. At lower steering angles the error is much less than 8%.

It follows from (1) that the motion of the vehicle is determined by $V$, $\Phi$ and the initial values of $X$, $Y$, $\Theta$. The processes $V$, $\Phi$ which drive the motion model are solutions to the SDEs

$$dV_s = q \, dB_s$$
$$d\Phi_s = -\alpha \Phi_s \, ds + \sigma \, dC_s \tag{2}$$

where $q$, $\alpha$, $\sigma$ are constants and $(B, C)^\top$ is a Brownian motion in $\mathbb{R}^2$ independent of the initial state $M_0$ of the vehicle. The tangential velocity $V$ is modelled by a scaled Brownian motion because this is the simplest choice. The steering angle $\Phi$ is modelled by an Ornstein-Uhlenbeck (OU) process rather than a Brownian motion. The reason is that, as noted above, the steering angle is restricted to a range depending on the geometry of the car. The SDE (2) for $\Phi$ contains a term $-\alpha \Phi_s \, ds$ which inhibits the occurrence of large values of $|\Phi|$ and tends to draw the paths of $\Phi$ towards the origin. The parameters $\alpha$, $\sigma$ are chosen such that the filter rarely produces an estimate of the steering angle outside the allowed range.

## 2.2 Advantages of Path Based Filtering

The task of filtering is to combine the information from the sequence of measurements and the motion model to i) obtain accurate estimates of the position and velocity of the vehicle; ii) predict future positions of the vehicle; and iii) make inferences about driver behaviour. In the usual approach to filtering, for example the extended Kalman filter [10], the information available at time $t$ is summarised by a conditional probability density function $\rho_t$ defined on the state space of the system. The expectation $E(\rho_t)$ is an estimate of the system state and the function $t \mapsto E(\rho_t)$ is an estimate of the path taken by the system state.

There are two major drawbacks to filtering based on $\rho_t$:

a) it is difficult to obtain a good estimate of $\rho_t$ if the state space has a high dimension (in this context 'high' means four or greater); and
b) the path $t \mapsto E(\rho_t)$ is usually physically impossible, in that it is incompatible with (1).

The incompatibility in *(b)* arises because the space of solutions to (1) is not linear. The expectation $t \mapsto E(\rho_t)$ need not be a solution to (1) and (2) for any realisation of $(B, C)$. An analogy may clarify things: let $q$ be a random variable taking values on the unit sphere $S^2$. The expected value, $E(q)$, need not be on $S^2$, even though all the possible values of $q$ are confined to $S^2$.

The drawbacks *(a)* and *(b)* are overcome by replacing the state space with the space of all paths taken by the system over a time $0 \leq s \leq t$. At time $t$ the noise process $(B, C)$ has a realisation at times prior to $t$ as a function $s \mapsto (b_s, c_s)$, $0 \leq s \leq t$. The functions $b$, $c$ are fixed (or deterministic), although they are of course unknown to the filter. For each possible realisation $(b, c)$ of $(B, C)$ and each initial value $M_0 = (x_0, y_0, \theta_0, v_0, \phi_0)$, (2) yields realisations $v$, $\phi$ of $V$, $\Phi$ and (1) yields realisations $x$, $y$, $\theta$ of $X$, $Y$, $\Theta$.

The possible realisations of $(V, \Phi)$ form an infinite dimensional function space. For the purposes of computation this space is approximated by a finite dimensional space $H$. The inclusion of the initial values $x_0$, $y_0$, $\theta_0$ leads to a space $\mathbb{R}^3 \times H$, each point of which corresponds to a realisation of $(X, Y, \Theta, V, \Phi)$. The space $\mathbb{R}^3 \times H$ carries a probability density function $p$ which depends in part on the Brownian motions $B$, $C$ and in part on the measurements. The location of the global maximum of $p$ is by definition the path of maximum likelihood (PML). It is chosen here as a good estimate of the path of the system conditional on the measurements.

Related work on the use of a PML for boundary finding and scene segmentation is described in [2].

## 2.3 The Cost of a Path

The approach to filtering sketched at the end of the preceding subsection relies on the following well known path based description of Brownian motion as given by [13], for example. For $t > 0$, let $\Omega$ be the space of all continuous functions $\omega : [0, t] \to \mathbb{R}$ such that $\omega(0) = 0$. Let $H$ be any finite dimensional subspace of $\Omega$ chosen such that all the elements of $H$ are absolutely continuous with derivatives in the space $L^2([0, t], \mathbb{R})$ of square integrable functions. Let $\langle ., . \rangle$ be the inner product defined on $H$ by

$$\langle h, g \rangle = \int_0^t \dot{h}_s \dot{g}_s \, ds$$

where $\dot{h} \equiv dh/ds$, and let $\|h\|^2 = \langle h, h \rangle$. The path with zero norm is constant, $h_s = 0$, $0 \leq s \leq t$. The norm $\|h\|$ is referred to as the cost of $h$. The mode of $\mu_H$ is the path in $H$ with the maximum likelihood, or equivalently the path in $H$ with the least cost.

Let $h_i$, $1 \leq i \leq n$ be an orthonormal basis for $H$, let $\lambda$ be Lebesgue measure on $H$ normalised such that the hypercube defined by the $h_i$ has unit measure and let $\mu_H$ be the Gaussian measure defined on $H$ by

$$\mu_H(dh) = (2\pi)^{-n/2} \exp\left(-\frac{1}{2}\|h\|^2\right) \lambda(dh)$$

It can be shown that there is a unique probability measure $\mu$ on $\Omega$ which induces on each subspace $H$ the measure $\mu_H$. The pair $\Omega$, $\mu$ is by definition a Brownian motion, or Weiner process.

The path based description of Brownian motion is readily extended to OU processes. An OU process $C$ is, by definition, a solution to an SDE

$$dC_s = -\alpha C_s \, ds + \sigma \, dB_s, \qquad 0 \le s \le t$$

where $\alpha > 0$ and $\sigma$ are constants. Let $H$ be a subspace of the path space of $C$, such that all the paths in $H$ are differentiable. If $h \in H$, then $\sigma^{-1}(\dot{h} + \alpha h)$ is the derivative of a realisation of the Brownian motion $B$. Let $\|.\|_o$ be the norm on $H$ defined by

$$\|h\|_o^2 = \int_0^t (\dot{h}_s + \alpha h_s)^2 \, ds$$

The Weiner measure $\mu$ for Brownian motion induces the measure $\sigma^{-1}\|.\|_o$ on $H$.

The path space for $(V, \Phi)$ is readily obtained from the path space for the driving Brownian motion $(B, C)$ of (2). Let $H$ be a finite dimensional approximation to the full, infinite dimensional path space of $(V, \Phi)$. Let $m = (x, y, \theta, v, \phi)^\top$ be a possible path for the car, chosen from the path space $\mathbb{R}^3 \times H$ introduced in §2.2. The cost $I(m)$ of $m$ arising from the choice (2) of models for $V$, $\Phi$ is

$$I(m) = \frac{1}{q^2}\|v\|^2 + \frac{1}{\sigma^2}\|\phi\|_o^2 \tag{3}$$

The contribution of the measurements to the cost of $m$ is obtained. The measurement function $f$ is

$$s \mapsto f_s \equiv (x_s, y_s, \theta_s) \qquad 0 \le s \le t$$

Let the measurements $z_i$ be obtained at times $t_i$, $1 \le i \le n$ where $0 \le t_1 < \ldots < t_n \le t$. It is assumed that the measurements are subject to zero mean Gaussian noise with covariance $R$. The cost of the measurements is

$$C(m) = \sum_{i=1}^n (z_i - f_{t_i})^\top R^{-1}(z_i - f_{t_i}) \tag{4}$$

The total cost $J(m)$ is the sum of (3) and (4),

$$J(m) = I(m) + C(m) \tag{5}$$

The probability density function associated with $J$ is $\exp(-J(m)/2)$.

The position of the global minimum of $J$ on $\mathbb{R}^3 \times H$ yields the PML. The PML depends on $H$ but it is likely to be close to the exact PML if the dimension of $H$ is large or the true motion of the car is simple.

## 2.4 PML in the Absence of Measurements

The PML in the absence of measurements is easily found. The norm $\|v\|$ attains a minimum of zero if and only if $v$ is constant, $v \equiv v_0$. The norm $\|\phi\|_o$ attains a minimum of zero if and only if

$$\frac{d\phi_s}{ds} + \alpha\phi_s = 0 \qquad\qquad 0 \leq s \leq t \qquad\qquad (6)$$

which yields $\phi_s = \phi_0 \exp(-\alpha s)$. As $s$ increases, $\phi_s$ tends towards zero and the PML is asymptotic to a straight line. Equation (6) is an example of an Euler-Lagrange equation from the calculus of variations (see [3]. The connections between the PML and the calculus of variations are explored further in §5.

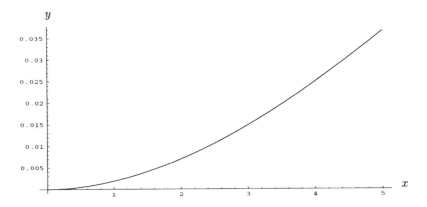

Figure 1. *Example of a path of maximum likelihood*

Let coordinates be chosen such that $x_0 = y_0 = 0$ and $\theta_0 = 0$. If $\alpha \neq 0$, then it follows from (1) and (6) that

$$\theta_s = v_0\phi_0(1 - \exp(-\alpha s))/(\alpha w)$$

$$x_s = v_0 \int_0^s \cos(\theta_u)\, du$$

$$y_s = v_0 \int_0^s \sin(\theta_u)\, du$$

Figure 1 shows the path $s \mapsto (x_s, y_s)$, $0 \leq s \leq 1$ of a PML with parameters $\alpha = 1\,\mathrm{s}^{-1}$, $q = 2\,\mathrm{ms}^{-1}$, $\sigma = 2\,\mathrm{s}^{-1}$, $x_0 = y_0 = 0\,\mathrm{m}$, $\theta_0 = 0\,\mathrm{rad}$, $w = 2.5\,\mathrm{m}$, $v_0 = 5\,\mathrm{ms}^{-1}$, $\phi_0 = 0.01\,\mathrm{rad}$. The orientation $\theta$ at $s = 1$ is $\approx 0.013\,\mathrm{rad}$.

# 3 An Approximation to the Cost Function

## 3.1 Taylor Expansion

The evaluation of the cost function $J$ of (5) is computationally expensive even if the path space $H$ has a low dimension. The cost can be reduced by expanding part of $(x_s, y_s)$ as a Taylor series in $s$.

Suppose firstly that $x_0 = y_0 = 0$, $\theta_0 = 0$ and that the time interval in which the approximation is required to hold is $[-t, t]$. The path space for $(V, \Phi)$ is approximated by a finite dimensional space $H$ of pairs of polynomials,

$$v_s = v_0 + v_1 s + \ldots v_d s^d$$
$$\phi_s = \phi_0 + \phi_1 s + \ldots \phi_d s^d \tag{7}$$

The space $H$ has dimension $2d + 2$ and coordinates $v_0, \ldots, v_d$, $\phi_0, \ldots, \phi_d$. It follows from (1), (7) that $\theta$ is a polynomial in $s$ of degree $2d + 3$. Let $f_s$, $g_s$ be the functions of order $O(s^2)$ defined by

$$f_s = \cos(\theta_s) - \cos\left(w^{-1}\phi_0 \int_0^s v_u \, du\right)$$

$$g_s = \sin(\theta_s) - \sin\left(w^{-1}\phi_0 \int_0^s v_u \, du\right)$$

The exact path in the $x$, $y$ plane is

$$x_s = w\phi_0^{-1} \sin\left(w^{-1}\phi_0 \int_0^s v_u \, du\right) + \int_0^s v_u f_u \, du \qquad 0 \le s \le t$$

$$y_s = -w\phi_0^{-1} \cos\left(w^{-1}\phi_0 \int_0^s v_u \, du\right) + \int_0^s v_u g_u \, du \qquad 0 \le s \le t \tag{8}$$

The coordinates $x_s$, $y_s$ are approximated by expanding the integrals

$$\int_0^s v_u f_u \, du \qquad \int_0^s v_u g_u \, du$$

on the right-hand side of (8) in Taylor series up to and including terms in $s^{3d}$.

If an approximation is required over a time interval $[t_1, t_2]$, then the time origin of (8) is changed by a substitution $s \mapsto s - t_0$ where $t_0 = (t_1 + t_2)/2$. If any of $x_0$, $y_0$, $\theta_0$ are non-zero, then appropriate expressions $x_s^{(n)}$, $y_s^{(n)}$, $\theta_s^{(n)}$ are obtained from $x_s$, $y_s$, $\theta_s$ as follows,

$$\theta_s^{(n)} = \theta_s + \theta_0$$
$$x_s^{(n)} = x_0 + \cos(\theta_0)x_s - \sin(\theta_0)y_s$$
$$y_s^{(n)} = y_0 + \sin(\theta_0)x_s + \cos(\theta_0)y_s$$

The resulting approximations are very accurate provided $t_2 - t_1$ is small. An example is shown in Figure 2 for $s \in [-1, 1]$, with $x_0 = y_0 = 0$ and $\theta_0 = \pi/4$. The tangential velocity is $v_s = 5 + s$, the steering angle is $\phi_s = 0.05 + 0.2s$

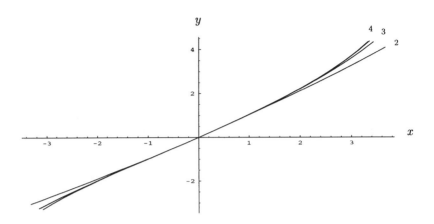

Figure 2. *Approximations to a path in the x, y plane.*

and the wheelbase is $w = 2.5$. The label $i = 2, 3, 4$ on a curve indicates that the error, or deviation from the exact path is $O(s^i)$, or equivalently that $f_s$, $g_s$ are approximated with an error of order $O(s^{i-1})$. In the case $i = 2$, $f_s$ and $g_s$ are approximated by zero. At the resolution of Figure 2 the path labelled 4 is indistinguishable from the exact path. The errors in the approximation at times $s = 0, 0.5, 1.0$ are given in the following table.

| $s$ | 2 | 3 | 4 |
|---|---|---|---|
| 0.0 | 0.000 | 0.000 | 0.000 |
| 0.5 | 0.047 | 0.006 | 0.001 |
| 1.0 | 0.421 | 0.104 | 0.038 |

Table 1. *Errors in the Taylor series approximation.*

It is apparent from Table 1 that the error in the Taylor series approximation can grow very rapidly. For example, in the case $i = 3$ the average rate of increase of the error in the range $0 \leq s \leq 0.5$ is 0.012, whilst the average rate of increase in the range $0.5 \leq s \leq 1$ is 0.196, which is more than 16 times greater than 0.012.

## 3.2 Gaussian Approximation and Driver Behaviour

To avoid a proliferation of symbols, '$J$' is used for the approximation to $J$ obtained in §3.1. The following notation is adopted: the measurements used in $J$ span an interval $[t_n - \Delta t, t_n]$, from the time $t_n - \Delta t$ of the first measurement to the time $t_n$ of the last measurement. The mid point is $t_0 = t_n - \Delta t/2$. The position and orientation of the car at time $t_0$ are $(x_0, y_0)$ and $\theta_0$, respectively.

The cost $J$ depends on the vector of $2d + 5$ parameters,

$$r = (x_0, y_0, \theta_0, v_0, \ldots, v_d, \phi_0, \ldots, \phi_d)$$

Let $r_0$ be the value of $r$ corresponding to the PML. It follows that

$$J = J(r_0) + \frac{1}{2}(r - r_0)^\top \left.\frac{\partial^2 J}{\partial r^2}\right|_{r_0} (r - r_0) + O(\|r - r_0\|^3)$$

The density $\exp(-J/2)$ is approximated by a Gaussian density with expectation $r_0$ and covariance

$$C_0 = \left(\left.\frac{1}{2}\frac{\partial^2 J}{\partial r^2}\right|_{r_0}\right)^{-1}$$

At time $t_n$ the driver may be accelerating, travelling with constant speed or decelerating. He or she may also be turning left, driving straight or turning right. The acceleration $\dot{v}_{t_n}$ and the steering angle $\phi_{t_n}$ are given by

$$\dot{v}_{t_n} = v_1 + \ldots dv_d t_n^{d-1}$$
$$\phi_{t_n} = \phi_0 + \ldots \phi_d t_n^d \tag{9}$$

The expectations and standard deviations of $\dot{v}_{t_n}$, $\phi_{t_n}$ are estimated from $r_0$, $C_0$. Let $p_1$, $p_2$ be the probabilities

$$p_1 = P(\dot{v}_{t_n} \geq 0) \qquad p_2 = P(\phi_{t_n} \geq 0)$$

and let $\tau_p$ be a threshold. The motion is classified as an acceleration if $p_1 \geq \tau_p$ and as a left turn if $p_2 \geq \tau_p$. It is classed as a deceleration if $p_1 \leq (1 - \tau_p)$ and as a right turn if $p_2 \leq (1 - \tau_p)$. The choice of $\tau_p$ depends on the noise level in the measurements and on the cost of making a wrong classification of driver behaviour.

## 4 Results

### 4.1 Data

The filter was applied to a sequence of measurements of the position and orientation of a car performing a three point turn. The motion is complicated and non-linear, making it difficult to track with conventional filters. The path of the car in the $x$, $y$ plane is shown in Figure 3, which is obtained by joining the points $(x, y)$ obtained from consecutive measurements. The numbers 1, 50, 100 ... on the path are indices to the list of measurements. The distances are in meters and the time between consecutive measurements is $2/25$ s.

The car halts at the first 'cusp' of the three point turn at about measurement number $n = 100$. It then reverses between n = 100 and n = 200, halts at the second cusp, and finally accelerates forward to leave the turn.

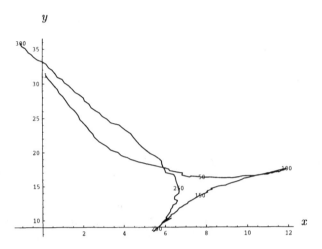

Figure 3. *Trajectory of the car in the x, y plane.*

## 4.2 Calculation of the PML

The filter is applied to every batch of 10 consecutive measurements with the parameters in (2) assigned the values $q = 1$, $\alpha = 4$, $\sigma = 1/4$. Each batch is obtained over an interval $I_n = [t_n - 18/25, t_n]$ where $t_n$ is the time of the last measurement in the batch. The path of the car during the interval $I_n$ is estimated with $d$ in (7) set to zero. The estimates of tangential velocity and the steering angle are thus held constant in $I_n$.

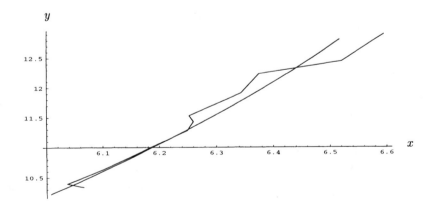

Figure 4. *Example of a PML*

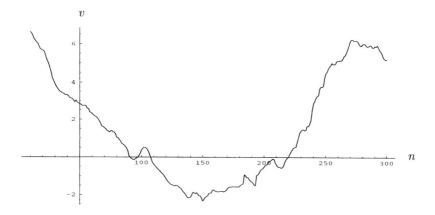

Figure 5. *Tangential velocity*

For each interval $I_n$, the covariance $R$ of the measurement noise is assumed to be diagonal with initial value $R_1 = \operatorname{diag}(0.0025, 0.0025, 0.0025)$. The PML $p^{(1)}$ is found by minimising the cost $J$ described in §3.2. A $\chi^2$ test with 25 degrees of freedom is applied to the estimated measurement errors. In most cases the $\chi^2$ value is too low indicating either that $R_1$ is too large or that there are too many variables fitted to the data.

The consistency of the estimates of the steering angle, as shown in Figure 6, indicates that there is no overfitting. Thus the covariance is varied on order to bring about a $\chi^2$ value of 0.5. Unfortunately, this cannot be done in a straightforward way because any change in the covariance affects the PML. To overcome this problem the following iterative procedure is adopted.

Let $z_i$ be the measurement vector obtained at time $t_i$ for $t_1 \leq \ldots \leq t_n$ and let $\Delta x_i, \Delta y_i, \Delta \theta_i$ be the estimated errors in $z_i$,

$$(\Delta x_i, \Delta y_i, \Delta \theta_i) = z_i - p^{(1)}_{t_i} \qquad 1 \leq i \leq n$$

A new covariance estimate of the form $R_2 = \{a^{-2}, a^{-2}, b^{-2}\}$ is calculated as follows. The normalised position errors are $a^{-1}(\Delta x, \Delta y)$ and the normalised angle errors are $b^{-1}\Delta \theta$. The variable $b$ is chosen such that the average of the squares of the normalised angle errors is equal to the average of the squares of the normalised position errors,

$$b = \frac{\sqrt{2}\, a \|\Delta \theta\|}{\sqrt{\|\Delta x\|^2 + \|\Delta y\|^2}}$$

Then $a$ is chosen such that the $\chi^2$ value is exactly 0.5. The PML $p^{(2)}$ for $R_2$ is found. The calculation is repeated, beginning with $R_2$, $p^{(2)}$, to yield a third pair of estimates $R_3$, $p^{(3)}$. For some intervals $I_n$ an estimated covariance $R_i$ has unusually low entries. In such cases $R_i$ is rejected and $p^{(i-1)}$ is chosen as the

PML. A diagonal entry of $R_i$ is low if it is less than $1/25$ of the corresponding entry of $R_1$.

The irregular path in Figure 4 is a section of the path in the $x$, $y$ plane for $236 \leq n \leq 245$. It contains the notch just below the label 250 in Figure 3. The smooth curve in figure 4 is the PML. The fit is good in spite of the notch.

## 4.3   Tangential Velocity and Steering Angle

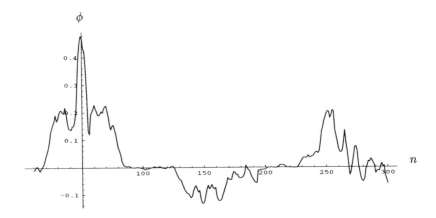

Figure 6. *Steering angle*

The estimates of tangential velocity and steering angle are shown in Figures 5 and 6 respectively. Both figures are clearly compatible with the three point turn displayed in Figure 3. Initially, the velocity falls, reaching zero at about measurement number $n = 100$. The velocity becomes negative as the car reverses, returns to zero again at about $n \approx 210$ and finally increases as the car moves away. The variations in the estimated velocity at the times when the car is known to be stationary suggest that the error in the velocity is no more than $0.5 \text{ ms}^{-1}$.

In Figure 6 the car first turns to the left ($\phi_s > 0$), then to the right during the reversing motion and finally to the left on leaving the three point turn. The value 4 assigned to $\alpha$ (see (2)) is large enough to ensure that the estimates of $\phi_s$ are confined to a range of approximately $\pm 0.46 \text{ rad}$. Figure 7 shows the probability that the car is making a left turn $P(\phi_s > 0)$, as a function of the measurement number $n$.

## 4.4   Lookahead

The PML for an interval $I_n = [t_n - 18/25, t_n]$ is used to predict the path of the car at times $s > t_n$. At time $t_n$ let the PML be at $(x^{(n)}, y^{(n)}, \theta^{(n)}, v^{(n)}, \phi^{(n)})$. It

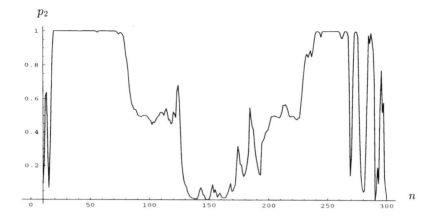

Figure 7. *Probability of a left turn*

is assumed, in line with §2.4, that

$$v_s = v^{(n)} \qquad\qquad s \geq t_n$$
$$\phi_s = \phi^{(n)} \exp(-\alpha(s - t_n)) \qquad\qquad s \geq t_n \qquad (10)$$

The PML is extrapolated beyond $s = t_n$, using (10). The extrapolated path $s \mapsto f_s \equiv (x_s, y_s)$ is compared with the measured path. Let $(z_i^{(x)}, z_i^{(y)})$ be the measurements of $(x_{t_i}, y_{t_i})$, $i \geq n$ and let $s \mapsto g_s$ be the path obtained from the $(z_i^{(x)}, z_i^{(y)})$ by linear interpolation,

$$g_{t_i} = (z_i^{(x)}, z_i^{(y)}) \qquad\qquad i \geq n$$
$$g_s = \frac{t_{i+1} - s}{t_{i+1} - t_i}(z_i^{(x)}, z_i^{(y)}) + \frac{s - t_i}{t_{i+1} - t_i}(z_{i+1}^{(x)}, z_{i+1}^{(y)}) \qquad s \in [t_i, t_{i+1}], i \geq n$$

Let $\tau_e$ be a fixed threshold. The lookahead time $e_n$ is defined such that $t_n + e_n$ is the first time after $t_n$ for which

$$\|g(t_n + e_n) - f(t_n + e_n)\| \geq \tau_e$$

In the experiments $\tau_e$ is a multiple of the nominal standard deviation, 0.05, of the errors in measuring $x$ or $y$, $\tau_e = (5\sqrt{2})0.05$.

The graph of $n \mapsto e_n$ is shown in Figure 8. The expected value of $e_n$ is 0.66.. and the standard deviation is 0.36...

## 4.5 Acceleration

To estimate accelerations the exponent $d$ in (7) must be at least one, but this increases the number of parameters fitted to each batch of ten measurements. Experiments show that there are too many parameters, overfitting occurs and

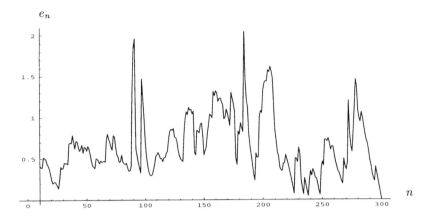

Figure 8. *Lookahead time*

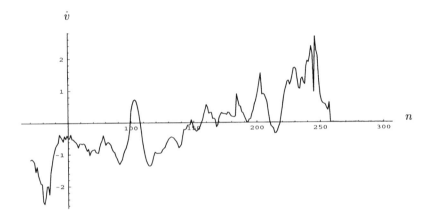

Figure 9. *Acceleration*

good estimates of the accelerations cannot be obtained. The solution is to apply the filter to batches of 20 measurements, with $d = 1$.

The estimated accelerations are shown in Figure 9. The estimates are for the most part negative in the range $1 \leq n \leq 150$. The point $n = 150$ is at the middle of the reversing motion. At this point the acceleration changes sign and the estimates are positive as the car moves towards the second cusp and then leaves the turn. The probabilities that the acceleration is positive are shown in Figure 10. The graph is consistent with Figure 9 except at the cusps of the motion ($n \approx 100$ and $n \approx 210$).

$p_1$

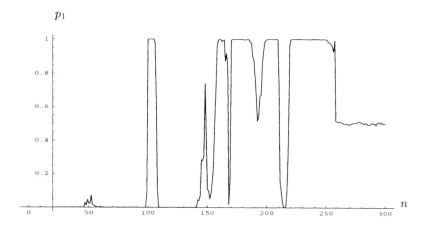

Figure 10. *Probability of positive acceleration*

# 5 Paths of Maximum Likelihood

Tracking is often difficult because of the variety of motions which the vehicle may undergo. A system model which works well for one particular motion may fail badly for other motions. One solution is to use a bank of filters (see [1]) running in parallel, each one with its own model. The output is chosen from the filter with a model closest to the observed motion. If the observed motion diverges too far from the current model, then a new filter is chosen. The filters are run in parallel to avoid the error prone initialisation stage when switching to a new filter.

The aim in this section is to develope an example to show how the path based approach might provide a framework for comparing different system models. The models may be allocated to separate filters or they may correspond to hypotheses about the future behaviour of the vehicle. From the probabilistic point of view, each model is obtained by conditioning the original path space.

The example chosen is the first leg of the three point turn illustrated in Figure 3.

## 5.1 Problem Statement

Suppose that at time $s = 0$ the car is at the origin with velocity $(u, 0)$ and with steering angle 0. It is hypothesised that the car will make the first leg a three point turn, reaching the cusp at time $t$, position $(x_t, y_t) = (a, b)$, orientation $\theta_t = \tan^{-1}(c)$ and steering angle 0. Under these constraints, what is the PML of the car in the time interval $[0, t]$?

It is convenient to use $\theta$, $v$ instead of $v$, $\phi$. The Lagrangian $V$ for the path is

$$V(\theta, v) = \frac{1}{q^2} \dot{v}_s^2 + \frac{w^2}{\sigma^2} \left( \frac{\ddot{\theta}}{v} - \frac{\dot{\theta}\dot{v}}{v^2} + \alpha\frac{\dot{\theta}}{v} \right)^2 + \lambda_1 v_s \cos(\theta_s) + \lambda_2 v_s \sin(\theta_s)$$

where $\lambda_1$, $\lambda_2$ are the usual analogues of Lagrange multipliers. The differential equations in $\theta$, $v$ are just the usual Euler-Lagrange equations, as given in [3] for example,

$$\frac{d^2}{ds^2}\left(\frac{\partial V}{\partial \ddot{\theta}}\right) - \frac{d}{ds}\left(\frac{\partial V}{\partial \dot{\theta}}\right) + \frac{\partial V}{\partial \theta} = 0$$

$$\frac{d}{ds}\left(\frac{\partial V}{\partial \dot{v}}\right) - \frac{\partial V}{\partial v} = 0 \tag{11}$$

It is difficult to find good solutions to (11), partly because they are of high order (four) in $\theta$ and partly because the constraints on the solution are given at two different times, namely the beginning of the path $(s = 0)$ and the end of the path $(s = t)$. The usual numerical methods for solving (11) assume that all the constraints are given at a single time.

## 5.2 Method for Approximating the PML

In view of the difficulties sketched in §5.1, the following approach is adopted. The path space is approximated by a finite dimensional space of pairs of polynomials,

$$x_\rho = x_0 + x_1\rho + \ldots + x_m\rho^n \qquad 0 \le \rho \le 1$$
$$y_\rho = y_0 + y_1\rho + \ldots + y_m\rho^n \qquad 0 \le \rho \le 1 \tag{12}$$

The parameter $\rho$ in (12) is not necessarily the time or the arc length.
The conditions on position and orientation at $\rho = 0, 1$ yield

$$x_0 = y_0 = y_1 = 0$$
$$x_1 = a$$
$$y_1 = b$$
$$\cos(c)\dot{y}_1 = \sin(c)\dot{x}_1$$

where $a$, $b$, $c$ are as given in §5.1. Zero curvature at $\rho = 0$ is imposed by setting $y_2 = 0$. Finally, zero curvature at $\rho = 1$ is imposed by rotating the path until $\dot{x}_1 = 0$ and then setting $\ddot{y}_1 = 0$. The pairs of polynomials satisfying all 8 constraints form a $2n - 6$ dimensional subspace $H$ of the $2n + 2$ dimensional space of pairs of polynomials defined by (12).

Let $(x, y)$ be a point of $H$. Suppose that the car moves with tangential velocity $v$ along the path described by $(x, y)$. The velocity $v$ is chosen such that the cost of traversing $(x, y)$ is minimised. It is convenient to parameterise the path by the arc length $\tau$. Let $l = l(x, y)$ be the total arc length and let $t$ be the time taken to traverse the path. Note that $v = d\tau/ds$. The cost $\tilde{J}$ of the path $(x, y)$ is

$$\tilde{J}(x, y) = \frac{1}{q^2}\int_0^t \dot{v}_s^2\, ds + \frac{1}{\sigma^2}\int_0^t (\dot{\phi}_s + \alpha\phi_s)^2\, ds$$

$$= \frac{1}{q^2}\int_0^l \left(\frac{dv}{d\tau}\right)^2 v_\tau\, d\tau + \frac{1}{\sigma^2}\int_0^l \left(\frac{d\phi}{d\tau} + \frac{\alpha\phi_\tau}{v_\tau}\right)^2 v_\tau\, d\tau \tag{13}$$

where the dot means, as usual, differentiation with respect to time $s$. The Lagrangian is

$$V(v) = \frac{1}{q^2}\left(\frac{dv}{d\tau}\right)^2 v_\tau + \frac{1}{\sigma^2}\left(\frac{d\phi}{d\tau} + \frac{\alpha\phi_\tau}{v_\tau}\right)^2 v_\tau$$

The Euler-Lagrange equation for variations in $v$ is

$$v\frac{d^2v}{d\tau^2} + \frac{1}{2}\left(\frac{dv}{d\tau}\right)^2 = \frac{q^2}{2\sigma^2}\left(\left(\frac{d\phi}{d\tau}\right)^2 - \frac{\alpha^2\phi^2}{v^2}\right) \tag{14}$$

Equation (14) is solved numerically subject to the constraints $v_0 = u$, $v_l = 0$.

Let $J(x, y)$ be the cost obtained by substituting the solution to (14) into (13). The PML is the global minimum of $J$ in $H$. Unfortunately, the computational cost of finding the PML is still too high. Instead, $H$ is searched first for the path $(x, y)$ at which

$$\int_0^1 \left(\frac{d\phi}{d\rho} + \alpha\phi_\rho\right)^2 d\rho$$

attains a global minimum. With this choice of $(x, y)$, (14) is solved numerically to obtain $v$. The exponent $m$ in (12) is assigned the value $m = 4$.

## 5.3   Results

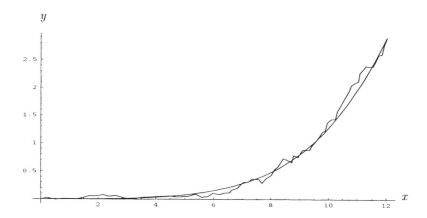

Figure 11. *First arc of the three point turn*

The measurements in the range $15 \le n \le 90$ comprise the first leg of the three point turn. The beginning is at $n = 15$ because Figure 6 shows that at this point the steering angle begins to increase away from zero. The end is at $n = 90$ because Figure 5 shows that at this point the velocity first reaches zero.

The $x$, $y$ coordinates of the measurements define points which are joined to make a polygonal arc. The arc is moved rigidly such that the first point is at $(0,0)$ and the initial section is (approximately) tangent to the $x$ axis. The final point is $(12.03, 2.90)$. The orientation of the car at $n = 90$ is estimated by taking the average value of the measured orientations for $86 \leq n \leq 90$. The result is $0.78\,\text{rad}$. The initial velocity (at $n = 15$) is estimated at $u = 6\,\text{ms}^{-1}$.

Figure 11 shows the polygonal arc with the PML superposed. The fit is surprisingly good, indicating that the model correctly describes the path traversed by the car in the $x$, $y$ plane. Figure 13 shows the steering angle as a function of time. This shows the same qualitative behaviour as the measured steering angle for $15 \leq n \leq 90$, except that the time interval is too short. The time taken by the car is $6\,\text{s}$, but the time taken by the PML is $3.51\,\text{s}$. This suggests that the true velocity diverges from the velocity predicted by the PML.

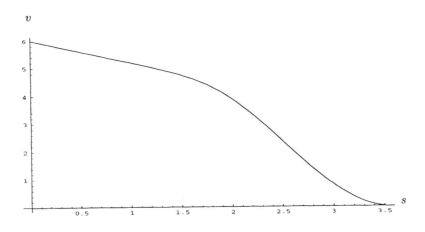

Figure 12. *Tangential velocity*

A closer examination of the true velocity in Figure 5 and the predicted velocity in Figure 12 bears out these observations. The optimal velocity decreases slowly at first and then much more rapidly towards the end of the motion, giving a concave graph. In the last quarter second the deceleration eases off and the curve becomes convex. In contrast, the graph of the observed velocity is concave. The deceleration is initially large and then gradually decreases towards the end of the trajectory.

## 5.4    A Change to the System Model

The results described in the preceding subsection suggest that the model for the steering angle is adequate to describe the behaviour of a human driver, but the model for the tangential velocity must be modified.

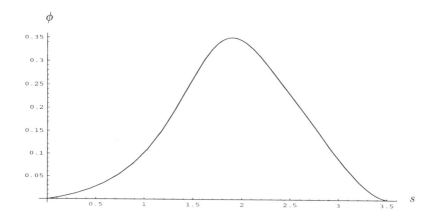

Figure 13. *Steering angle*

Equation (13) suggests that $\alpha$ may be the cause of the problem. If $\alpha \neq 0$, then the contribution of

$$\frac{1}{\sigma^2} \int_0^l \alpha^2 \frac{\phi_\tau^2}{v_\tau} \, d\tau$$

to the cost of the path favours high velocities. If $\alpha = 0$, then the contribution of the steering angle process to the cost of the path is

$$\frac{1}{\sigma^2} \int_0^l \left( \frac{d\phi}{d\tau} \right)^2 v_\tau \, d\tau$$

which favours low velocities. On setting $\alpha = 0$ the tangential velocity of the estimated PML becomes a concave function of time as shown in Figure 14.

Unfortunately, on setting $\alpha = 0$ large and erratic estimates of $\phi$ are obtained from the data displayed in Figure 3. At present it is not clear how to resolve the conflict.

## 6   Conclusion

A new path based filter has been developed and applied to car tracking in sequences of video images. The experimental results are good: the steering angle of the car can be estimated after taking measurements for about 1 s, and the acceleration can be estimated after 2 s.

The path based approach gives a framework in which to compare human driver behaviour with the optimal behaviour predicted by the motion model in the filter. A good agreement is found for steering but there are significant discrepancies for acceleration. The resolution of these discrepancies may in future lead to changes in the system model.

**Acknowledgement:** this work is funded in part by the SECURE project, with partners DERA, Lucas Industries and Jaguar.

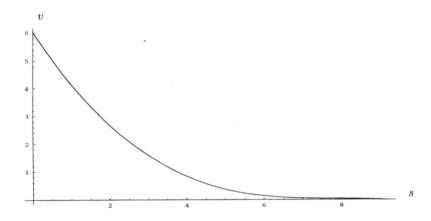

Figure 14. *Tangential velocity for* $\alpha = 0$

# References

1. Bar-Shalom, Y., Fortmann, T.E.: Tracking and Data Association. Mathematics in Science and Engineering Series **179**. Academic Press: San Diego, CA (1988).
2. Blake, A., Zisserman, A.: Visual Reconstruction. MIT Press (1987).
3. Bolza, O.: Lectures on the Calculus of Variations. New York: Dover Publications Inc (1961).
4. Heisler, H.: Advanced Vehicle Technology. Hodder and Stoughton (1989).
5. Hull, J.: An Introduction to Futures and Options Markets. Prentice Hall International Editions (2nd edition, 1995).
6. Kee, R.J., Irwin, G.W.: Investigation of trellis based filters for tracking. IEE Proc. Radar, Sonar and Navigation **141** (1994) 9–18.
7. Maybank, S.J., Worrall, A.D., Sullivan, G.D.: A filter for visual tracking based on a stochastic model for driver behaviour. In Buxton, B., Cipolla, R. (eds.) Computer Vision-ECCV'96, Lecture Notes in Computer Science **1065** 540–549. Springer-Verlag: Berlin, Heidelberg, New York (1996).
8. Maybank, S.J., Worrall, A.D., Sullivan, G.D.: Filter for car tracking based on acceleration and steering angle. In Fisher, R.B., Trucco, E. (eds.) British Machine Vision Conf. 1996, **2** (1996) 615–624.
9. Maybeck, P.S.: Stochastic Models, Estimation and Control - Volume 1. Mathematics in Science and Engineering Series **141-1**. San Diego, CA, USA: Academic Press (1979).
10. Maybeck, P.S.: Stochastic Models, Estimation and Control - Volume 3. Mathematics in Science and Engineering Series **141-3**. London, UK: Academic Press (1982).
11. Øksendal, B.: Stochastic Differential Equations: an introduction with applications. Springer-Verlag: Berlin, Heidelberg, New York (Third edition 1992).
12. Press, W.H. et al. Numerical Recipes. Cambridge University Press (1986).
13. Stroock, D.W.: Gaussian measures in traditional and not so traditional settings. Bulletin of the American Mathematical Society **33** (1996) 135–155.
14. Sullivan, G.D.: Model-based vision for traffic scenes using the ground plane constraint. In Terzopoulos, D., Brown, C.: (eds) Real-time Computer Vision. Cambridge University Press (1994).

15. Wolfram, S.: The Mathematica book. Cambridge University Press (Third edition 1996).
16. Worrall, A.D., Sullivan, G.D., Baker, K.D.: A simple, intuitive camera calibration tool for natural images. In Hancock, E.: (ed.) British Machine Vision Conf. 1994, **2** (1994) 781–790.

# Bottom-up Derivation of the Qualitatively Different Behaviors of a Car Across Varying Spatio-Temporal Scales: A Study in Abstraction of Goal-Directed Motion

Leo Dorst

RWCP Novel Functions: SNN Laboratory Amsterdam,
Dept. of Mathematics, Computer Science, Physics and Astronomy,
University of Amsterdam, Kruislaan 403, NL-1098 SJ Amsterdam, The Netherlands

**Abstract.** Driving a car involves considering it at different spatio-temporal scales, and somehow leads to behavior such as the parallel parking maneuver, the three-point turn, free Euclidean driving in a desert, following a road, and translationally passing other vehicles at high speed. In the study of autonomous systems, it is desirable to find a representation in which such different behaviors of a single system can be related to each other, and to find precisely how and under what conditions a change of representation and corresponding choice of motions occurs.

In this paper, we formulate an abstraction mechanism based on approximations of flows of commutators of vector fields. We apply it to the goal-directed motion of a car and show how the environmental constraints induce, through this abstraction mechanism, a recognizable hierarchy of descriptions of the car's motion.

## 1  Levels of Motion Description

Imagine that you are planning to visit some friends that live in $B$. You decide to go from $A$ (your town) to $B$ by car. At you kitchen table, you decide which route to take, using maps of rather coarse scale, choosing the quickest or most scenic route. You then go to the garage, maneuver the car onto the driveway and turn onto the road. You take your favorite route through town to the beginning of the freeway or scenic route, and drive along, passing other cars. When you arrive at $B$, some planning through that town is required to find your friends' house. You pass it, make a three point turn, and parallel-park in front of their house. You straighten out the wheels, thereby ending a typical car trip.

Now consider at what level of detail you think of your car's motions at each of these stretches. On the map, your car is a 'directed point', and as you mentally move it over the map it behaves like a particle pushed through a tube (the road on the map), taking turns as the tube leads it. When you are actually on the freeway, passing cars, you consider your car as a translational device: the steering moves you sideways relative to the car in front of you, the accelerator moves you forward. Your heading is still determined by the road. During the local trips through the towns $A$ and $B$, the turning of your car is essential, as it is for going around the corner through the maze of streets. In the street, during the slow maneuvers of getting into the parking spot, you are very aware of the fact that your car can also go backwards, and of the limitations of its turning radius (about 10 meter). The straightening out of the wheels, your final act, implies that at this stage you are also aware of the wheel-angle directly, rather than as a means to make turns.

It is intriguing that all these views of the car exist in the driver's mind, that they are consistent, and that the switch between views appears to correspond to switches in the behavior of car-and-driver (viewed as a single autonomous system) and the types of motions making up a maneuver. A full understanding of these issues is required if we are to generate autonomous behavior automatically.

In this paper, we attempt a full analysis of these transitions in representation for the car-and-driver system, within a single representational framework and with a single abstraction mechanism based on flows and commutators. In doing so, we are beginning to bridge the gap between continuous representations as used in control theory and more discrete representations used in AI when describing or planning motion. Our goal is to pinpoint precisely how the abstraction is done and how the relationship between the different levels is established.

## 2 Fields, Flows and Commutators

In control theory, there is a well developed formalism to describe motions and analyze the controllability of complex systems. We give a brief introduction, immediately applying the concepts to the car. For a more general and formal treatment, the reader is referred to introductory works such as [3],[4].

### 2.1 Parametrization

A car is parametrized, at the lowest level of description that we will consider, by 4 parameters, 3 related to the position and orientation of the vehicle, and 1 to the steering angle. As a reference point on the car we choose the middle of the front axle, and in our figures we will draw a 3-wheeled car with a wheel at the same point (which is obviously equivalent to a 4-wheeled car with steering), see Fig. 1a. We call the positional variables $x$ and $y$, the orientation $\phi$, and the angle of the steering wheel $\theta$. We can thus represent a car as a point in a 4-dimensional *configuration space* $\mathcal{C}$, with each point in $\mathcal{C}$ parametrized by $(x, y, \phi, \theta)$.

### 2.2 Vector Fields of Control

For the car, there are two controls: we may turn the steering wheel, or move forward or backward. We describe the motions that are a consequence of these controls by means of two vector fields in $\mathcal{C}$, which we call **Steer** and **Drive** (following the terminology in [3]). From straightforward geometry using Fig.1 we obtain:

$$\textbf{Steer} \equiv \mathbf{e}_\theta \tag{1}$$

$$\textbf{Drive} \equiv \cos(\theta + \phi)\mathbf{e}_x + \sin(\theta + \phi)\mathbf{e}_y + \frac{1}{\ell}\sin\theta\,\mathbf{e}_\phi \tag{2}$$

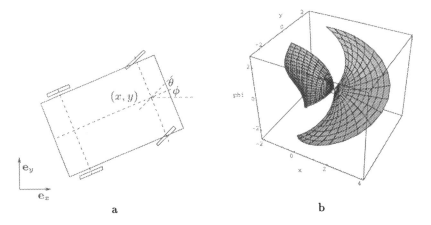

**Fig. 1.** *a) The parametrization of a car by $(x, y, \phi, \theta)$; b) A part of the flow of* **Drive** *in $(x, y, \phi)$-space.*

where $\ell$ is the length of the car, as measured between front axle and rear axle. This gives the basic motion of the car; all higher level abstractions should be consistent with it.

## 2.3 Basic Motions: the Flows

When the car moves under the influence of one of these vector fields, the point in $C$ representing the car moves along *flow lines* of those fields. These flow lines can be computed by integration; for the **Steer** field of the equation: $\frac{\partial}{\partial t}\theta(t) = 1$; and for the **Drive** field of the coupled equations:

$$
\begin{cases}
\frac{\partial}{\partial t}x(t) = \cos(\theta + \phi(t)) \\
\frac{\partial}{\partial t}y(t) = \sin(\theta + \phi(t)) \\
\frac{\partial}{\partial t}\phi(t) = \frac{\sin(\theta)}{\ell}
\end{cases} . \tag{3}
$$

Thus the flow for the field **Steer** acts on a point $(x_0, y_0, \phi_0, \theta_0)^T$ of $C$-space to produce a new point which is along a straight line in the $\theta$ direction:

$$
\Phi_{\textbf{Steer}}(t)\left(x_0, y_0, \phi_0, \theta_0\right)^T = \left(x_0, y_0, \phi_0, \theta_0 + t\right)^T, \tag{4}
$$

while the flow for **Drive** is:

$$
\Phi_{\textbf{Drive}}(t)\begin{pmatrix} x_0 \\ y_0 \\ \phi_0 \\ \theta_0 \end{pmatrix} = \begin{pmatrix} x_0 - \ell\sin(\theta_0 + \phi_0)/\sin\theta_0 + \ell\sin(\theta_0 + \phi_0 + \frac{t}{\ell}\sin\theta_0)/\sin\theta_0 \\ y_0 + \ell\cos(\theta_0 + \phi_0)/\sin\theta_0 - \ell\cos(\theta_0 + \phi_0 + \frac{t}{\ell}\sin\theta_0)/\sin\theta_0 \\ \phi_0 + \frac{t}{\ell}\sin\theta_0 \\ \theta_0 \end{pmatrix}
$$
$$\tag{5}$$

which is a family of helices in $(x, y, \phi)$-space, with $\theta_0$ indexing the family members. These helices are indicated in the plot of Fig.1b. The turning radius of the front wheel when set at an angle $\theta_0$ follows from eq.(5), and is: $R(\theta) = \ell/\sin\theta_0$.

Feasible paths of the car consist of a concatenation of basic motions of the point representing it in $\mathcal{C}$-space along the flow lines, and hence of pieces of straight $\theta$-lines and pieces of $(x, y, \phi)$ helices (each possibly infinitesimally long).

## 2.4 Controllability and Commutators

If the reference point of the car is permitted to move only very little, at each elementary move (steer a little, and/or drive a little) the representation of the car in $\mathcal{C}$-space will only reach a small part of that space, namely the set of points just along the flow lines in the **Steer** and **Drive** flows. If the temporal scale is long enough, there is time to concatenate these elementary motions. It is not obvious that by doing so one could reach arbitrary orientations and positions (within the prescribed space) – however, that turns out to be the case. The tools to demonstrate this are those of *Lie-algebra*, which describes the concatenation of infinitesimal motions.

The most direct way of showing that the two fields **Drive** and **Steer** could fill the entire 4-dimensional $\mathcal{C}$-space with their flows is by considering the vector fields themselves. Each vector field can be interpreted as a derivative acting on scalar functions, with as its value at each point: the value of the derivative of that function in the direction of the local vector, see e.g. [3]. Thus if the components of the vector field $A$ in a coordinate system $\{x_i\}$ are $A_i$, then the action of $A$ on a scalar function $f$, denoted $Af$, is: $Af = \sum_i A_i \frac{\partial}{\partial x_i} f$. These derivatives can be concatenated, and if this is done as a *Lie-product* (which is a commutator of two derivatives), then the result is again a derivative, and hence a vector field. For two vector fields $A$ and $B$ with components $A_i$ and $B_i$ on the coordinate basis $\{x_i\}$, the Lie product has the components: $[A, B]_j = \sum_i (A_i \frac{\partial}{\partial x_i} B_j - B_i \frac{\partial}{\partial x_i} A_j)$. These new vector fields are in general linearly independent of the old; and a system is *controllable* if the set of all vector fields followed by the Lie products *form a basis for the space*, since one can then use combinations of the fields to move in arbitrary directions in the configuration space. In fact, the elementary interpretation of the flow $\Phi_{[A,B]}$ of such a new field $[A, B]$, constructed as the Lie product of two vector fields $A$ and $B$, is found by expressing it in the flows of $A$ and $B$, as follows (see [5], [3]):

$$\Phi_{[A,B]}(t)p = \lim_{n \to \infty} \left( \Phi_B(\sqrt{t/n}) \cdot \Phi_A(\sqrt{t/n}) \cdot \Phi_B(-\sqrt{t/n}) \cdot \Phi_A(-\sqrt{t/n}) \right)^n p. \quad (6)$$

In words, eq.(6) means that one can move in the field $[A, B]$ by concatenating infinitesimal moves in $A$ and $B$ in the order: first against (the flow of) $A$, then against $B$, then with (the flow of) $A$, then with $B$.

## 2.5 Car Control

Let us verify the controllability of a car. By computing the Lie products of the fields **Steer** and **Drive** using eq.(2.4), recursively, one finds new fields which we dub **Wriggle** and **Slide** (following [3]):

$$[\textbf{Steer}, \textbf{Drive}] = -\sin(\theta + \phi)\, \mathbf{e}_x + \cos(\theta + \phi)\, \mathbf{e}_y + \frac{\cos\theta}{\ell}\, \mathbf{e}_\phi \equiv \textbf{Wriggle}, \quad (7)$$

$$[\textbf{Steer}, \textbf{Wriggle}] = -\cos(\theta + \phi)\mathbf{e}_x - \sin(\theta + \phi)\mathbf{e}_y + \frac{\sin\theta}{\ell}\,\mathbf{e}_\phi = -\textbf{Drive}, \qquad (8)$$

$$[\textbf{Wriggle}, \textbf{Drive}] = \frac{1}{\ell}(-\sin\phi\,\mathbf{e}_x + \cos\phi\,\mathbf{e}_y) \equiv \frac{1}{\ell}\,\textbf{Slide}, \qquad (9)$$

(Note that **Slide** is a translational motion vector perpendicular to the driving direction, without any rotational component, hence its name. The name **Wriggle** is slightly unfortunate since it describes the motion rather than its result, which is mostly a change of orientation; we maintain it mainly to be consistent with [3].) Since computation shows that: $[\textbf{Slide}, \textbf{Steer}] = [\textbf{Slide}, \textbf{Drive}] = [\textbf{Slide}, \textbf{Wriggle}] = 0$, there are no other fields that can be constructed by taking Lie products. So we obtain a *Lie module* of actions, as 4 independent vector fields, which thus span the 4-dimensional $(x, y, \phi, \theta)$-configuration space.

The flows of the new fields **Wriggle** and **Slide** are found by integration as:

$$\Phi_{\textbf{Wriggle}}(t)\begin{pmatrix} x_0 \\ y_0 \\ \phi_0 \\ \theta_0 \end{pmatrix} = \begin{pmatrix} x_0 - \ell\cos(\theta_0 + \phi_0)/\cos\theta_0 + \ell\cos(\theta_0 + \phi_0 + \frac{t}{\ell}\cos\theta_0)/\cos\theta_0 \\ y_0 - \ell\sin(\theta_0 + \phi_0)/\cos\theta_0 + \ell\sin(\theta_0 + \phi_0 + \frac{t}{\ell}\cos\theta_0)/\cos\theta_0 \\ \phi_0 + \frac{t}{\ell}\cos\theta_0 \\ \theta_0 \end{pmatrix}$$

$$(10)$$

(a family of helices with radius $\ell/\cos(\theta_0)$), and $\Phi_{\textbf{Slide}}(t)\left(x_0, y_0, \phi_0, \theta_0\right)^T = \left(x_0 - t\sin\phi_0, y_0 + t\cos\phi_0, \phi_0, \theta_0\right)^T$ (a straight line perpendicular to the car).

An intuitive demonstration of eq.(6) may be given by sketching a finite approximation for the Lie products of **Steer** and **Drive**, for large $n$ (say $n \approx 10$). Fig.2a indicates the flow of **Wriggle** = [**Steer**, **Drive**]. For fixed $\theta$, it is a turning motion along a circle with radius $\ell/\cos(\theta)$. Note that the motion along the circle is an order of magnitude smaller than the distance the car travels, in agreement with eq.(6). Fig.2b sketches the flow of **Slide** = [**Wriggle**, **Drive**]. Since it is a commutator of commutators, it consists of motions at two levels of magnitude: the internal **Wriggle** motions need to be implemented by iterations of flows as in Fig.2a. The resulting flow of **Slide** consists of a single line perpendicular to the car, independent of the value of $\theta$. It will be clear from these examples that the basic principles are sound, but that implementation of motions by taking eq.(6) literally is very time consuming and far from optimal.

But although actual motions in the commutator fields are impractical, the availability of those fields simplifies the description of arbitrary motions. This description has some continuous and discrete aspects: *discrete*, or *symbolic*, in the name of the motion primitive giving the *kind* of flow that is followed, and *continuous* in the *amount* that this flow is followed.

## 3 Motions at Different Spatio-Temporal Scales

### 3.1 The Constrained Flows of Goal-Directed Motion

Not all flows generated by the fields **Drive** and **Steer** and their infinitesimal concatenation are legitimate behaviors of the car: there are *physical constraints* to motion which exclude many flows. These constraints include limits on values

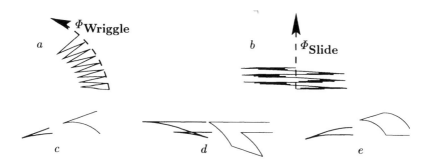

**Fig. 2.** *Flows of commutators, see text.*

of variables, such as the maximum steering angle of the actual car, or obstacles (which interact with the shape of the car to make certain values of $(x, y, \phi)$ unattainable, see e.g. [1]). But there are also constraints which relate to the dynamic behavior of a car, such as a maximum acceleration and a maximum permissible centripetal acceleration. In this way, the car's own motion, its construction, and the environment limit the flow in $\mathcal{C}$-space. Such limited flows can still be described exactly, using the full formalism of infinitesimal moves and commutators of section 2.5, but it is often more insightful to approximate them by much simpler flows, of much simpler fields.

This approximation of flows must be related to some measure of similarity between flows, and this should be based on a notion of 'reasonable' motion of the car, which is a distinguishing characteristic between goal-directed motions and random motions. Mathematically, it is most analyzable when we demand strict *optimality* of motion, expressed as minimum length paths in an appropriately chosen metric. Since strictly optimal motions do not behave smoothly under scaling, we are in this paper already satisfied with close-to-optimal motions, at the level of spatio-temporal resolution considered.

### 3.2 Optimality Induces Macroscopic Manifestation of Commutators

Motion in the pure flows of the fields is not the only action we can perform on a real car; it is possible to drive and steer simultaneously, or at the very least follow in an interleaved manner very short portions of each of their flows. Because this is possible, we may approximate the flow of the commutator in eq.(6) by very few terms, since we will be able to perform corrections while driving to arrive at, for instance, the exact parallel displacement that the **Slide** flow demands. Indeed, we conjecture that very reasonable approximate results for small $t$ are obtained by taking $n = 1$. We have not been able to find (a source for) a proof of this conjecture; though the derivation in [5] based on exponential operators might be modified to yield this proof.

We are led to consider such simplified flows by the *optimality principle*, which asks that goal-directed paths be reasonably short. Indeed, eq.(6) demands that the total 'distance' of moves in the flows of $A$ and $B$ to construct the flow of $[A, B]$ is proportional to $4\sqrt{t/n} \cdot n = 4\sqrt{tn}$. Thus the 'distance' along an infinitesimal move into the flow of a commutator field, performed in this way, is infinite. Obviously, to reach the same distance in the flow field with a shorter motion, $n$ *should be as small as possible*. Thus we desire to execute simplified moves in the contributing flows (where, as we said, we hope to correct possibly inaccuracies by the freedom that simultaneous control offers). However, even when taking $n = 1$, we find that the distance to effectively travel a distance of $t$ in the desired flow field is proportional to $4\sqrt{t}$. For small $t$, this is an unfortunate ratio. So apart from simplifying the individual moves (by taking $n = 1$), we can come closer to optimal paths by *chosing t as large as possible*. We assume that the proof of our earlier conjecture will put convergence limitations on the arguments in eq.(6), and that these demand that we cannot go much beyond comparable flow distances in each of the constituent flows. But for now, it is enough to note that the square root in combination with the optimality criterion forces us to want to travel as long as possible in each of the flows. Thus *the paths will tend to fill the available room for maneuvering*: positional fields such as **Drive** will go to the limits of the obstacle free space, and the field **Steer** to the limits of the steering angle.

In terms of finite flows with $n = 1$ we can write eq.(6) in shorthand (for convenience in notation ignoring the arguments, which are of similar magnitude by the reasoning above) as:

$$\Phi_{[A,B]} \approx \Phi_B \Phi_A \Phi_{\overline{B}} \Phi_{\overline{A}} \qquad (11)$$

(the bar denoting a negative move in the flow).

Fig.2c sketches the finite **Wriggle** motion $\Phi_{\textbf{Wriggle}} = \Phi_{[S,D]} \approx \Phi_D \Phi_S \Phi_{\overline{D}} \Phi_{\overline{S}}$ (with $S = $ **Steer** and $D = $ **Drive**), moving far in the flows concerned. In the figure, we indicate the motion both for the front wheel (thin curves) and for the middle of the rear axle (thick curves). The latter sketch is somewhat more easy to interpret since $\phi$ is everywhere equal to the tangent direction of the track. Fig.2d sketches the **Slide** = [[**Steer**, **Drive**], **Drive**] motion, using: $\Phi_{[[S,D],D]} \approx \Phi_D \Phi_{[S,D]} \Phi_{\overline{D}} \Phi_{\overline{[S,D]}} \approx \Phi_D \Phi_D \Phi_S \Phi_{\overline{D}} \Phi_{\overline{S}} \Phi_{\overline{D}} \Phi_S \Phi_D \Phi_{\overline{S}} \Phi_{\overline{D}}$, (where we used $\Phi_{\overline{[A,B]}} = \Phi_{[B,A]}$). Note that we do not quite recognize this motion as a reasonable maneuver for the car; however, since the Lie product is anti-symmetric, we may write the same field in several ways. Some of these lead to a cancellation of flows, and hence to a more efficient move: when the commutator is written in the equivalent form [**Drive**, [**Drive**, **Steer**]] we obtain: $\Phi_{[D,[D,S]]} \approx \Phi_{[D,S]} \Phi_D \Phi_{\overline{[D,S]}} \Phi_{\overline{D}} \approx \Phi_S \Phi_D \Phi_{\overline{S}} \Phi_{\overline{D}} \Phi_D \Phi_D \Phi_S \Phi_{\overline{D}} \Phi_{\overline{S}} \Phi_{\overline{D}} = \Phi_S \Phi_D \Phi_{\overline{S}} \Phi_D \Phi_S \Phi_{\overline{D}} \Phi_{\overline{S}} \Phi_{\overline{D}}$ This shorter description (because of the cancellation of terms) is sketched in Fig.2e and readily recognizable as an approximation of the sequence of actions that humans take to move a parked car closer to the curb.

We thus have the interesting property that the Lie module of the vector fields, which was purely infinitesimal, determines the behavior of a hemmed-in vehicle to a fairly large macroscopic scale. And not only have we given a macroscopic meaning to the controllability of the car, but we have done so in a highly structured way which 'lumps' basic motions in certain patterns of behavior, which we can sensibly give names such as **Slide**. Now that these motions are actually executable in finite time, we can use them to give a description of a motion at a higher level, *which can be translated into executable basic motions*. Thus we might describe the parking of a car as: **Drive** through the street till you find a parking spot, then **Slide** into the spot till you hit the curb. This is a good and understandable description of the motion at the street level; which 'opens up' at the level of resolution of the parking spot into the constituent commutators.

### 3.3 The Proposed Abstraction Mechanism

Based on these observations, we can formulate the steps in our proposed abstraction mechanism for goal-directed motion:

1. **Vector fields of control**: Form the vector fields at level $i$.
2. **Flows**: Compute their flows.
3. **Constraints**: Consider the constraints (including the optimality criterion) that apply to level $(i+1)$, and use those to approximate the close-to-optimal flows. If useful, apply a coordinate transformation to simplify the flows.
4. **New fields**: Differentiate those flows to obtain vector fields for level $(i+1)$.
5. **Commutators**: Form the Lie products for the level $(i+1)$ fields to derive the full set of fields required for total controllability at that level.
6. **Descriptive terms**: Use those fields from the Lie module at level $(i+1)$ to describe motions at that level, giving meaningful names to their function.
7. **Planning**: Plan a motion on level $(i+1)$ using the new fields and their flows, and the new terms for the description of the resulting motions.
8. **Specification**: When executing the motion, use the expression eq.(6) to implement a higher level commutator field using the lower level fields that composed it, recursively. Use this equation approximately: $\Phi_{[A,B]} \approx \Phi_B \Phi_A \Phi_{\overline{B}} \Phi_{\overline{A}}$, using the freedom of control at level $i$ to correct deviations.

In the remainder of this chapter, we apply these principles to the car, under various constraints in space, time and speed.

### 3.4 Motions in a Tight Spot

Let us consider the case of a car moving in a tight spot. This *spatial constraint* will force the flows of the optimal paths to assume a particular form.

For this case, we have already shown in section 3.2 how to apply the abstraction mechanism to the underlying infinitesimal level to obtain the relevant maneuvers. The result is that we may turn a car by repeatedly performing a **Wriggle**, implemented as the sequence: $\Phi_D \Phi_S \Phi_{\overline{D}} \Phi_{\overline{S}}$ and move it perpendicular to itself by a **Slide**, which is a complicated motion derived from a commutator of commutators, resulting in: $\Phi_S \Phi_D \Phi_{\overline{S}} \Phi_D \Phi_S \Phi_{\overline{D}} \Phi_{\overline{S}} \Phi_{\overline{D}}$ These terms **Wriggle** and **Slide** are useful qualitative descriptive terms of arbitrary motions; and the corresponding flows $\Phi_{\textbf{Wriggle}}(t)$ and $\Phi_{\textbf{Slide}}(t)$ can be used to quantify the amounts required. Thus we have reached a proper abstraction, although still very similar to the infinitesimal description on which it was based.

## 3.5 Beyond the Turning Radius: the Euclidean Regime

It can be shown that the number of commutator type maneuvers to achieve an arbitrary orientation in a circle with radius $r$ is approximately $\pi R/r$. Therefore, at a spatial scale of about $\pi R$ the approximately optimal flows change: they now reach the desired orientation simple by turning while driving to the perimeter.

We can see this effect in the equations of the flows. Eq.(10) shows that the additive part of the spatial component of the flow of **Wriggle** is of order $\ell/\cos(\theta_0)$, and when this becomes negligible relative to the spatial changes, we may write instead of eq.(10):

$$\Phi_{\textbf{Wriggle}}(t)\left(x_0, y_0, \phi_0, \theta_0\right)^T \approx \left(x_0, y_0, \phi_0 + \tfrac{t}{\ell}\cos\theta_0, \theta_0\right)^T \qquad (12)$$

This is the flow of a vector field which we dub **Turn**, defined by differentiation as: $\textbf{Turn} \equiv \frac{\ell}{\cos\theta_0}\, \mathbf{e}_\phi$. At the longer distances, the required optimality in distance or time will lead the optimal flow connecting two states to consist of straight lines in the direction of their connection:

$$\Phi_{\textbf{Drive}}(t)\left(x_0, y_0, \phi_0, \theta_0\right)^T \approx \left(x_0 + t\cos\phi, y_0 + t\sin\phi, \phi_0, \theta_0\right)^T \qquad (13)$$

This is the flow of a field we dub **Straight**, defined by taking its derivative: $\textbf{Straight} = \cos\phi\, \mathbf{e}_x + \sin\phi\, \mathbf{e}_y$. Note that it is simply implemented as the **Drive** field with $\theta = 0$, hence its name.

Those basic fields **Turn** and **Straight** generate a Lie-module, since their commutator equals: $[\textbf{Turn}, \textbf{Straight}] = \frac{\ell}{\cos\theta}(-\sin\phi\, \mathbf{e}_x + \cos\phi\, \mathbf{e}_y) = \frac{\ell}{\cos\theta}\,\textbf{Slide}$ (which generates motions perpendicular to the car). The steering of the car, governed by the variable $\theta$, has now uncoupled from the other variables, since using it in commutators does not produce new fields: $[\textbf{Steer}, \textbf{Turn}] = -\frac{\ell\sin\theta}{\cos^2\theta}\, \mathbf{e}_\theta = -\ell\tan\theta\,\textbf{Turn}$, and all other commutators are zero. Thus the two fields **Turn** and **Straight** are enough to generate all motions. The variable $\theta$ has become a parameter of the system, and we will usually take it equal to $\theta_{max}$ since this leads to optimal motions (at least in the sense of minimum distance traveled).

In practice, one prefers to incorporate some of the **Wriggle** motion leading to the **Turn** into the large motion in the flow of **Drive**, which implies just going forward along a circle of turning radius $R$ till the correct orientation has been achieved but not reversing after that, so $\Phi_{\textbf{Straight}}\Phi_{\overline{D}}\Phi_{\overline{S}}\Phi_D\Phi_S \approx \Phi_{\textbf{Straight}}(\Phi_{\overline{S}}\Phi_D\Phi_S)$.

Humans would describe motions at this scale mostly as 'driving straight there' (in any language), and would see the final re-orientation as trivial ('and then make it face west'), and might even forget about specifying the initial aiming of the car in the right direction as a separate maneuver. We would interpret this as the flows of **Straight** and **Turn**, and the fact that we follow the former a lot further at this spatial scale as an explanation of its perceived higher importance. The **Turn** is really of a different spatial scale, though macroscopically important for the orientation.

## 3.6 Following a Road

When the car follows a road, its orientation is to a large extent determined by the local orientation of the road. In a way, the road enforces most of the steering. When passing other vehicles, or obstacles, the driver obviously still has some influence. Let us see how this affects the description of motion.

We take a local coordinate system $(x', y', \phi')$ at $(x, y)(s)$ ($s$ is the parameter along the road) in the direction of the road $\phi(s)$, and denote the position of the car by the *center of the rear axle*. We make two approximations: first, we only permit rather small deviations of the orientation of the car relative to $\phi(s)$ (which is $\phi'_0 = 0$ in the new coordinates). Secondly, we take the approximation for high speeds, which means: small $\theta_0$.

$$
\Phi_{\mathbf{Drive}}(t) \begin{pmatrix} 0 \\ 0 \\ 0 \\ \theta_0 \end{pmatrix}' \overset{small\ \Delta\phi}{\approx} \begin{pmatrix} t \\ \frac{1}{2}(t)(\frac{t}{\ell}\tan\theta_0) \\ \frac{t}{\ell}\tan\theta_0 \\ \theta_0 \end{pmatrix}' \overset{high\ speed}{\approx} \begin{pmatrix} t \\ 0 \\ 0 \\ \theta_0 \end{pmatrix}' \tag{14}
$$

where the final transition is made in a metric to which the degrees of freedom $(x, y, \phi)$ contribute similarly. Thus the **Drive** flow with non-zero but small $\theta_0$ is in this approximation indistinguishable from that with $\theta_0 = 0$. It is the flow of a field **Straight**, defined through differentiation of eq.(14) with $\theta_0 = 0$ as: **Straight** $= \mathbf{e}_{x'}$ (in car-centered coordinates). This is just straight driving.

The maneuver **Wriggle** becomes, in the same approximations for small angles and high speeds, $\Phi_{\mathbf{Wriggle}}(t)\left(0, 0, 0, \theta_0\right)^{'T} \approx \left(0, 0, \frac{t}{\ell}\cos\theta_0, \theta_0\right)^{'T}$ Thus the **Wriggle** consists mostly of a rotational part in $\phi$, parametrized by the steering angle $\theta_0$. It is the flow of a field **Turn** defined (by differentiating) as: **Turn** $= \frac{\cos\theta_0}{\ell}\,\mathbf{e}_{\phi'}$. Some remarks on the interpretation of this derivation:

- The reason for the coordinate transformation to the middle of the rear axle is given by the desire to make the flows, in particular $\Phi_{\mathbf{Wriggle}}$, as simple as possible. Apparently, this kind of trivial re-representation is a useful part of the abstraction mechanism, so we included it under point 3 of section 3.3.
- Neither of the fields **Straight** and **Turn** appear to affect the $y'$-coordinate, but this does not imply that the car does not move sideways in the corresponding commutator [**Straight**, **Turn**]. After a **Turn**, the car has changed orientation, so a subsequent **Straight** does change the $y'$ *relative to the original coordinates*. Proper computation of the vector fields now needs to take into account the covariant derivative, describing the differential coordinate transformation (this is standard differential geometry, see e.g. [2]).
- It may seem strange that we invoke **Wriggle**, which involves driving in the reverse direction. However, when we execute it *simultaneously* with the fast **Drive**, we can show by a proper change of coordinates to a frame moving with the car we are passing that moving positively in the flow of **Drive** involves moving with a slightly higher speed than that car, and hence it is implemented by applying the accelerator; moving against the flow is done by deceleration; and the commutator **Slide** becomes a pure lateral displacement relative to the car that is being passed.

Humans driving along a freeway do not refer to the absolute angle of the car into their description, and so appear to make the same coordinate adjustment by $(x, y, \phi)(s)$. In a sense, because *the road steers the car*, there is no need for explicit representation of $\phi$ since this parameter is imposed, not governed by one's own actions. Within that representation, the motion **Straight** is of course recognizable as just driving along (with an orientation determined by the road). The part $\Phi_{\overline{S}}\Phi_D\Phi_S$ of the **Wriggle** maneuver, with small rotations and small displacement in a short time, is recognizable as the elementary way of changing the lateral position on the road, used for instance in passing cars, by a little jerk to the steering wheel (maybe **Jerk** is a better name than **Turn**). The maneuver **Slide**, which consists of two quick and opposite jerks to the steering wheel, is similar to the correction one makes to center oneself in the lane.

# 4 The General Principles of Abstraction of Motion

We have seen that the different levels of representation of a car which were sketched in the introduction can indeed be related to the lowest level representation of the kinematics of the car mechanism. The various levels appear in an interaction between the constraints imposed by the environment and the flows of these basic motions. The advantage of using the car as an example is that the levels which we find are recognizable; but the disadvantage is that we may have used extraneous knowledge to create those levels, so that it is not surprising that they turn out so familiar. In this section, we attempt to indicate explicitly the principles that govern this abstraction process, in order to understand how objective it is, and to what extent we might expect it to be automated.

- **Grounding:**
  A basic level of representation must be chosen. For the study of motion we have chosen the kinematic level. This may not be the best choice when dynamic effects are important. If it is true that levels of motion description can be formed objectively by consistent application of abstraction principles, then the issue of grounding is less important; but for the car we should at least show that the kinematics level is indeed one of those levels.
- **Moving across abstraction levels:**
  We have moved up in levels of description using the abstraction mechanism indicated in section 3.3, based on the approximation of optimal flows under constraints, with the commutator playing the important role of generating new maneuvers from basic motions.
- **Formation of symbols:**
  This abstraction mechanism leads to symbols for the description of motion at the different levels, which are grounded in the lower levels through the operation of commutation. Note the role the constraints play in their formation (in step 3 in section 3.3): they govern the simplification which leads to the new fields. Thus *symbolic motion description is induced by environmental constraints*, and that seems very reasonable.

- **Optimality:**

  Not *all* flows are considered in the abstraction of level $i$, but only those corresponding to approximately optimal paths, using the optimality at that level. At level $(i + 1)$, a different optimality criterion will apply. The relationship between these criteria should be investigated.

- **Automatic abstraction:**

  In doing the abstraction for the car we have worked towards representations that we knew, *explaining* them rather than *deriving* them from first principles. We would like to understand better whether there is an inevitability to the levels found. Since they do depend on the environmental constraints, this would seem hard to generalize for arbitrary devices. Yet there is a sense of reasonableness to the approximations we made, as if the basic motions and their flows dictate how to classify the environment at different scales.

- **Proofs:**

  The essential part of our reasoning (the approximation of flows of commutators by a much smaller number of terms than the exact expression eq.(6) demands, with smaller motions in each of them) lacks a proof. We need to know the conditions under which this step is permitted.

- **Basis of Lie module:**

  One the reviewers paper remarked that our derivation of the motion abstractions depends on the original basis we choose to express the Lie module: in this case **Steer** and **Drive**; another basis might give a different abstraction. This is a good point; and for now we can only say that the basis chosen seems 'natural' since these are the actual controls of the car...

In conclusion, although we do not claim to have uncovered *the* universal abstraction mechanism, we nevertheless hope to have convinced the reader that the mechanism described is an important principle in the automated formation of well-founded levels of description of goal-directed motion.

# References

1. L. Dorst, I. Mandhyan, K.I. Trovato: The Geometrical Representation of Path Planning Problems, Robotics and Aut. Syst., Elsevier, vol.7, 1991, pp.181-195.
2. B. O'Neill: Elementary Differential Geometry, Academic Press, 1966.
3. E. Nelson: Tensor Analysis, Princeton U. & U. of Tokyo Press, 1967.
4. E. Nelson: Topics in Dynamics, I: Flows, Princeton U. & U. of Tokyo Press, 1969.
5. H.F. Trotter: On the product of semi-groups of operators, Proc. AMS 10(1959), pp.545-551.

# Neural Network Approaches for Perception and Action

Helge Ritter

Technical Faculty, Bielefeld University, Bielefeld D-33501, Germany

**Abstract.** We argue that endowing machines with perception and action raises many problems for which solutions derived from first principles are unfeasable or too costly and that artificial neural networks offer a worthwhile and natural approach towards a solution. As a result, learning methods for creating mappings or dynamical systems that implement a desired functionality replace the implementation of algorithms by programming. We discuss two neural network approaches, the Local Linear Maps (LLMs) and the Parametrized Self-Organizing Maps (PSOMs), that are well suited for the rapid construction of mapping modules and illustrate some of their possibilies with examples from robot vision and manipulator control. We also address some more general issues, such as the need for mechanisms of attention control and for the flexible association of continuous degrees of freedom, together with an ability for handling varying constraints during the selection of actions.

## 1 Introduction

Neural networks were evolved by nature to enable perception and action. Therefore, artificial neural networks seem as an appropriate approach for building robots that can perceive and that can meaningfully coordinate their actions according to their sensory perception.

Before the advent of artificial neural networks, traditional robotics research had focused on the use of sophisticated, explicit world models, which have to be be built from sensory data and on which then algorithms operate to derive the robot's actions [1]. While this is a clear-cut and very powerful approach whenever it can be carried through, it may be simply too ambitious for many tasks a robot has to carry out. In the domain of robot vision, the emphasis of explicit models has attracted a lot of research with the goal of scene reconstruction. While a 3d representation of its environment undoubtedly is very useful for a robot, many useful behaviors can be carried out with much less information that may be much easier to extract from images than first going through a full geometric representation. This has been demonstrated impressively by the approach of "behavior based robotics", which in its more radical forms even denies the necessity any explicit world models and instead attempts to engineer the behavior of a robot from a suitably interacting set of simple sensory reflexes or reactive behaviors, possibly augmented by some simple memory scheme (for a collection of representative papers, see, e.g. [15]). In addition, by viewing vision as intrinsically embedded in a perception-action cycle, many vision tasks become significantly changed and often easier, due to the availability of more data and the possibility to actively control them.

If we abandon the traditional, algorithm-oriented approach and instead use neural networks as our main tool, several new issues arise. First, there is a shift from constructing algorithms to the construction of *mappings* and of *dynamical systems* that solve a particular task or that contribute to a desired behavior.

At the lowest level, we need mappings that can form representations from the multitude of sensory signals that must be processed by any perceiving robot. In biological brains, the bulk of these operations are carried out in the primary sensory cortices. These are organized topographically, forming dimension-reduced representations of the sensory signals in the form of activity patterns on two-dimensional arrays of feature selective neurons. The spatial structure of these topographically organized "feature maps" can be modeled [20,3]. to a surprising extent with Kohonen's Self-Organizing Map ("SOM") algorithm ([12,25]; for brief surveys see, e.g., [11,28]) Using this approach, it has become possible not only to model many intriguing properties of topographic brain maps, but also to create artificial feature maps for a variety of computational tasks in pattern recognition, data analysis and visualization, process control, data mining and robotics ( [12] gives comprehensive references).

Originally, SOMs were developed to model some initial steps in the organization of the perceptual apparatus in the brain, namely the adaptive creation of dimension-reduced mappings of various sensory spaces. However, it is equally possible to include in the mapped stimuli also features that are related to action, such as parameters of motor commands. This leads to "sensory-motor-maps" that link perception with action [23]. Such maps are known to exist at least in some places in the brain, the best-known example being the collicular map where maps of retinal and of auditory space are in register with a topographic map of motor signals required to perform a saccade to a stimulus that elicited activity in one of the sensory maps.

Very similar maps can be created with the SOM algorithm [25] and work along these lines has shown that sensori-motor SOMs can, e.g., be used to learn hand-eye coordination for robots [24].

Unlike traditional algorithms, most knowledge that is represented in neural networks is encoded in a non-symbolic form, distributed over a large number of "synaptic weights". As a consequence, both our burden and our freedom to specify a desired behavior in terms of a symbolic program is greatly diminished. Programming largely becomes replaced by *learning algorithms* for constructing the necessary mappings or the desired dynamical behavior.

On the one side, this is an appealing feature, since it allows us to create systems that require only a very modest amount of explicit knowledge. Most information can be provided in *implicit* form, by providing a sufficient number of training examples. On the other hand, we then have to face a very important new issue, namely the task to develop methods for the *rapid construction of neural modules* from limited sets of examples. This is particularly important in robotics, since robot actions are rather costly and, say, a few ten thousand training examples are in most cases a limit beyond which robot learning usually becomes unattractive.

Below, we will present two approaches that have both been motivated by the SOM and that adress this issue. The first approach, the *Local Linear Map* (LLM-) networks, improves the learning capabilities of the basic SOM for smooth mappings by associating with each neuron a locally valid, linear mapping [27]. In differential geometric terms, this can be viewed as constructing for a manifold that is approximated by a discrete SOM a cross section of its tangent bundle.

The second approach, the *Parametrized Self-Organizing Map* [26,32], goes one step further and replaces the discrete representation of the SOM by a differentiable manifold that is constructed with the help of a set of basis functions. A significant amount of flexibility of this approach comes from a generalization of the discrete SOM bestmatch search into the continuous domain, yielding the functionality of a *continuous associative memory*.

Both approaches provide us with convenient functional building blocks for the construction of larger systems. However, before we can use them to build more comprehensive systems, we must consider as an important third issue the question of a suitable *architecture* that can coordinate the cooperation of multiple modules.

One important principle of coordination apparently is *focal attention*. The ability of focal attention is a characteristic of most living cognitive systems, and its implementation in artificial robots is an essential means for coping with the high demands of processing real world multimodal sensory data in real time. Besides the economic aspect of concentrating and thereby making more efficient use of processing resources, a further important aspect of of focal attention is the ability *to ignore the unimportant*. In the overwhelming majority of tasks, the meaningful coordination of action usually requires the coordination of a comparably small number – if compared to the dimensionality of the sensory input – of degrees of freedom. It is the task of our perception system to extract these few relevant degrees of freedom from the almost infinite dimensional visual, auditory, tactile and proprioceptive sensory inputs and to maintain a stable binding to these relevant degrees of freedom during the execution of a task. This, however, requires also to reliably ignore the influence of the majority of sensory inputs. Below, we will describe one particular approach how visual focal attention can be learnt within a hierarchical system of neural networks by an approach that may be described as "neural zooming". The architecture behind this approach is, however, purely computationally motivated and has not been developed with any deeper cognitive modeling in mind.

# 2  Local Linear Maps

While the discrete nature of a SOM poses no difficulty for tasks such as classification or data analysis and visualization, it becomes a problem when we want to use SOMs to represent sensory-motor mappings. Here, we usually require smoothness, and, in addition, the underlying spaces are frequently higher than just two-dimensional. While we can increase the degree of smoothness by using more nodes, and it is in principle also straightforward to extend the SOM algorithm to higher-dimensional situations, both measures together very rapidly exhaust any reasonable amount of computational resources (providing only 10 discretization steps for a mapping with 6 DOFs would require already the substantial number of $10^6$ nodes!). A simple though effective way to deal with this problem is to attach to each SOM node $\mathbf{r}$ a locally valid linear mapping that is used to compute a correction to the node's output reference vector $\mathbf{w}_{\mathbf{r}}^{(out)}$ that is linear in the deviation of the input $\mathbf{x}$ from the prototype vector $\mathbf{w}_{\mathbf{r}}^{(in)}$ of the best-matching node $\mathbf{s}$. We then obtain for the output $\mathbf{y}_{\mathbf{r}}$ computed by a node $\mathbf{r}$ the equation

$$\mathbf{y}_{\mathbf{r}} = \mathbf{w}_{\mathbf{r}}^{(out)} + \mathbf{A}_r(\mathbf{x} - \mathbf{w}_{\mathbf{r}}^{(in)}). \tag{1}$$

where $\mathbf{A}$ is a $M \times L$ matrix and $L$ and $M$ denote the dimensionalities of the input and of the output space, resp.

Mathematically, the mapping $\mathbf{r} \mapsto \mathbf{A}_{\mathbf{r}}$ can be viewed as a "cross section" through a "fiber bundle" with its base manifold given by the manifold that is represented (in discretized form) by the set of reference vectors $\mathbf{w}_{\mathbf{r}}^{(in)}$ of the SOM. In particular, if $M = L$, this fiber bundle can be identified with the tangent bundle of the SOM manifold. For $M \neq L$, we can represent general fiber bundles, for instance, the fiber bundle of inertia tensor fields on the configuration manifold of a manipulator (note that the "affine" term $\mathbf{w}_{\mathbf{r}}^{(out)}$ in (1) could be easily absorbed in $\mathbf{A}_{\mathbf{r}}$ by augmenting all input reference vectors $\mathbf{w}^{(in)}$ with an additional, constant component).

To complete our definition, we still have to specify how to use the node outputs $\mathbf{y}_{\mathbf{r}}$. In the simplest case, we just minimally extend the best-match step of the original SOM algorithm and define the output $\mathbf{y}(\mathbf{x})$ of the LLM-network as the response $\mathbf{y}_{\mathbf{s}}(\mathbf{x})$ of the best-match unit $\mathbf{s}$ that satisfies $d_{\mathbf{s}}(\mathbf{x}) = \min_{\mathbf{r}} d_{\mathbf{r}}(\mathbf{x})$, where $d_{\mathbf{r}}(\mathbf{x})$ denotes some distance function, such as $d_{\mathbf{r}}(\mathbf{x}) = \|\mathbf{x} - \mathbf{w}_{\mathbf{r}}^{(in)}\|$ ("winner-take-all"-network). However, this introduces discontinuities at the borders of the Voronoi tesselation cells defined by the $\mathbf{w}_{\mathbf{r}}^{(in)}$ vectors. Such discontinuities can be avoided by introducing a "soft-max"-function

$$g_{\mathbf{s}}(\mathbf{x}) = \frac{\exp(-\beta d_{\mathbf{s}}(\mathbf{x}))}{\sum_{\mathbf{r}} \exp(-\beta d_{\mathbf{r}}(\mathbf{x}))} \tag{2}$$

with an "inverse temperature" $\beta > 0$ and blending the contributions of the individual nodes according to

$$\mathbf{y}(\mathbf{x}) = \sum_{\mathbf{r}} g_{\mathbf{r}}(\mathbf{x}) \cdot [\mathbf{w}_{\mathbf{r}}^{(out)} + \mathbf{A}_{\mathbf{r}}(\mathbf{x} - \mathbf{w}_{\mathbf{r}}^{(in)})]. \tag{3}$$

The prototype vectors $\mathbf{w}_{\mathbf{r}}^{(in)}$ can be determined in a variety of ways. Closest in spirit to the original SOM is to use the SOM algorithm with some neighborhood function that is chosen according to the expected topology of the input manifold in order to find a suitable arrangement of the input prototypes $\mathbf{w}_{\mathbf{r}}^{(in)}$. Often, it even suffices to choose a random subset of the training data and, optionally, to refine the positions of the chosen $\mathbf{w}_{\mathbf{r}}^{(in)}$ prototypes by some LBG-type vector quantization procedure (see, e.g., [12]). Together with a gradient based adaptation rule for the output prototypes $\mathbf{w}_{\mathbf{r}}^{(out)}$ and for the Jacobian matrices $\mathbf{A}_{\mathbf{r}}$, this leads to the following adaptation equations for training a LLM network from a set of input-output pairs $(\mathbf{x}, \mathbf{y})$:

$$\Delta\mathbf{w}_{\mathbf{r}}^{(in)} = \epsilon_1(\mathbf{x} - \mathbf{w}_{\mathbf{r}}^{(in)}) \tag{4}$$

$$\Delta\mathbf{w}_{\mathbf{r}}^{(out)} = \epsilon_2(\mathbf{y} - \mathbf{y}^{(net)}(\mathbf{x})) + \mathbf{A}\Delta\mathbf{w}_{\mathbf{r}}^{(in)} \tag{5}$$

$$\Delta\mathbf{A}_{\mathbf{r}} = \epsilon_3(\mathbf{y} - \mathbf{y}^{(net)}(\mathbf{x}))\frac{\mathbf{x}^T}{\|\mathbf{x}\|^2} \tag{6}$$

Here, the learning rule for output values $\mathbf{w}_{\mathbf{r}}^{(out)}$ differs from the equation for $\mathbf{w}_{\mathbf{r}}^{(in)}$ by the extra term $\mathbf{A}_{\mathbf{r}}\Delta\mathbf{w}_{\mathbf{r}}^{(in)}$ to compensate for the shift $\Delta\mathbf{w}_{\mathbf{r}}^{(in)}$ of the input prototype vectors. The adaptation rule for the Jacobians is the perceptron learning rule, but restricted to those input-output pairs $(\mathbf{x}, \mathbf{y})$ for which $\mathbf{x}$ is in the Voronoi cell of center $\mathbf{w}_{\mathbf{r}}^{(in)}$ (in practice, one should decrease the learning rates $\epsilon_{1...3}$ slowly towards zero, with $\epsilon_3$ decreasing more slowly than the other two in order to leave any fine-tuning to the matrices $\mathbf{A}_{\mathbf{r}}$).

A variant of this scheme weights the adaptation steps (4) by the node activities $g_{\mathbf{r}}(\mathbf{x})$. In this case is important to note that for finite temperature parameter $\beta$ the positions of the centers to which the $\mathbf{w}_{\mathbf{r}}^{(in)}$ will converge under (4) are then subject to a bifurcation scenario that forces the centers to become clustered. For $\beta$ below some critical limit $\beta_0$, there exists only a single cluster (a weighted average of all input data points) into which all $\mathbf{w}_{\mathbf{r}}^{(in)}$ coalesce. For increasing values of $\beta$, this single cluster and its descendants become repeatedly split by a cascade of bifurcation steps. Usually, one is interested in a sufficiently high value of $\beta$ that admits at least as many clusters as there are nodes. This value is distribution dependent, and it may differ from the value of $\beta$ that optimizes the blending of the different $\mathbf{y}_{\mathbf{r}}$ contributions according to (3. This may make it necessary to use different values of $\beta$ for training and for recall.

From a different perspective, a LLM-network can also be viewed as a "gated expert net", with the gating achieved by the SOM layer, and the

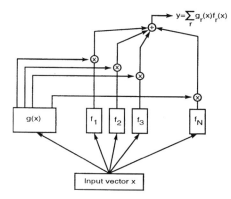

$$y = \sum_r g_r(x) f_r(x)$$

**Fig. 1.** LLM network as a mixture of "linear experts". The output is a weighted superposition of the outputs of locally valid linear maps $f_1 \ldots f_N$ ("expert nets"). The superposition coefficients $g_r(\mathbf{x})$ are determined by a shared "gating network" $\mathbf{g}(\mathbf{x})$, described by Eq.(2).

expert nets chosen as the local linear mappings defined by (1). This is depicted schematically in Fig.1, showing a number of linear "expert modules" that process a common input vector $\mathbf{x}$ and that contribute to the output according to a weighted superposition with weighting coefficients determined by the gating function $\mathbf{g}(\mathbf{x})$. From this point of view, LLM networks can also be seen as a natural extension of a single linear model, with the nonlinearity nicely "encapsulated" in the gating net equation (2) that determines how strongly the individual local linear maps contribute.

## 3 Robot Vision with LLM Networks

We have found the approach of LLM networks very suitable to learn both continuous and discrete mappings in various vision tasks, such as pose identification of deformable objects, e.g., hands [16] or object classification [4,5].

In both cases, a very important first step is a suitable preprocessing of the input image to obtain a manageably sized feature vector which also should already have some basic invariance properties, such as (at least approximate) translational invariance to facilitate the subsequent classification or identification task.

The first preprocessing operation usually is a figure ground segmentation, which can be color based (for the use of LLM-networks to learn a classifier for color-based segmentation, see e.g., [14]). For non-overlapping objects, this results in a number of "blobs" that can then be taken as the centers of an equal number of regions of interest (ROIs) which then are processed separately.

Within each ROI we choose a "fixation point" which becomes the center of a "jet" of Gabor filters that are convolved with the image intensity to obtain a low-dimensional feature vector. This choice of filter functions is motivated by the response properties of cells in the primary visual cortex, which can often be reasonably well described by Gabor filter profiles [2]. In the simplest case, the fixation point is taken as the centroid of the input intensity. For a more sophisticated scheme, based on the evaluation of local symmetries, see

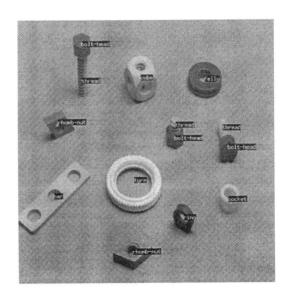

**Fig. 2.** *(a,top):* optimized Gabor filter mask. *(b,right):* some wooden toy pieces with their classification by the LLM-network (taken from ref. [4]).

e.g. [7]. Often, it is also advantageous to equalize the intensity histogram by a logarithmic intensity transformation, optionally followed by a convolution with a laplacian to suppress the DC component of the image.

While this scheme works already sufficiently well for many tasks, we can still obtain some further improvement by optimizing the chosen Gabor filter masks for an improved separation of the different object classes in feature space. We have investigated this approach for the task of classifying a set of wooden toy pieces in the context of a robot vision task [4]. This leads to optimized spatial filter functions such as the filter mask that is shown in Fig.2a, which usually show no longer any obvious symmetries. It also turns out that the parameter space in which the optimization takes place exhibits numerous local minima that correspond to filter sets that look strikingly different to the human eye (this, however, should not be too surprising, since a similar phenomenon also occurs when correlation filters belonging to different eigenfunctions, but the same eigenvalue, of a correlation matrix are linearly combined).

Figure 2b shows a selection of the toy objects that have been classified with this approach. The training set contains for each of the objects multiple views, so that recognition is also only weakly view-dependent. The same approach can be extended to the recognition of entire aggregates and to the recognition of parts within aggregates. This requires, however, to train a specialized recognition network for each aggregate, and also for the recognition of constituents within an aggregate. Figures 3 and 4 gives some impression of the degree of complexity that still is manageable within such a "holistic" approach. An attractive feature of this method is its speed. We have implemented the preprocessing steps on a MaxVideo200 Datacube. The re-

**Fig. 3.** Holistic identification of toy aggregates by specialized LLM-nets.

**Fig. 4.** Identification of individual constituents within an aggregate.

maining classification can then be done at about 10 frames per second on a workstation.

Of course, such "holistic" recognition approach cannot scale up to the recognition of more complex objects that are composed of many different parts. Therefore, we pursue the strategy to use the holistic object recognition networks a "holistic preprocessors" for a subsequent semantic network, which then draws upon symbolic world knowledge in order to guide a decomposition of more complex objects by positioning ROIs on subregions where particular, holistically recognizable object parts are expected. Work along these lines is currently under way, for some initial results, see e.g. [6].

## 4  Focal Attention by "Neural Zooming"

The previous section indicated already the importance for some control mechanism to guide a decomposition of complex objects into simpler constituents. An important role in this regard is played by *focal attention*, and semantic networks [18] appear to be very well suited for implementing a knowledge-driven, top-down attention control mechanism that can assist a collection of more specialized, neural object recognizers in decomposing object aggregates into more easily recognizable constituents [17].

However, one can also devise mechanisms that are based on simpler, and more general principles and that may even be implemented within an entirely neural architecture. One such approach is based on the observation that an object often can be quite easily identified when a particular set of characteristic object "landmarks" has been found. The detection of any individual landmark (such as a finger tip when a hand shape is to be recognized) depends on both local and global information, the latter providing the context for the correct interpretation (e.g., whether the same local pattern belongs to the middle or to the index finger) of the former.

364

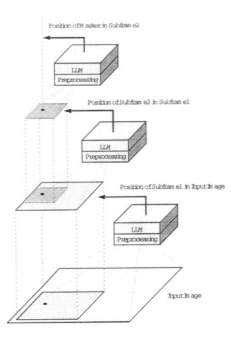

**Fig. 5.** Hierarchical architecture of LLM-networks to successively locate and "zoom in" on target feature in an image. Each network determines a smaller image region that then serves as input for the next network. The final network outputs the position estimate for the target feature.

The proper weighting of global and local context can be achieved within a hierarchical architecture of recognition networks, operating on a nested sequence of subimages whose spatial resolution and diameter vary inversely. Fig. 5 shows such architecture for the case of three hierarchically arranged LLM networks. The first network processes a coarse view of the entire image and passes to the second network a decision about the position of a smaller image rectangle which is then processed by the next network in the hierarchy, using a correspondingly higher spatial resolution. This step can be repeated several times, until the last network in the hierarchy, analyzing only a comparably small surround of the putative landmark location, makes a final decision of the precise landmark location in the image.

In this way, the initial networks in the cascade can first eliminate peripheral visual input and provide increasingly focused "regions of attention" for their successor networks. Since the feature vectors in the smaller input regions of the successor networks are computed with correspondingly downscaled versions of the gabor wavelets used in the first processing stage, these networks automatically become trained, and subsequently operate, at a finer spatial scale. This allows the final network to achieve a high spatial resolution

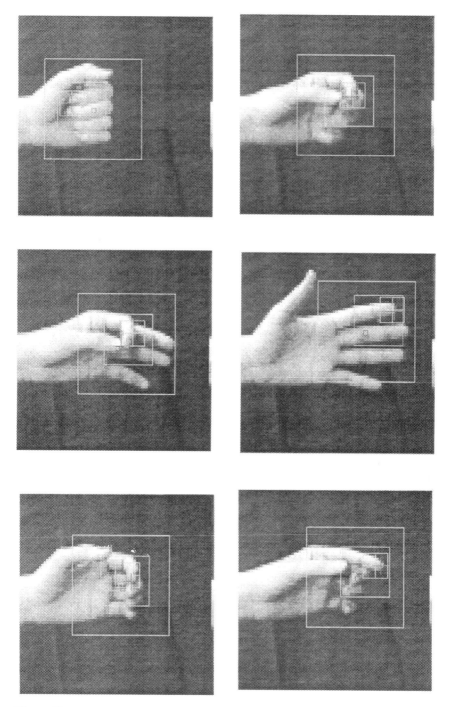

**Fig. 6.** Detection of index finger tip in hand images, using "neural zooming" with the hierarchical architecture depicted in Fig.5. Innermost bright plus mark indicates position estimate by final LLM network in the hierarchy, dark plus mark indicates correct target position.

while still more global visual information can be taken into account by the processing of the initial networks in the hierarchy.

This approach might be termed "neural zooming", since each network that is lower in the hierarchy determines a smaller subarea in its own image subregion into which then the successor network in the hierarchy can "zoom in".

Training of the network hierarchy can be achieved by providing training images with labeled landmark points (suitable target values for the centers of the intermediate image subregions can be obtained by simple interpolation between the final target point and, e.g., the image center).

We have been investigating this approach for the identification of finger tip locations in images of human hands. Fig.6 shows a number of recognition examples, obtained for identifying the location of the index finger tip for various hand postures before a black background (to simplify preprocessing, which was similar as explained in the previous section).

We used a three-stage recognition hierarchy (as depicted in Fig.5, with a monochrome input image of $80 \times 80$ pixels and successive processing stages of $40 \times 40$ and $20 \times 20$ pixels size, respectively). A feature vector was computed for each image, using a set of 36 gabor wavelets, centered at the points of a quadratic $3 \times 3$ lattice (using for each point the same spatial resolution and 4 different orientations, separated by angles of $45^0$), covering a central region of the image area.

Using a set of only 50 training images, we can locate the index finger tip in a similar test set to mostly within a few pixels deviation from its true location. The depicted recognition examples are some typical cases from an independent test set of 50 further images (showing the same hand and using the same illumination conditions; at present, we did not yet provide any measures for normalizing illumination or the size of the hand). We are currently extending this approach for recognizing multiple finger tips simultaneously, including a more robust preprocessing and segmentation stage and considering the evaluation of context information that becomes available when several finger tip positions are known or when continuous motion sequences are considered. For a fuller account with more quantitative results see [19].

## 5   Parametrized Self-Organizing Maps

The LLM networks were motivated by the desire to overcome the discrete nature of the SOM manifold and to realize a smooth input-output mapping, even in higher-dimensional spaces. However, instead of achieving smoothness by attaching tangent planes to a coarsely discretized manifold, we also can try to construct the entire manifold in a parametrized form, using a set of topologically ordered reference vectors of a coarse SOM as support points. This is the basic idea behind the *Parametrized Self-Organizing Map* ("PSOM"). Its construction proceeds in two steps. The first step has to provide a set of topologically ordered reference vectors $\mathbf{w_r}$, $\mathbf{r} \in \tilde{A}$, through which the desired manifold shall pass. As in the SOM, $\tilde{A}$ is usually some cartesian grid of the same dimensionality as the intrinsic dimensionality of the desired PSOM manifold $S$. Each grid point of $A$ has "attached" a reference vector $\mathbf{w_r}$ that represents a discretization point on $S$, with the understanding that the labeling is such that neighboring grid points $\mathbf{r}, \mathbf{r}'$ belong to points $\mathbf{w_r}, \mathbf{w_{r'}}$ that are "close" to each other on the manifold (this is what we want to understand by "topologically ordered reference vectors"; at present, a mathematically rigorous definition encounters the same problems as the definition of "topological order" for higher dimensional SOMs; currently, no unversally accepted definition is available for dimensions larger than one). $S$ itself is represented as an embedding of a $d$-dimensional hypersurface in a $D$-dimensional cartesian space and is given explicitly by a vector-valued function

$$\mathbf{w}(\mathbf{s}) = \sum_{\mathbf{r} \in \tilde{A}} H(\mathbf{r}, \mathbf{s}) \mathbf{w_r}. \tag{7}$$

Here, $H(\mathbf{r}, \mathbf{s})$ is a set of scalar valued basis functions (one for each $\mathbf{r} \in \tilde{A}$) that must obey the two conditions

$$H(\mathbf{r}, \mathbf{r}') = \delta_{\mathbf{r}, \mathbf{r}'} \text{ for } \mathbf{r}, \mathbf{r}' \in A \text{ and} \tag{8}$$

$$\sum_{\mathbf{r} \in A} H(\mathbf{r}, \mathbf{s}) = 1 \tag{9}$$

The first of these two conditions ensures that the manifold defined by (7) indeed passes through the given "prototypes" $\mathbf{w_r}$. The second condition ensures that the shape of $S$ is not altered under a global translation $\mathbf{w_r} \mapsto \mathbf{w_r} + \mathbf{t}$ of all prototypes.

Equation (7) is the continuous counterpart of the discrete SOM. In order to make the generalization into the continuous domain complete, we also have to specify an analogue of the SOM best-match search. As in the case of the SOM, we require here, too, minimization of some distance measure $d(\mathbf{x}, \mathbf{w}(\mathbf{s}))$ between input $\mathbf{x}$ and a general prototype $\mathbf{w}(\mathbf{s})$ on the PSOM, i.e., we have to find the "best match location" $\mathbf{s}(\mathbf{x})$ according to

$$\mathbf{s}(\mathbf{x}) = \arg \min_{\mathbf{s}} \sum_{i=1}^{D} c_i \cdot (x_i - w_i(\mathbf{s}))^2. \tag{10}$$

Here, we have specialized the distance measure $d(\mathbf{x}, \mathbf{w})$ to a weighted euclidean distance with weighting coefficients $c_i \geq 0$. In contrast to the standard SOM, where the determination of the discrete best match location $\mathbf{s}$ requires only a discrete search, we arrive in the case of a PSOM at a non-linear least-squares problem in continuous variables $\mathbf{s}$. While in principle we might use simple gradient descent to obtain a local solution, a more efficient method for this type of problem is the Levenberg-Marquardt algorithm [22] (simple gradient descent turns out to require a rather delicate step-size control).

The output of the PSOM is represented by the "best match prototype" $\mathbf{w}(\mathbf{s}(\mathbf{x}))$, which is the continuous equivalent of the discrete best match proto-type $\mathbf{w_s}$ in the discrete SOM. If the distance could be reduced to zero (and all $c_i$ are positive), we have the rather uninteresting result that $\mathbf{w}(\mathbf{s}(\mathbf{x})) = \mathbf{x}$. A much more interesting situation arises, when only a subset (say, $k$) of the $c_i$ are positive and the rest vanishes. In this case, minimization of the weighted distance $d(\mathbf{x}, \mathbf{w}(\mathbf{s}))$ is unaffected by those components of $\mathbf{x}$ that belong to dimensions $i$ for which $c_i = 0$. In other words, we now can accept as input an only *partially specified input vector* $\mathbf{x}$, provided that the index set $\tilde{I}(\mathbf{x})$ of its specified components comprises at least all those dimensions for which $c_i > 0$. If $k$ (the number of nonvanishing $c_i$'s ) equals the intrinsic dimen-sionality $d$ of the manifold $S$ or is larger, specification of $\mathbf{x}$ will (except for singular cases) uniquely specify a best match point $\mathbf{w}(\mathbf{s}(\mathbf{x}))$ on $S$. The map-ping $\mathbf{x} \mapsto \mathbf{w}(\mathbf{s}(\mathbf{x}))$ then becomes a non-linear projection from the embedding space $I\!\!R^D$ into $S$, and the image $\mathbf{w}(\mathbf{s}(\mathbf{x}))$ of any partially specified input $\mathbf{x}$ (such "partially specified inputs" actually being linear subspaces) under this mapping can be viewed as an *associative completion* of a fragmentary input.

Therefore, the PSOM acts as an *associative memory* that can complete fragmentary input vectors that are only specified along a subset $\tilde{I}(\mathbf{x}) \subset \{1 \ldots D\}$ of all embedding dimensions. All that then is required is to choose the weighting coefficients $c_i$ according to the simple rule

$$c_i = \begin{cases} 1, & \text{if } i \in \tilde{I}(\mathbf{x}) \\ 0 & \text{otherwise} \end{cases} \tag{11}$$

and to use the bestmatch step of the PSOM to obtain values for the missing components of $\mathbf{x}$.

The PSOM has some interesting similarities with the well-known spin-glass type associative memory (see, e.g. [8]). In both cases, retrieval occurs by offering a fragmentary input vector as initial condition, letting some dy-namics (in the case of the PSOM implemented by the iterative Levenberg-Marquard algorithm) do its "associative completion". However, an important difference is the dimensionality of the attractors: a spin-glass type attractor network usually contains a large number of discrete point attractors (as a consequence of the symmetry of the coupling coefficients), while a PSOM contains a single, *continuous attractor manifold* of some intrinsic dimension-ality $d > 0$. This is precisely what we need in many situations in robotics: here, we usually want to represent smooth relationships between continuous valued degrees of freedom instead of isolated and discrete patterns. Therefore, PSOMs are very well suited for robotics (continuous attractors can in prin-ciple also be obtained with recurrent multilayer perceptrons; however, these are very difficult to train [for a survey of algorithms, see, e.g., [21]], while the PSOM manifold can be rather directly constructed by simply specifying a reasonably dense set of prototype vectors $\mathbf{w_r}$ that sufficiently constrain the shape of the desired manifold).

So far, we did not specify a particular set of basis functions $H(\mathbf{r}, \mathbf{s})$. The two conditions (8,9) do not yet specify them uniquely. A particularly simple choice becomes possible if we assume that the sampling grid $\tilde{A}$ for the prototype vectors $\mathbf{w_r}$ is a (possibly distorted) $d$-dimensional cartesian lattice (cf. (7)). Then we can always construct a set of basis functions with the required properties (8,9) as products of one-dimensional Lagrange polynomials in the single axis variables $s_i$. For technical details, see [32,31] and also Eq.(12) below.

# 6    Robot Control with PSOMs

Many of the properties of PSOMs are particularly convenient in robotics, where different coordinate systems abound and training data are limited, since each training sample usually requires a rather time-consuming movement of the robot.

We first illustrate a typical use of PSOMs with the example of learning the inverse kinematics for a six-axis PUMA robot arm (for a fuller account of this work, see [26]). Simplified versions of this task have been considered earlier in the literature, for instance, [33,24].

The configuration space of a six-axis robot is a six-dimensional manifold. We can use a PSOM to represent this manifold in any higher-dimensional embedding space and we can choose for the coordinates of this space any set of coordinates that we find useful or convenient to relate with the arm configurations. A straightforward choice is a 15-dimensional embedding space, spanned by the set of the six joint angles $\theta_1 \ldots \theta_6$, the cartesian world coordinates $(x, y, z)$ of the end effector, together with its approach vector $(a_x, a_y, a_z)$ and the normal vector $(n_x, n_y, n_z)$ to its "palm". Note that the last nine coordinates are just a different coordinate system to specify an arm posture, and that we have taken the freedom to specify manipulator orientation (3 DOFs) with the non-minimal, but more convenient set of six coordinates $a_x \ldots n_z$. We could easily extend this choice of an embedding space beyond $D = 15$ by including further coordinates, such as homogeneous transforms, Euler angles or polar coordinates. Moreover, some of these additional coordinates could encode information beyond the mere location of the end effector, e.g., we might include the configuration dependent arm Jacobian, the inertia tensor or homogeneous transforms describing the location of individual arm segments in world coordinates, or even relative to other arm segments. Extending the embedding space in this way increases the range of parameter associations that can be formed after the PSOM has been constructed (for instance, we might then retrieve the joint angles that produce a given matrix value of the inertia tensor). Note that this increases the computational cost only linearly in the number of additionally included parameters, whereas the main cost factor, the *intrinsic* dimensionality of the PSOM manifold remains unchanged at its value of $d = 6$.

Having set up our embedding space, we have to decide on a suitable grid $\tilde{A}$ for the prototypes $\mathbf{w_r}$ that we use to define the PSOM manifold. In the present case, the most parsimonious choice is a cartesian $3 \times 3 \times 3 \times 3 \times 3 \times 3$-grid[1], requiring us to provide $729 = 3^6$ training prototypes $\mathbf{w_r}$ to construct the expression (7) for $\mathbf{w_r}$. The simplest way to generate these training vectors is to identify $\tilde{A}$ with a suitable hypercubical volume in the joint variables $\theta_1 \ldots \theta_6$ of the robot and to compute the remaining cartesian coordinates with the forward kinematics of the arm. The only remaining step is some choice for the basis functions $H(\mathbf{r}, \mathbf{s})$. In the present case we choose them as products $\prod_{i=1}^{6} H_i^{r_i}(\theta_i, s_i)$ of the simple one-dimensional Lagrange polynomials

$$H_i^j(\theta_i, s_i) = \prod_{k, k \neq j} \frac{s_i - \theta_i^k}{\theta_i^j - \theta_i^k} \tag{12}$$

along the six coordinate axes of our hypercubical joint space volume. As support points $\theta_i^{j=0,1,2}$ along axes $i = 1 \ldots 6$ we choose $\theta_i^0 = \theta_i^* - 45^0$, $\theta_i^1 = \theta_i^*$ and $\theta_i^2 = \theta_i^* + 45^0$, where $\theta_i^*$ denotes the angle of joint $i$ when the robot is in its "home position".

Fig.7 shows the accuracy of the inverse kinematics mapping that is obtained from the resulting PSOM by depicting the robot's positioning error with an error cross for a set of 200 randomly chosen target positions in its workspace. The resulting dimensionless NRMS positioning error is 0.06, which compares very favorably with the earlier approach of [33] (these authors obtain a similar accuracy with a specially structured multilayer perceptron, but for the simplified task that remains when manipulator orientation is neglected; moreover, they use several thousands of training examples instead of just 729).

Considering the 3d-positioning task alone and staying with the 3 discretization points per manifold dimension reduces the required number of training vectors to only 27 and the mean root square positioning error to 2.6 cm (corresponding to a NMRSE of 0.05). By using five instead of three nodes per dimension (requiring 125 data samples), the error can be reduced to 5mm (NMRSE of 0.0097). By making use of the further improvements considered in [30], the error can be reduced still further. For a more recent discussion, including extensions to combining several PSOMs, see, e.g., [29,31].

This example indicates that PSOMs compare very favorably with other neural network approaches when it comes to construct smooth and highly non-linear mappings for robot control from limited numbers of data examples. Here we only stress that this comes with the additional benefit of a great freedom in the choice of input and output variables: even without extending the embedding space as indicated at the outset, we could as well use the

---

[1] a corresponding $2 \times 2 \times 2 \times 2 \times 2 \times 2$ grid would appear as even more parsimonious, but the resulting basis functions would then be linear in each single coordinate and we would lose the ability to obtain a good approximation for functions that possess a local extremum in the interior of their domain.

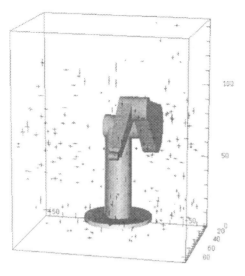

**Fig. 7.** Positioning errors of six-axis PUMA robot arm when the inverse kinematics transform is computed with a six-dimensional PSOM with $3^6$ nodes, representing the configuration manifold of the manipulator embedded in a 15-dimensional $r, a, n, \theta$-space (for details, see text).

present PSOM to find an arm configuration for which, say, target values for three arm angles and three cartesian parameters are given.

Instead of supporting this assertion with additional simulation results, we illustrate as a further capability the use of a PSOM to flexibly satisfy different constraints to a manipulator with excess degrees of freedom. This is a frequent situation in robotics, and the use of the associative completion capability of the PSOM to solve this type of task was first pointed out in [31].

A common way to determine a robot configuration in the presence of excess degrees of freedom is the formulation of one or several additional cost functions that express additional constraints that are to be met by the manipulator. For instance, if the cost function is a weighted sum of squared arm velocities, we arrive a the well-known pseudo-inverse control for redundant manipulators (for a review, see e.g., [10]). By including each desired cost function in the set of embedding coordinates of the PSOM manifold we can use the PSOM to *associate manipulator configurations with desired values of particular cost functions.* For each query, we can thus "activate" a different cost function, simply by setting its associated distance weighting coefficient $c_i$ to a positive value. If we choose this value to be large we make the associated constraint mandatory. If we specify only a small value, we establish

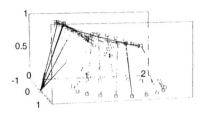

**Fig. 8.** *(a,left)* Positioning errors (magnified by a factor of 100 to become visible) for a redundant 4-link manipulator, evaluated at a cartesian grid of $3 \times 4 \times 5$ test target points it its workspace (+ and x-marks indicate target and actual positions, resp., with the deviation of the latter from the former magnified by a factor of 100). *(b,right)* Stroboscopic rendering of the arm when traversing a circular path in the ground plane, resolving the redundancy with an additional constraint feature maximizing the similarity of the last (distal) two joint angles.

a "soft constraint" that can be violated if it conflicts with other constraints that have larger $c_i$ coefficients.

To illustrate these possibilities with a concrete example, let us consider a robot arm based at the coordinate origin and composed of three coplanar links in a vertical plane that can be rotated about the vertical $z$-axis (angle $\theta_0$). The base link of the robot can be rotated about an axis passing through the origin and perpendicularly to the plane of the arm $(\theta_1)$, with the remaining two distal arm segments attached via two further revolute joints (angles $\theta_2$ and $\theta_3$, resp.) (see Fig.8b). The entire arm thus has four DOFs and, therefore, one excess degree of freedom with respect to the task of positioning its end effector at a specified location. For the construction of a PSOM for the configuration manifold of this arm, we choose as the lattice $\tilde{\mathbf{A}}$ a $5 \times 5 \times 5 \times 5$ grid covering the joint angle hypercube $[0^0, 180^0] \times [0^0, 120^0] \times [-120^0, 0^0] \times [-120^0, 0^0]$ for $\theta_{0...3}$, resp. (segment lengths were 1.0, 0.8 and 0.7 from base to distal end). For the embedding space, we use the four joint angles and the three cartesian end effector coordinates. To later allow resolving the arm redundancy in different ways, we augment the embedding space by three further coordinate parameters, namely the variance of the last two joint angles, the angle of the middle arm segment against the horizontal, and the angle of the most distal arm segment with the vertical. This yields a 10-dimensional embedding space. If we activate the first constraint by specifying a target value of 0 and its associated weighting coefficient as 0.01 ("soft minimization of joint angle variance"), we obtain for the inverse kinematics (i.e., $(x, y, z) \mapsto (\theta_0 \ldots \theta_3)$) a positioning error of less than 0.008 for all indicated test target points within the work space depicted in Fig.8a (NOTE: to make the positioning errors visible, the length all error vectors depicted in Fig.8a has been scaled by a factor of 100). Fig.8b illustrates how in this case the redundancy is resolved when the end effector traverses a circular path in the ground plane. We can now readily change this behavior by activating a different constraint. Activating

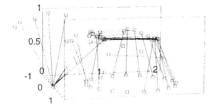 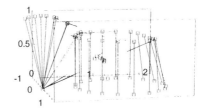

**Fig. 9.** Tracing of a horizontal circle with the same arm and the same PSOM mapping shown already in Fig.8, but activating different constraint features. *(a,left:* require middle arm segment to stay horizontal. Note that for the given geometry this constraint cannot be fulfilled for some target points on the path near the base the robot. Here, the PSOM still correctly positions the manipulator and selects arm configurations that try to violate the constraint only weakly. *(b,right):* Similar as *b)*, but now with a constraint specifying a vertical direction for the last arm segment. This time, the given constraint cannot be fulfilled for the most distal points of the circle, but the positioning again is correct along the entire path.

the constraint for the direction of the middle arm segment and specifying a target value of 0 (horizontal direction) results in a different sequence of arm configurations depicted in Fig.9a. As can be seen, the traversed path has remained the same, but the middle arm segment is now kept horizontal for all those target points for which this constraint can be met. For some points close to the base of the robot, meeting this constraint is impossible. Having, however, specified only a "soft constraint", allows the PSOM to violate it without (significantly) sacrificing accuracy in the tracking of the target trajectory. Finally, Fig.9b shows the result when instead the constraint for the vertical direction of the distal arm segment is activated, with a similar "graceful degradation" for some points, this time along the most distal points on the path.

Therefore, if we have some foresight which cost functions will be useful, we can augment our embedding space in advance, enabling us to construct very flexibly configurable "mapping modules" which have as their simultaneous benefits that we can (*i*) construct them readily from examples, (*ii*) apply them to flexibly associate partial state specifications, without being rigidly committed to particular mapping directions and (*iii*) use them as dynamically reconfigurable optimizers for a previously selected set of cost functions that have become associated with the stored configuration manifold.

## 7 Recognition of Hand Postures

The flexible mapping capabilities of PSOMs are also useful in other domains. We conclude this paper with a brief report of ongoing work in applying PSOMs to the identification of object pose from images. In Sec.4 we presented an approach to identify finger tip locations in video images. A natural next step is to use this information to estimate the three-dimensional posture of the hand. Since a human hand has a large number of degrees of freedom, an accurate estimate will probably require the inclusion of a number of further "landmark points" on the hand; for the time being we bypass this issue by restricting the set of allowed hand postures to a subset that comprises those finger joint angle configurations that can be expressed as a linear superposition of a small number (currently three) of basis postures (such as a "fist", the open hand and a "precision grip", in which only the index finger and the thumb are flexed to touch each other at their tips). If we allow in addition some rotation about the arm axis and about a second axis that is perpendicular to the arm and to the line of sight (we left out rotations about the line of sight itself, since these correspond to a simple image rotation, while the other two change the appearance of the image itself) we end up with a five-dimensional configuration manifold for the described, restricted set of hand postures.

Our goal is to estimate postures from this set on the basis of a number of observed finger tip locations. To study how well this task can be solved with PSOMs, we are currently using a (simplified) articulated hand model that can be rendered on a computer screen. Using this model, we can generate training examples for the "forward transform", i.e, the mapping from joint coordinates to projected finger tip coordinates in the image. Presently, we have a 13-dimensional embedding space (five parameters to specify a hand posture and its orientation [subject to the restrictions explained previously], and eight parameters to specify the 2d-locations of four finger tips) for a configuration manifold of intrinsic dimensionality $d = 5$. For the construction of the PSOM we use 243 training points, obtained by computing the projected finger tip positions for all points of a $3 \times 3 \times 3 \times 3 \times 3$ grid of the configuration/orientation parameters of the synthetic hand model. Estimation of the hand posture that is associated with an observed set of finger tip locations can then be achieved by using the PSOM in the reverse mapping direction. Fig.10 gives some impression of the identification accuracy that can be obtained with this approach. The cross marks indicate observed finger tip locations (these were generated by randomly selecting a hand posture from the range of posture parameters that is covered by the training grid). The depicted hand posture is the estimate obtained by the PSOM, with its finger tip positions surrounded by the gray squares. As can be seen, the estimated postures very accurately fit the observed finger tip locations. The average accuracy is in the range of a few pixel positions. There is, however, as small fraction of cases for which the PSOM responds with a grossly incorrect posture (an example being given by the last picture in Fig.10). We attribute the occurence of these outliers to the occurrence of a local minimum for the best match search. Fortunately, the percentage of these cases is only small (about 3%) and they can probably be eliminated rather easily, e.g., by training several PSOMs with different data sets, making it less likely that more than a single PSOM gets simultaneously trapped in one of these rare local minima.

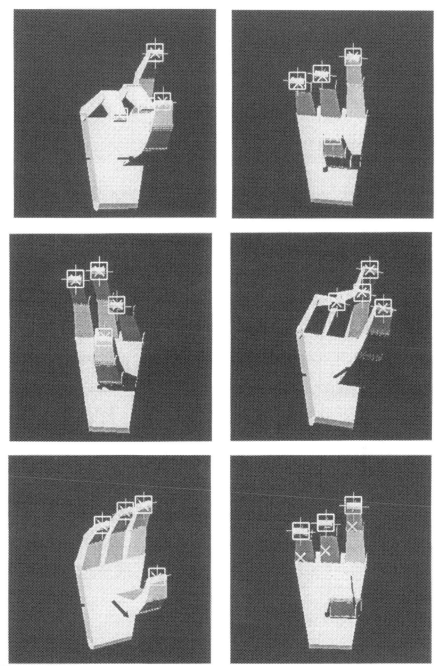

**Fig. 10.** Hand postures, estimated from finger tip locations (cross marks), using a PSOM network trained with 243 examples of the forward transform (posture parameters ↦ finger tip locations). The first five images illustrate the typical performance of the PSOM. However, a small percentage of cases results in grossly wrong responses, such as depicted in the lower right image.

# 8 Discussion

The complexities involved in linking perception with action challenge our traditional ways of constructing systems in a purely analytic way, working from basic principles upwards to cleanly specified algorithms that can deal transparently with all the complexities of real world tasks. While such an approach will always remain desirable, it is most likely unfeasable for the majority of tasks a robot has to solve. Therefore, we must develop alternative methods, with an emphasis on the adaptive construction of mappings and of dynamical systems that can achieve a desired behavior. Neural networks offer a natural framework for meeting this challenge, and we have identified as some of the important questions that have to be solved along this way the need for algorithms that allow the *rapid construction of mapping modules* from limited amounts of data, the development of strategies for *attention control*, both for the efficient allocation of processing resources as well as to support a tight man-machine cooperation, and finally, the necessity for exploring ways to implement additional functional building blocks, such as *continuous associative memories* with an intrinsic "multiway-mapping" capability in conjunction with an ability to deal with dynamically changing and possibly conflicting *constraints*.

We have presented examples of concrete approaches that attempt to address some of these issues, including *LLM* and *PSOM networks* for the rapid construction of mappings and continuous associative memories, a hierarchical architecture to implement a simple type of focal attention that can be learned from examples, and work showing how these approaches can be used in the context of various robot tasks, such as object classification, the identification of hand postures of a human user or the control of robot manipulators with the inclusion of excess degrees of freedom.

In all cases, the emphasis was on learning approaches with an ability to extract the necessary world knowledge in implicit form from modest numbers of training examples. Undoubtedly, there is still a long way to go until robots even remotely reach the learning competence that we routinely unfold when we learn to carry out a new task. However, compared with the long history of biological neural nets, artificial neural networks are still in their infancy and the rapid progress in their development encourages us to expect significant further progress ahead.

## Acknowledgement
Part of this work has been supported by the German Research foundation (DFG) in the SFB 360 project.

# References

1. Brady M. (ed.): Robotics Science. MIT Press, Cambridge Massachusetts (1989)

2. Daugman, J.G.: Two-dimensional spectral analysis of cortical receptive field profiles, Vision Research 20 (1980) 847-856
3. Erwin E., Obermayer K., Schulten K.: Models of Orientation and Ocular Dominance Columns in the Visual Cortex: A Critical Comparison. Neural Computation 7 (1995) 425-468
4. Heidemann G., Ritter H.: A Neural 3-D Object Recognition Architecture Using Optimized Gabor Filters. Proc. 13th Int. Conf. Patt. Recog., Vol. IV, (1996) 70-74, IEEE Computer Society Press.
5. Heidemann G., Ritter H.: A neural recognition architecture for composed objects. DAGM Symposium Mustererkennung (B. Jähne, P. Geißler, H. Haußecker, F. Hering eds.), (1996) 475-482, Springer Verlag Berlin Heidelberg.
6. Heidemann, G., Kummert F., Ritter H., Sagerer G.: A Hybrid Object Recognition Architecture, ICANN 96, Springer Lecture Notes in Computer Science 1112, (1996) 305-310. Springer Berlin Heidelberg.
7. Heidemann, G., Nattkemper T., Menkhaus G., Ritter H.: Blicksteuerung durch präattentive Fokussierungspunkte. Proceedings in Artificial Intelligence (B. Mertsching ed.) (1996) 109-116 Infix Verlag, St. Augustin (in German).
8. van Hemmen, J.L., Kühn R.: Collective Phenomena in Neural Networks. In: Models of Neural Networks (E. Domany et al. eds.) 1-106, Springer Berlin Heidelberg 1991.
9. Hunt K.J., Sbarbaro D., Zbikowski R., Gawthrop P.J.: Neural Networks for Control Systems: A Survey. Automatica 28 (1992) 1083ff.
10. Klein C.A., Huang C-H.: Review of Pseudoinverse Control for Use with Kinematically Redundant Manipulators. IEEE Trans. Sys. Man and Cybern. SMC-13, (1983) 245-250.
11. Kohonen, T.: The Self-Organizing Map, Proc. IEEE 78 (1990) 1464-1480
12. Kohonen, T.: Self-Organizing Maps, Springer Series in Information Sciences, Springer Berlin Heidelberg 1997.
13. Littmann E., Drees A., Ritter H.: Neural Recognition of Human Pointing Gestures in Real Images. Neural Processing Letters (1996) 61-71, Kluwer Academic Publishers.
14. Littmann E., Ritter H.: Adaptive Color Segmentation – A comparison of Neural and Statistical Methods. IEEE Trans. Neural Networks 8 (1997) 175-185
15. Maes P. (ed.): Designing Autonomous Agents (Special Issue of Robotics and Autonomous Systems), MIT Press Cambridge (1990)
16. Meyering A., Ritter H.: Learning 3D-Shape Perception with Local Linear Maps. Proc. Int. Joint Conf. on Neural Networks (1992) IV:432-436, Baltimore
17. Moratz R., Heidemann G., Posch S., Ritter H., Sagerer G.: Representing Procedural Knowledge for Semantic Networks using Neural Nets. Proc. 9th Scand. Conf. Image Anal. (1995) 819-828 Uppsala.
18. Niemann H., Sagerer G., Schröder S., Kummert F.: ERNEST: A Semantic Network for Pattern Understanding. IEEE Transactions on Pattern Analysis and Machine Intelligence 12 (1990) 883-905
   Nölker, C., Ritter, H.:
19. Nölker, C., Ritter, H.: Detektion von Fingerspitzen in Videobildern DAGM Mustererkennung (F. Wahl et al. ed.) (1997) (accepted) Springer Verlag Berlin Heidelberg.
20. Obermayer K., Ritter H., Schulten K.: A Model for the Development of the Spatial Structure of Retinotopic Maps and Orientation Columns. IEICE Trans. Fundamentals E75-A, (1992) 537-545.

21. Pearlmutter B.: Gradient calculation for dynamic recurrent neural networks: a survey. IEEE Trans. Neural Networks 6 (1995) 1212-1228
22. Press, W.H., Flannery, B.P., Teukolsky, S.A., Vetterling, W.T.: Numerical Recipes in C, Cambridge University Press (1990).
23. Ritter H., Schulten K.: Extending Kohonen's Self-Organizing Mapping Algorithm to Learn Ballistic Movements. In: Neural Computers (Eckmiller R., v.d.Malsburg Ch. eds.) (1987) 393-406 Springer Heidelberg
24. Ritter, H., Martinetz, T., Schulten, K.: Topology Conserving Maps for Learning Visuomotor-Coordination. Neural Networks 2 (1989) 159-168.
25. Ritter, H., Martinetz, T., Schulten, K.: Neural Computation and Self-organizing Maps, Addison-Wesley, Reading, MA. (1992)
26. Ritter, H. Parametrized Self-Organizing Maps. ICANN 93 Proceedings (S.Gielen and B.Kappen eds.), (1993) 268-273, Springer Berlin Heidelberg.
27. Ritter H.: Learning with the Self-Organizing Map. In: Artificial Neural Networks (T. Kohonen et al. ed.) (1991) 379-384.
28. Ritter H.: Self-Organizing Feature Maps: Kohonen Maps. In: The Handbook of Brain Theory and Neural Networks (M.A. Arbib ed.) 846-851 MIT Press, Cambridge (1995)
29. Walter J., Ritter H.: Investment Learning with Hierarchical PSOM. In: Advances in Neural Information Processing Systems 8 (NIPS*95) (D. Touretzky, M. Mozer and M. Hasselmo eds.) (1996) 570-576, MIT Press
30. Walter J., Ritter H.: Local PSOMs and Chebychev PSOMs – Improving the Parametrized Self-Organizing Maps. ICANN95 Conf. Proceedings (1995) 95-102
31. Walter J.: Rapid Learning in Robotics. Cuvillier Verlag Göttingen 1996.
32. Walter, J., Ritter, H., Rapid Learning with Parametrized Self-organizing Maps. Neurocomputing 12, (1996) 131-153
33. Yeung D., Bekey G.: On Reducing Learning Time in Context Dependent Mappings, IEEE Transactions on Neural Networks 4 (1993) 31-42

# Geometric Neural Networks

Eduardo Bayro-Corrochano and Sven Buchholz
Computer Science Institute, Cognitive Systems Group
Christian Albrechts University, Kiel, Germany
email: edb,sbh@informatik.uni-kiel.de

**Abstract.** The representation of the external world in biological creatures appears to be defined in terms of geometry. This suggests that researchers should look for suitable mathematical systems with powerful geometric and algebraic characteristics. In such mathematical context the design and implementation of neural networks will be certainly more advantageous. This paper presents the generalization of feedforward neural networks in the Clifford or geometric algebra framework. The efficiency of the geometric neural nets indicate a step forward in the design of algorithms for multidimensional artificial learning.

**Categories**: Clifford algebra; geometric algebra; feedforward neural networks; hyper-complex neural networks; RBF geometric neural networks.

## 1 Introduction

Biological creatures interact with their environment in order to survive. This activity is triggered by different needs which should be satisfied. The most important ones are nutrition and conservation. As soon a creature moves secure its internal activity may switch on higher cognition levels to satisfy other sophisticated needs, e.g. the joy during playing. If we are interested to build artificial intelligent systems which should autonomously perceive and act in their surroundings we should first of all ask how the machine should build its internal representation of the world. Nowadays there is the believe that the brain might be seen as a geometric engine [14, 16]. A general hypothesis of the geometric interpretation of the brain may relay on the assumption that the mapping between the external world and the brain is certainly a result of the perception and action activities within a time cycle. These activities controlled by the central nervous system might be seen in the context of learning by experience as a basic way to build the internal geometric representation.

In mathematical terms we can formalize the relations of the physical signals of world objects with the creature ones by using extrinsic vectors coming from the world and intrinsic vectors which depict the internal representation. We can also assume that the external world and the internal world have different reference coordinate systems. If we see the acquisition of knowledge as a distributed process it is imaginable that there exist various domains of representation with different vectorial basis and obeying different metrics. How is it possible that nature through the evolution has acquired such tremendous representation power for dealing with such geometric vector representations. In a stimulating series of articles Pellionisz and Llinàs [16, 17] claim that the formalization of the geometrical representation seems to be the dual expression of extrinsic physical cues by intrinsic central nervous system vectors. They quoted that these vectorial representations related to reference frames intrinsic to the creature are co-variant for perception analysis and contra-variant for action synthesis. These authors explain that the geometric mapping between these two vectorial spaces is implemented by a neural network which performs as a metric tensor [17]. The Clifford algebra in the geometric interpretation of Hestenes [10] appears to be an alternative to the tensor analysis employed since 1980 by Pellionisz and Llinàs for the perception action cycle theory. Since tensor calculus is co-variant, in other words it requires of transformation laws for getting coordinate-independent relations, Clifford or geometric algebra appears more attractive as it is not only essentially a coordinate free or invariant system but also includes spinors which tensor theory does not. The computational efficiency of geometric algebra has been shown in various challenging areas of mathematical physics [6]. Preliminary attempts for applying the geometric algebra in neural geometry have been already done by Hestenes and in the fields of computer vision, robotics and neural nets can be found in [11, 12, 3, 4, 2]. Analyzing other approaches for neural computation we see that the mostly used is the matrix algebra. Geometric algebra and matrix algebra both are associative algebras, yet geometric algebra captures the geometric characteristics of the problem better independent of a coordinate reference system and offers also other computational mechanisms that matrix algebra has not, e.g. the geometric product using hypercomplex, double and dual entities.

In this paper we will specify a geometric algebra $\mathcal{G}_n$ of the n-dimensional space by $\mathcal{G}_{p,q}$, where p and q stand for the number of basis vectors which squares to 1 and -1 respectively and fulfill n=p+q. See [10, 2, 3] for a more complete introduction in geometric algebra . The next section reviews the computing principles of feedforward neural networks underlining their most important characteristics. Section three deals with the extension of the multilayer perceptron (MLP) to complex and quaternionic MLPs. Section four presents the generalization of the feedforward neural networks in the geometric algebra system. Section five describes the generalized learning rule across different geometric algebras. Section six presents various comparative experiments of geometric neural networks with real valued MLPs. The last section discusses the suitability of the geometric algebra system for neural computing.

## 2   Real Valued Neural Networks

The approximation of nonlinear mappings using neural networks is useful in various areas of signal processing like pattern classification, prediction, system modelling and identification. This section reviews the fundamentals of standard real valued feedforward architectures.

Cybenko [5] used for the approximation of a given continuous function $g(\mathbf{x})$ the superposition of weighted functions:

$$y(\mathbf{x}) = \sum_{j=1}^{N} w_j \sigma_j(\mathbf{w}_j^T \mathbf{x} + \theta_j), \tag{1}$$

where $\sigma(.)$ is a continuous discriminatory function like a sigmoid, $w_j \in \mathcal{R}$ and $\mathbf{x}, \theta_j, \mathbf{w}_j \in \mathcal{R}^n$. The finite sums of the form of Eq. (1) are dense in $C^0(I_n)$ if $|g(\mathbf{x}) - y(\mathbf{x})| < \varepsilon$ for a given $\varepsilon > 0$ and all $\mathbf{x} \in [0,1]^n$. This is called a *density theorem* and is a fundamental concept in approximation theory and nonlinear system modelling [5, 13].

A structure with k outputs $y_k$ having several layers using logistic functions is known as the Multilayer Perceptron (MLP) [22]. The output of any neuron of a hidden layer or of the output layer are represented in similar way,

$$o_j = f_j\left(\sum_{i=1}^{N_i} w_{ji} x_{ji} + \theta_j\right) \qquad y_k = f_k\left(\sum_{j=1}^{N_j} w_{kj} o_{kj} + \theta_k\right), \tag{2}$$

where $f_j(\cdot)$ is logistic and $f_k(\cdot)$ is logistic or linear. Linear functions at the outputs are often used for pattern classification. In some tasks of pattern classification suffices one hidden layer whereas in some tasks of automatic control it may be required two hidden layers. Hornik [13] showed that standard multilayer feedforward networks are able to approximate accurately any measurable function to a desired degree. Thus they can be seen as *universal approximators*. In case of a training failure we should rather attribute to an inadequate learning, incorrect number of hidden neurons or a poor deterministic relation between input and output patterns.

Poggio and Girosi [19] developed the Radial Basis Function (RBF) network which consists of a superposition of weighted Gaussian functions as follows

$$y_j(\mathbf{x}) = \sum_{i=1}^{N} w_{ji} G_i\big(\mathbf{D}_i(\mathbf{x} - \mathbf{t}_i)\big) \tag{3}$$

where $y_j$ is the $j$-output, $w_{ji} \in \mathcal{R}$, $G_i$ is a Gaussian function, $\mathbf{D}_i$ a $N \times N$ dilatation diagonal matrix and $\mathbf{x}, \mathbf{t}_i \in \mathcal{R}^n$. The vector $\mathbf{t}_i$ is a translation vector. This architecture is supported by the regularization theory.

## 3 Complex MLP and Quaternionic MLP

A MLP is extended in the complex domain when its weights, activation function and outputs are complex valued. Yet, the selection of the activation function is a non-trivial matter. For example, the extension of the sigmoid function from $\mathcal{R}$ to $\mathcal{C}$, i.e.

$$f(z) = \frac{1}{(1 + e^{-z})} \qquad (4)$$

where $z \in \mathcal{C}$, is not allowed as this function is analytic and unbounded [7]. Similar is the case of $\tanh(z)$ and $e^{-z^2}$. This kind of activation functions troubles the convergence during training due to its singularities. The necessary conditions that a complex activation $f(z) = a(x,y) + ib(x,y)$ has to fulfill are: $f(z)$ non-linear in $x$ and $y$, partial derivatives $a_x$, $a_y$, $b_x$ and $b_y$ exist (where $a_x b_y \not\equiv b_x a_y$) and $f(z)$ is not entire. Accordingly Georgiou and Koutsougeras [7] proposed

$$f(z) = \frac{z}{c + \frac{1}{r}|z|} \qquad (5)$$

where $c, r \in \mathcal{R}^+$. These authors extended the usual real back-propagation learning rule for the Complex MLP (CMLP).

Arena et al.[1] introduced the Quaternionic MLP (QMLP) which is an extension of the CMLP. The weights, activations functions and outputs of this net are represented in terms of quaternions [8]. They choose the following non-analytic bounded function

$$\begin{aligned} f(q) &= f(q_0 + q_1 i + q_2 j + q_3 k) \\ &= (\frac{1}{1 + e^{-q_0}}) + (\frac{1}{1 + e^{-q_1}})i + (\frac{1}{1 + e^{-q_2}})j + (\frac{1}{1 + e^{-q_3}})k, \end{aligned} \qquad (6)$$

where $f(\cdot)$ is now the function for quaternions. These authors proved that superpositions of such functions approximate accurately any continuous quaternionic function defined in the unit polydisc of $\mathcal{C}^n$. The extension of the training rule along the lines of the CMLP was done straightforwardly [1].

## 4 Geometric Algebra Neural Networks

Real, complex and quaternionic neural networks can be further generalized in the Clifford or geometric algebra framework. The weights, the activation functions and the outputs will be now represented using multivectors. In the real valued neural networks of section 3, the vectors are multiplied with the weights using the scalar product. For geometric neural networks the scalar product will be substituted by the Clifford or geometric product.

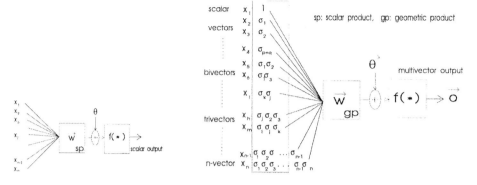

Fig. 1. *McCulloch-Pitts Neuron and Geometric Neuron*

## 4.1 The activation function

The activation function of Eq. (5) used for the CMLP was extended by Pearson and Bisset [15] for a type of Clifford MLP for different Clifford algebras including the quaternion algebra. In this paper we propose an activation function which affects each multivector basis element. This function was introduced independently by the authors [2] and is in fact a generalization of the function of Arena et al [1]. The function for a n-dimensional multivector $\boldsymbol{m}$ reads

$$\boldsymbol{f}(\boldsymbol{m}) = \boldsymbol{f}(m_0 + m_i\sigma_i + m_j\sigma_j + m_k\sigma_k + ... + m_{ij}\sigma_i \wedge \sigma_j + ... +$$
$$m_{ijk}\sigma_i \wedge \sigma_j \wedge \sigma_k + ... + m_n\sigma_1 \wedge \sigma_2 \wedge ... \wedge \sigma_n)$$
$$= f(m_0) + f(m_i)\sigma_i + f(m_j)\sigma_j + f(m_k)\sigma_k + .. + f(m_{ij})\sigma_i \wedge \sigma_j + ... +$$
$$f(m_{ijk})\sigma_i \wedge \sigma_j \wedge \sigma_k + ... + f(m_n)\sigma_1 \wedge \sigma_2 \wedge ... \wedge \sigma_n, \qquad (7)$$

where $\boldsymbol{f}(\cdot)$ is written in bold to be distinguished from the one used for a single argument $f(\cdot)$. The values of $f(\cdot)$ can be of the type sigmoid or Gaussian.

## 4.2 The geometric neuron

The McCulloch-Pitts neuron uses the scalar product of the input vector and its weight vector [22]. The extension of this model to the geometric neuron requires the substitution of the scalar product with the Clifford or geometric product, i.e.

$$\boldsymbol{w}^T\mathbf{x} + \theta \qquad \Rightarrow \qquad \boldsymbol{wx} + \boldsymbol{\theta} = \boldsymbol{w} \cdot \boldsymbol{x} + \boldsymbol{w} \wedge \boldsymbol{x} + \boldsymbol{\theta} \qquad (8)$$

Figure 1 shows in detail the McCulloch-Pitts neuron and the geometric neuron. This figure also depicts the way how the input pattern is formated in a specific geometric algebra. The geometric neuron outputs a more rich kind of pattern, let us illustrate this with an example in $\mathcal{G}_{3,0}$

$$
\begin{aligned}
o &= f(wx + \theta) \\
&= f(s_0 + s_1\sigma_1 + s_2\sigma_2 + s_3\sigma_3 + s_4\sigma_1\sigma_2 + s_5\sigma_1\sigma_3 + s_6\sigma_2\sigma_3 + s_7\sigma_1\sigma_2\sigma_3) \\
&= f(s_0) + f(s_1)\sigma_1 + f(s_2)\sigma_2 + f(s_3)\sigma_3 + f(s_4)\sigma_1\sigma_2 + \ldots + \\
&\quad\; f(s_5)\sigma_1\sigma_3 + f(s_6)\sigma_2\sigma_3 + f(s_7)\sigma_1\sigma_2\sigma_3,
\end{aligned} \tag{9}
$$

where $f$ is the activation function defined in Eq. (7) and $s_i \in \mathcal{R}$. Using the McCulloch-Pitts neuron in the real valued neural networks the output is simply a scalar given by

$$
o = f\left(\sum_i^N w_i x_i + \theta\right). \tag{10}
$$

The geometric neuron outputs a signal with more geometric information

$$
o = f(wx + \theta) = f(w \cdot x + w \wedge x + \theta) \tag{11}
$$

which on the one hand has the scalar product like the McCulloch-Pitts neuron, i.e.

$$
f(w \cdot x + \theta) = f(s_0) \equiv f\left(\sum_i^N w_i x_i + \theta\right) \tag{12}
$$

and on the other hand the outer product expressed by

$$
\begin{aligned}
f(w \wedge x + \theta - \theta) &= f(s_1)\sigma_1 + f(s_2)\sigma_2 + f(s_3)\sigma_3 + f(s_4)\sigma_1\sigma_2 + \ldots + \\
&\quad\; f(s_5)\sigma_1\sigma_3 + f(s_6)\sigma_2\sigma_3 + f(s_7)\sigma_1\sigma_2\sigma_3.
\end{aligned} \tag{13}
$$

Note that the outer product supplies the scalar cross-products between the individual components of the vector which are nothing else as the multivector components of higher grade like point or lines (vectors), planes (bivectors) and volumes (trivectors). This characteristic will be used in section 7.2 for the implementation of the embedded geometric processing in the extended geometric neural networks. These kind of neural networks resemble to the higher order neural networks, however the extended geometric neural networks use not only scalar products of higher order but all the necessary scalar cross-products for carrying out a geometric cross-correlation. That is why a geometric neuron can be seen as a sort of geometric correlator which for the interpolation offers in contrast to the McCulloch-Pitts neuron not only points but also higher grade multivector components like planes, volumes,...,hyper-volumes.

## 4.3  Feedforward geometric neural networks

Figure (2) depicts the standard neural and network structures for function approximation in the geometric algebra framework. Here the inner vector product

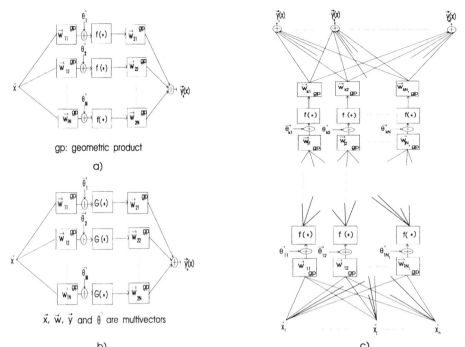

gp: geometric product

a)

$\vec{x}$, $\vec{w}$, $\vec{y}$ and $\theta$ are multivectors

b)

c)

**Fig. 2.** *Geometric Network Structures for Approximation: (a) Cybenko's (b) GRBF network (c) GMLP$_{p,q}$*

has been extended to the geometric product and the activation functions are according (7).

The equation (1) of the Cybenko's model in geometric algebra reads

$$y(x) = \sum_{j=1}^{N} w_j f(w_j \cdot x + w_j \wedge x + \theta_j). \qquad (14)$$

The extension of the MLP is straightforward. The equations using the geometric product of the outputs of hidden and output layers read

$$o_j = f_j(\sum_{i=1}^{N_i} w_{ji} \cdot x_{ji} + w_{ji} \wedge x_{ji} + \theta_j)$$

$$y_k = f_k(\sum_{j=1}^{N_j} w_{kj} \cdot o_{kj} + w_{kj} \wedge o_{kj} + \theta_k) \qquad (15)$$

In the radial basis function networks, the dilatation operation (via the diagonal matrix $D_i$) can be implemented by means of a geometric product with a dilation $\boldsymbol{D}_i = e^{\alpha \frac{ii}{2}}$ [10], i.e.

$$D_i(\mathbf{x} - \mathbf{t}_i) \Rightarrow \boldsymbol{D}_i(\boldsymbol{x} - \boldsymbol{t}_i)\tilde{\boldsymbol{D}}_i \qquad (16)$$

$$y_k(x) = \sum_{j=1}^{N} w_{kj} G_j(D_j(x_{ji} - t_j)\tilde{D}_j) \tag{17}$$

Note that in the case of the geometric RBF we are also using an activation function according the Eq. (7).

# 5 Learning Rule

This section presents the multidimensional generalization of the gradient descent learning rule in the geometric algebra framework. This rule can be used for the Geometric MLP (GMLP) and for tuning the weights of the Geometric RBF (GRBF). Previous learning rules for the real valued MLP, complex MLP [7] and the quaternionic MLP [1] are simply especial cases of this extended rule.

## 5.1 Generalized multi-dimensional back-propagation training rule

The norm of a multivector $x$ for the learning rule reads

$$|x| = (x|x)^{\frac{1}{2}} = \left(\sum_A [x]_A^2\right)^{\frac{1}{2}}. \tag{18}$$

The geometric neural network with $n$ inputs and $m$ outputs is supposed to approximate the target mapping function

$$\mathcal{Y}_t : (\mathcal{G}_{p,q})^n \rightarrow (\mathcal{G}_{p,q})^m, \tag{19}$$

where $(\mathcal{G}_{p,q})^n$ is the n-dimensional module over the geometric algebra $\mathcal{G}_{p,q}$ [15]. The error at the output of the net is measured according the metric

$$E = \frac{1}{2} \int_{x \in X} ||\mathcal{Y}_w - \mathcal{Y}_t||^2, \tag{20}$$

where $X$ is some compact subset of the Clifford module $(\mathcal{G}_{p,q})^n$ involving the product topology derived from equation (18) for the norm and $\mathcal{Y}_w$ and $\mathcal{Y}_t$ are the learned and target mapping functions respectively. The back-propagation algorithm [22] is a procedure for updating the weights and biases. This algorithm is a function of the negative derivative of the error function (Eq. (20)) with respect to the weights and biases themselves. The computing of this procedure is straightforward and here we will only give the main results. The updating equation for the multivector weights of any hidden $j$−layer is

$$w_{ij}(t+1) = \eta\left[\left(\sum_{k}^{N_k} \delta_{kj} \otimes \overline{w_{kj}}\right) \odot F'(net_{ij})\right] \otimes \overline{o_i} + \alpha w_{ij}(t), \tag{21}$$

for any $k-$output with a non-linear activation function

$$\boldsymbol{w}_{jk}(t+1) = \eta\big[(\boldsymbol{y}_{k_t} - \boldsymbol{y}_{k_a}) \odot \boldsymbol{F}'(\boldsymbol{net}_{jk})\big] \otimes \overline{\boldsymbol{o}_j} + \alpha \boldsymbol{w}_{jk}(t), \qquad (22)$$

and for any $k-$output with a linear activation function,

$$\boldsymbol{w}_{jk}(t+1) = \eta(\boldsymbol{y}_{k_t} - \boldsymbol{y}_{k_a}) \otimes \overline{\boldsymbol{o}_j} + \alpha \boldsymbol{w}_{jk}(t), \qquad (23)$$

where $\boldsymbol{F}$ is the activation function defined in equation (7), $t$ is the update step, $\eta$ and $\alpha$ are the learning rate and the momentum respectively, $\otimes$ defined for clearness is the Clifford or geometric product, $\odot$ is a scalar component by component product and $\overline{(\cdot)}$ is a multivector antiinvolution (reversion or conjugation).

In the case of the non-Euclidean geometric algebra $\mathcal{G}_{0,3}$ $\overline{(\cdot)}$ corresponds to the simple conjugation. Each neuron now consist of p+q units, each for a multivector component. The biases are also multivectors and are absorbed as usual in the sum of the activation signal called here $\boldsymbol{net}_{ij}$. In the learning rules, Eqs. (21)-(23), the way how the geometric product and the antiinvolution are computed varies according the geometric algebra being used [20]. As illustration we give the conjugation required in the learning rule for the quaternion algebra or $\mathcal{G}_{0,2}$, where the index indicates the number of the basis vector, i.e. 0 for 1, 1 for $\sigma_1$, 2 for $\sigma_2$ and 3 for $\sigma_1\sigma_2$. The conjugation reads: $\bar{x} = x_0 - x_1\sigma_1 - x_2\sigma_2 - x_3\sigma_1\sigma_2$, where $\boldsymbol{x} \in \mathcal{G}_{0,2}$.

The reversion in case of non-Euclidean $\mathcal{G}_{0,3}$ is given by $\bar{x} = x_0 + x_1\sigma_1 + x_2\sigma_2 + x_3\sigma_3 - x_4\sigma_1\sigma_2 - x_5\sigma_2\sigma_3 - x_6\sigma_3\sigma_1 - x_7i$.

## 5.2 Simplification of the learning rule using the density theorem

Given $\boldsymbol{X}$ and $\boldsymbol{Y}$ as compact subsets belonging to $(\mathcal{G}_{p,q})^n$ and to $(\mathcal{G}_{p,q})^m$ respectively and considering $\mathcal{Y}_t : \boldsymbol{X} \to \boldsymbol{Y}$ a continuous function, there are some coefficients $w_1, w_2, w_3, ..., w_{N_j} \in \mathcal{R}$ and some multivectors $\boldsymbol{y}_1, \boldsymbol{y}_2, \boldsymbol{y}_3, ..., \boldsymbol{y}_{N_j} \in \mathcal{G}_{p,q}$ and $\boldsymbol{\theta}_1, \boldsymbol{\theta}_2, \boldsymbol{\theta}_3, ..., \boldsymbol{\theta}_{N_j} \in \mathcal{G}_{p,q}$ so that the following inequality $\forall \epsilon > 0$ is valid

$$E(\mathcal{Y}_t, \mathcal{Y}_w) = sup\big[|\mathcal{Y}_t(\boldsymbol{x}) - \sum_{j=1}^{N_j} w_j \boldsymbol{f}_j(\sum_{i=1}^{N_i} \boldsymbol{w}_i \boldsymbol{x} + \boldsymbol{\theta}_i)|\boldsymbol{x} \in \boldsymbol{X}\big] < \epsilon, \qquad (24)$$

where $\boldsymbol{f}_j$ is the multivector activation function of Eq. (7). Here the approximation given by

$$S = \sum_{j=1}^{N_j} w_j \boldsymbol{f}_j(\sum_{i=1}^{N_i} \boldsymbol{w}_i \boldsymbol{x} + \boldsymbol{\theta}_i) \qquad (25)$$

is the subset of the class of functions $C^0(\mathcal{G}_{p,q})$ with the norm $|\mathcal{Y}_t| = sup_{\boldsymbol{x} \in \boldsymbol{X}} |\mathcal{Y}_t(\boldsymbol{x})|$. Since equation (24) is true we can finally say that $S$ is dense in $C^0(\mathcal{G}_{p,q})$. The density theorem presented here is the generalization of the one used for the quaternionic MLP by Arena et al. [1]. Its complete prove will be published else-where.

The density theorem shows that for the training of geometric feedforward networks the weights of the output layer could be real values. Therefore the training of the output layer can be simplified, i.e. the output weight multivectors could be scalars of $k$-grade. This $k$-grade element of the multivector is selected by convenience.

## 5.3 Learning using the appropriate geometric algebras

The main motivation to process signals in the geometric algebra framework is to have access to representations with a strong geometric character and to take advantage of the geometric product. An important question arises regarding the type of geometric algebra we should use for a specific problem. For some appli-cation the modelling of the suitable space would be straightforward. However in other cases it would be somehow difficult unless some a priori knowledge of the problem is available. In case we do not have any clue we should then explore various network topologies in different geometric algebras. This will require some orientation about the different geometric algebras we could use. Since each geo-metric algebra is either isomorphic to a matrix algebra of $\mathcal{R}$, $\mathcal{C}$ or $\mathcal{H}$ or simply a product of these algebras a great care has to be taken for choosing the algebras. Porteous [20] showed the isomorphisms

$$\mathcal{G}_{p+1,q} = \mathcal{R}_{p+1,q} \cong \mathcal{G}_{q+1,p} = \mathcal{R}_{q+1,p} \tag{26}$$

and presented the following expressions for completing the universal table of geometric algebras

$$\mathcal{G}_{p,q+4} = \mathcal{R}_{p,q+4} \cong \mathcal{R}_{p,q} \otimes \mathcal{R}_{0,4} \cong \mathcal{R}_{0,4} \cong \mathcal{H}(2)$$
$$\mathcal{G}_{p,q+8} = \mathcal{R}_{p,q+8} \cong \mathcal{R}_{p,q} \otimes \mathcal{R}_{0,8} \cong \mathcal{R}_{0,8} \cong \mathcal{R}(16), \tag{27}$$

where $\otimes$ stands for the real tensor product of two algebras. The last equation is known as the periodicity theorem [20]. Examples of this table are R$\cong \mathcal{G}_{0,0}$, $\mathcal{R}_{0,1} \cong \mathcal{C} \cong \mathcal{G}_{0,1}$, $\mathcal{H} \cong \mathcal{G}_{0,2}$ and $\mathcal{R}_{1,1} \cong {}^2\mathcal{R} \cong \mathcal{G}_{1,1}$, $\mathcal{C}(2) \cong \mathcal{C} \otimes \mathcal{R}(2)\cong \mathcal{G}_{3,0} \cong \mathcal{G}_{1,2}$ for the 3D space and $\mathcal{H}(2) \cong \mathcal{G}_{1,3}$ for the 4D space. The two later examples correspond to the geometric algebras mentioned in section 2.

## 6 Experiments

In this section the GMLP using the geometric algebra $\mathcal{G}_{p,q}$ will be denoted as GMLP$_{p,q}$. Firstly we test the component-wise activation function and then investigate new ways of geometric preprocessing embedded in the first network layer. Finally we identify the key role of the geometric learning in 3D object recognition and prediction in chaotic processes.

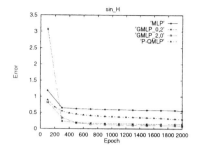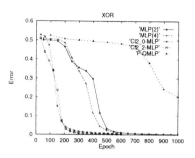

**Fig. 3.** *a) Learning $sin_H$ (left) b) XOR (right) using the MLP(2 and 4), $GMLP_{0,2}$ , $GMLP_{2,0}$ and P-QMLP.*

## 6.1 Inputs without any preprocessing

First the performance of the geometric net was tested using the sigmoid multi-vector activation function (7) against the real valued one and the one of Person (P-QMLP) [15]. The function to be approximated is $sin_H = sin(q)$, where $q \in \mathcal{H}$. All nets had the same 4-D input and 4-D output. The MLP uses 4 inputs and the GMLPs one 4-D input. The MLP had 3 hidden nodes and the GMLPs 3 geometric hidden nodes. Figure 3.a shows the performance of MLP, $GMLP_{0,2}$, $GMLP_{2,0}$ and the P-QMLP. In this figure the vertical axis indicates the total error per epoch. We can see clearly that the $GMLP_{0,2}$ and the $GMLP_{2,0}$ have a better performance than the P-QMLP. The reason is probably that during learning the activation function of the P-QMLP suffers of the consequence a zero division which slows down the convergence rate. In order to verify the generalization capability using different amount of hidden neurons after the training the nets MLP, $GMLP_{0,2}$ and $GMLP_{2,0}$ were tested using 50 before unseen patterns. For this evaluation the mean error was considered. Table 1 shows that the GMLPs have almost a similar generalization capability using 4 hidden nodes than the MLP, better using 3 and much better using just 2. The P-QMLP is not considered due to its poor performance. These experiments show so far that the complex neuron of the GMLPs process compactly the information better than the real neuron of the MLP. That is because the GMLPs have more weights due to their multivector valued nodes and presumably in this example the GMLPs profit of the geometric product based processing.

|  | 4 | 3 | 2 |
|---|---|---|---|
| MLP | 0.0827 | 0.5014 | 1.0145 |
| $GMLP_{0,2}$ | 0.0971 | 0.1592 | 0.2243 |
| $GMLP_{2,0}$ | 0.0882 | 0.1259 | 0.1974 |

Table 1. Mean error by different amount of hidden nodes.

A dramatic confirmation that the component-wise activation function (7) works much better than the one of Pearson (P-QMLP) is observed when testing the XOR problem. Figure 3.b shows that geometric nets $GMLP_{0,2}$ and $GMLP_{2,0}$ have a faster convergence rate than the MLP with 2- and 4- dimensional inputs and by far than the P-QMLP. Since the MLP(4) working even in 4D can not beat the GMLP, it can be claimed that the better performance of the geometric neural network is not only due to the higher dimensional quaternionic inputs but rather due to the algebraic advantages of the geometric neurons of the net.

## 6.2 Embedded geometric preprocessing

This subsection shows experiments with geometric neural networks with an embedded geometric preprocessing in the first layer. This has been done in order to capture the geometric cross-products of the input data much better. This proved to improve the convergence. In this paper these kind of nets will be called $EGMLP_{p,q}$. Figure 2 shows a geometric network with its extended first layer. The function $sin_H$ is again employed to test the performance of the MLP,

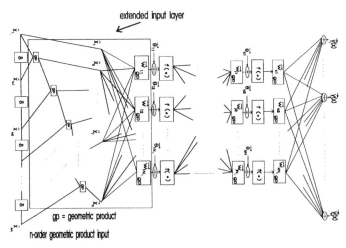

**Fig. 4.** *Geometric neural network with extended input layer*

$GMLP_{0,2}$ and $GMLP_{2,0}$. Here we use as first input $x_i$ and as second one the geometric product $x_i x_{i+1}$ (second order) or $x_i x_{i+1} x_{i+2} x_{i+3}$ (fourth order). Figure 3 shows that the performance of the three networks improves. It is clear that the extension of the first layer of the geometric network helps. Since the MLP uses the same type of inputs, its improvement relays also on this kind of geometric preprocessing.

## 6.3 Geometric RBF networks for 3D object recognition

In this section we explore the potential of the geometric RBF (GRBF) net to deal with the geometric nature of the task in question. The net is supposed to

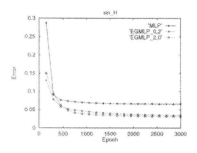
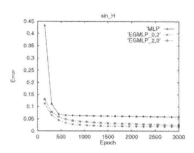

**Fig. 5.** *Learning $sin_H$ using MLP , $EGMLP_{0,2}$ and $EGMLP_{2,0}$ with second (left) and fourth order geometric product inputs (right).*

recognize a wire–frame 3D object from any of its perspective views. The object attributes are its N feature points, N-1 lengths and N-2 angles, see Figure 4a. The nets are trained on several random views and should map any view of the same object into a standard view. We trained real valued RBF neural networks and GRBF nets using the embedded geometric processing.

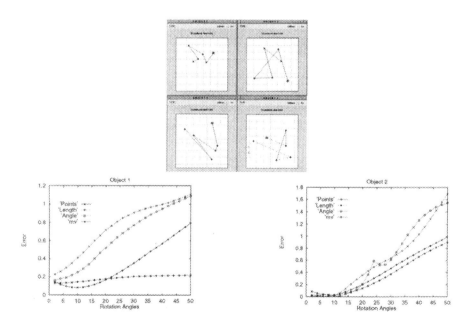

**Fig. 6.** *a) Wire frames b-c) Recognition error by RBF net using points or lines or angles and by $GRBF_{0,2}$ using $x_i x_{i+1}$ (mv). The two rotation angles are equal.*

In Figure 4b-c we can see for two of the objects that the $GRBF_{0,2}$ with $x_i x_{i+1}$ (second order embedded preprocessing depicted as *mv*) in a range from 0 to 50 degrees performs the generalization better than a real valued RBF using points or lengths or angles. This results should encourage researchers to apply this kind of geometric RBF networks with higher order of geometric products for various tasks like recognition of objects and the recover of pose.

## 6.4 Recognition of geometry in chaotic processes

This experiments shows that the geometric neural networks are well suited to distinguish the geometric information in a chaotic process.

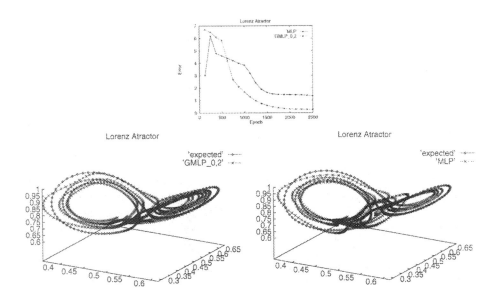

**Fig. 7.** a) Training error b) Prediction by $GMLP_{0,2}$ and expected trend c) Prediction by MLP and expected trend.

For that we used the well known Lorenz attractor ($\sigma=3$, r=26.5 and b=1) with the initial conditions [0,1,0] and sample rate 0.02 sec. A 3-12-3 MLP and a 1-4-1 $GMLP_{0,2}$ were trained in the interval [12, 17] sec. to perform a 8 $\tau$ step ahead prediction. The next 750 samples unseen during training were used for the test. The Figure 7.a show the error during training, note that the $GMLP_{0,2}$ converges faster than the MLP. Interesting is how they behave by the prediction. The figures 7b-c shows that the $GMLP_{0,2}$ predicts better than the MLP. Analyzing the covariance parameters of the MLP [0.96815, 0.67420, 0.95675] and of the $GMLP_{0,2}$ [0.9727, 0.93588, 0.95797] we can see that the MLP requires longer to get the geometry involved in the second variable, that is why the convergence is slower. As a result of that the MLP loses control to predict well in the other side of the looping. On contrast the geometric net from early stage captures the geometric characteristics of the attractor so it can not fail if it has to predict in the other side of the looping.

# 7  Conclusions

This paper started with basic reflections regarding geometry in biological creatures. The complexity of the mapping between the external and the internal world demands that the representation and calculus has to be carried out in a coordinate-free mathematical system with a strong algebraic and geometric characteristic. According the literature there are basically two mathematical systems used in neural computing: the tensor algebra and the matrix algebra. This paper chooses the coordinate-free system of Clifford or geometric algebra. The authors use this general and powerful mathematical system for the analysis and design of multidimensional feedforward neural networks. The reader can see that real-, complex- and quaternionic–valued neural networks are simple particular cases of the geometric algebra multidimensional neural networks. This work shows that the component-wise activation function defeats the activation function used in the complex neural nets [7] and in the hypercomplex nets by Pearson [15]. In case of the XOR problem the MLP using a 2-D or 4-D coded inputs can not perform as well as the GMLPs. The authors show also how the embedded geometric processing in the first layer helps to capture the geometric correlation of the data. The algebraic character of the nets is due to the activation function of the geometric neurons and the operations through the layers. The GA algebra is a coordinate free approach for neural nets. This can be seen by the experiment with GRBF net where the geometric products capture the geometric relations of the lines using the bivector between points liberating in this way the coordinate dependency existing in the point manifold. The ability of the geometric neural networks to recognize the geometric characteristics during the dynamic evolution of a chaotic process exemplifies the power of geometric neural learning for prediction.

**Acknowledgment**

This work was financed by the Deutsche Forschungsgemeinschaft project SO 320-2-1 Geometrische Algebra ein Repräsentationsrahmen für den Wahrnehmungs-Handlungs-Zyklus.

# References

1. Arena P., Caponetto R., Fortuna L., Muscato G. and Xibilia M.G. 1996. Quaternionic multilayer perceptrons for chaotic time series prediction. IEICE Trans. Fundamentals. Vol. E79-A. No. 10 October, pp. 1-6.
2. Bayro-Corrochano E., Buchholz S., Sommer G. 1996. Selforganizing Clifford neural network *IEEE ICNN'96 Washington, DC*, June, pp. 120-125.

3. Bayro-Corrochano E., Lasenby J., Sommer G. Geometric Algebra: A framework for computing point and line correspondences and projective structure using n-uncalibrated cameras *IEEE Proceedings of ICPR'96 Viena, Austria*, Vol. I, pages 334-338, August, 1996.

4. Bayro-Corrochano E., Daniilidis K. and Sommer G. 1997. Hand-eye calibration in terms of motion of lines using Geometric Algebra. To appear in SCIA'97, June 9-11,Lappeenranta, Finland.

5. Cybenko G. Approximation by superposition of a sigmoidal function. *Mathematics of control, signals and systems*, Vol. 2, 303:314, 1989.

6. Doran C.J.L. 1994. Geometric algebra and its applications to mathematical physics. *Ph.D. Thesis*, University of Cambridge.

7. Georgiou G. M. and Koutsougeras C. Complex domain backpropagation. *IEEE Trans. on Circuits and Systems*, 330:334, 1992.

8. Hamilton W.R. 1853. Lectures on Quaternions. Hodges and Smith, Dublin.

9. Hestenes D. 1966. Space-Time Algebra. *Gordon and Breach*.

10. Hestenes D. and Sobczyk G. 1984. Clifford Algebra to Geometric Calculus: A unified language for mathematics and physics. *D. Reidel*, Dordrecht.

11. Hestenes D. 1993. Invariant body kinematics I: Saccadic and compensatory eye movements. Neural Networks, Vol. 7, 65-77.

12. Hestenes D. 1993. Invariant body kinematics II: Reaching and neurogeometry. Neural Networks, Vol. 7, 79-88.

13. Hornik K. 1989. Multilayer feedforward networks are universal approximators. Neural Networks, Vol. 2, pp. 359-366.

14. Koenderink J. J. 1990. The brain a geometry engine. Psychological Research, Vol. 52, pp. 122-127.

15. Pearson J. K. and Bisset D.L. . 1992. Back Propagation in a Clifford Algebra. *Artificial Neural Networks, 2, I. Aleksander and J. Taylor (Ed.)*, 413:416.

16. Pellionisz A. and Llinás R. 1980. Tensorial approach to the geometry of brain function: cerebellar coordination via a metric tensor. Neuroscience Vol. 5, pp. 1125-1136

17. Pellionisz A. and Llinás R. 1985. Tensor network theory of the metaorganization of functional geometries in th central nervous system. Neuroscience Vol. 16, No. 2, pp. 245-273.

18. Perantonis S. J. and Lisboa P. J. G. 1992. Translation, rotation, and scale invariant pattern recognition by high-order neural networks and moment classifiers. IEEE Trans. on Neural Networks, Vol. 3, No. 2, March, pp 241-251.

19. Poggio T. and Girosi F. Networks for approximation and learning. *IEEE Proc.*, Vol. 78, No. 9, 1481:1497, Sept. 1990.

20. Porteous I.R. 1995. Clifford Algebras and the Classical Groups. *Cambridge University Press*, Cambridge.

21. Sommer G., Bayro-Corrochano E. and Bülow T. 1997. Geometric algebra as a framework for the perception–action cycle. Workshop on Theoretical Foundation of Computer Vision, Dagstuhl, March 13-19, 1996, Springer Wien.

22. Rumelhart D.E. and McClelland J. L.. 1986. Parallel Distributed Processing: Explorations in the Microstructure of Cognition. 2 Vols. Cambridge: MIT Press.

# Author Index

# Springer
# and the
# environment

# Lecture Notes in Computer Science

For information about Vols. 1–1229

please contact your bookseller or Springer-Verlag